Probabilistic Methods in Discrete Mathematics

PROBABILISTIC METHODS IN DISCRETE MATHEMATICS

PROCEEDINGS OF THE FOURTH INTERNATIONAL PETROZAVODSK CONFERENCE

Petrozavodsk, Russia, June 3 - 7, 1996

Editors:
V.F. Kolchin, V.Ya. Kozlov,
Yu.L. Pavlov and *Yu.V. Prokhorov*

///VSP///

Utrecht, The Netherlands, 1997

VSP BV
P.O. Box 346
3700 AH Zeist
The Netherlands

First published in 1997

ISBN 90-6764-245-2

Printed in The Netherlands by Ridderprint bv, Ridderkerk.

CONTENTS

distribution, is found:

$$\mathbf{P}\{\eta_1(n) = n_1\} = \prod_{j=0}^{n_1-1}(b+jc)\prod_{j=0}^{n_2-1}(r+jc)\left(\prod_{j=0}^{n-1}(b+r+jc)\right)^{-1} \tag{1}$$

Hence it follows that if $c \neq 0$, then

$$\begin{aligned}\mathbf{P}\{\eta_1(n) = n_1\} &= \binom{b/c+n_1-1}{n_1}\binom{r/c+n_2-1}{n_2}\binom{(b+r)/c+n-1}{n}^{-1} \\ &= \binom{-b/c}{n_1}\binom{-r/c}{n_2}\binom{-(b+r)/c}{n}, \end{aligned} \tag{2}$$

where $n_1+n_2 = n$. Remark that (1) for $c = 0$ is the binomial distribution $\mathrm{Bi}(n, b/(b+r))$ of the number of successes (the success means drawing a black ball) in the scheme of random sampling with replacement, and the second formula in (2) gives the hyper-geometric distribution which appears in the scheme of random sampling without replacement (in this case we suppose that $n \leq b + r$).

Presenting this model, Feller referred to the paper [9] published by Eggenberger and Pólya at 1923 as a pioneer work on this scheme. In that paper the authors obtained the explicit distribution of the random variable $\eta_1(n)$ given by (1) and (2), derived the factorial moments of this random variable and proved two limit theorems on the convergence of the distribution of $\eta_1(n)$ to the negative binomial distribution $\overline{\mathrm{Bi}}$ and the Poisson distribution Π as $n \to \infty$ and the parameters b, r, and c have a prescribed behaviour. Later many papers devoted to the study of this model and its generalizations were published where this model received deservedly the name Pólya urn model.

However, as it is pointed out in [10], six years before Eggenberger and Pólya publication [9] in 1917 Markov published the paper [11] in the *Izvestiya of the Petersburg Academy of Science*, where he introduced and investigated in detail exactly the same urn model.

Markov carried out the exhaustive asymptotic analysis of this model and obtained the complete list of limit distributions of the random variable $\eta_1(n)$ in the array scheme. It appeared that this list of limit distributions contained in addition to the distributions mentioned above the normal, beta-, and gamma-distributions (see section 5 below). Moreover, a particular case of this scheme for $c = 1$ was considered by Markov even earlier in 1906 as an example of the model where the law of large numbers is not valid [12].

Apparently these Markov's papers were known neither to Pólya and Eggenberger (they do not cite them), nor to Feller who also does not mention Markov in this connection.

Taking the situation into account it seems to us that it would be natural and justified from the historic point of view to call the scheme under consideration by the Markov–Pólya urn model (the abridged notation MP-model).

We do not know other papers devoted to Laplace's model, although it is obvious that the cases considered by Markov do not cover all possible variants of the behaviour of parameters of the model if we include the scheme of arrays as is usually done in the modern asymptotic investigations.

3. The next model considered by Markov became known in the modern notation as the scheme of allocating particles to cells by groups. A separate chapter of the book [4] is devoted to this problem (see also [7]). Therefore we give only a short description of some its parts.

Let us have N cells and let groups consisting of m particles be allocated to these cells. The particles of one group are allocated in such a way that each particle occupies a separate cell and all $\binom{N}{m}$ possible allocations are equiprobable. The main characteristics of the scheme are the random variables $\mu_r(n, N, m)$ equal to the number of cells containing exactly r particles each after n groups are allocated, $r = 0, 1, \ldots, n$, and the random variables $\nu_m(N, k)$ equal to the number of groups allocated until no less than k cells become occupied, $m \le k \le N$. This model is mentioned in Markov's book [5], where, in particular, the explicit distribution of the random variable $\mu_0(n, N, m)$ is given:

$$\mathbf{P}\{\mu_0(n, N, m) = s\} = \frac{\binom{N}{s}}{\binom{N}{m}} \sum_{i=0}^{N-s} (-1)^{N-s-i} \binom{N-s}{i} \binom{i}{m}^n.$$

This model and its various generalizations were studied by many mathematicians. Essential results in this area were obtained by Pólya, Békéssy, Sevastyanov, Ivchenko, Medvedev, Ivanov, Vatutin, Mikhailov, Savchuk (the corresponding bibliography can be found in [4, 6, 7]). From what has been said it is clear that the main contribution to the study of this important in applications model was made by the Russian mathematicians.

4. Our main concern is with the following well-known in the literature urn model given in Feller's book under the name Pólya urn model: an urn initially contains b black and r red balls, the balls are sequentially drawn at random from the urn in such a way that at each draw the probability of obtaining a specific ball is the same for all balls, after each draw we replace the drawn ball together with c of the same colour. The parameter c can take any integer value, in particular, if $c = 0$, then we have the scheme of simple random sampling (with replacement), if $c = -1$, then we obtain the scheme of random sampling without replacement (in the last case the process is stopped after $b + r$ draws). For $c > 0$, the model possesses an important property of after-effect: if we obtain a ball of some colour, then the probability to draw a ball of the same colour at the next draw increases. As Feller notes, this scheme can be taken as an approximate model for epidemic phenomena where a realization of some events increases the probability of their repetition.

The analytical properties of this model are investigated in detail in [1], where the distribution of the number $\eta_1(n)$ of black balls appeared during n drawings, the Pólya functions, theory of differential equations. As concerns to the probabilistic researches it is sufficient to recall that Markov was the first who gave a rigorous proof of the main limit theorem of probability theory and extended the result to sequences of dependent random variables. This extension led him to the notion of the general scheme of trials forming a chain which is now referred to as a Markov chain. Apart from these well-known facts, Markov obtained pioneer results concerning some classical urn models such as Laplace's urn model with ball exchange between two urns, allocating groups of balls to urns, the model of sequential sampling of balls from an urn with addition of a fixed number of balls of the same colour as the ball drawn (without sufficient reasons the last model is called the Pólya urn model).

In this paper we try to compensate to some extent this gap in the presentation of the history of probability theory (its discrete problems) and to bring to the attention of the mathematical society some problems on priority of the Russian mathematicians in this area of science.

2. Following the chronology, we begin with a short description of Laplace's scheme of ball exchange between two urns and present Markov's contribution to this area.

Let one of the two urns contain n balls and the other urn contain n_1 balls. Each ball is either white or black, the total number of white balls is $(n + n_1)p$ and the total number of black balls is $(n + n_1)q$, $p + q = 1$.

It is assumed that the number of white balls in the first urn is a random variable with a given probability distribution. The process of ball exchanging consists of sequential simultaneous exchanges of two balls from the two urns, one ball from each of the urn is replaced to another in a moment. The main characteristic of this process is the number $X(r)$ of white balls in the first urn after r exchanges. This problem was first formulated by Laplace in 1812 who gave its partial solution for $p = q = 1/2$, $n = n_1$ (see [1]).

An in-depth analysis of this problem in the general case was made by Markov in 1912 and 1915 [2, 3]. He showed the connection of this problem with the scheme of trials forming a chain (Markov chain), found in the explicit form the distribution and the moments of the random variables $X(r)$. Markov used a method which consisted in setting up and analyzing a difference-differential equation for the probabilities

$$z_{x,r} = \mathbf{P}\{X(r) = x\}.$$

An asymptotic analysis of the distribution of $X(r)$ as $n, n_1, r \to \infty$ was carried out by the method of moments with the use of the Chebyshov-Hermite polynomials. Namely, he showed that if $r(1/n + 1/n_1) \to \infty$, then the moments of the normalized random variable

$$\mu = \frac{X(r) - np}{\sqrt{2pqnn_1/(n + n_1)}}$$

converges to the corresponding moments of the standard normal distribution. If $r(1/n + 1/n_1)$ tends to a constant, then the limit distribution is some generalized normal distribution.

Probabilistic Methods in Discrete Mathematics, pp. 1–9
V. F. Kolchin et al. (Eds.)

THE CONTRIBUTION OF THE RUSSIAN MATHEMATICIANS TO THE STUDY OF URN MODELS

G. I. IVCHENKO and YU. I. MEDVEDEV

Moscow

Abstract — We discuss the history of development of some parts of probabilistic combinatorics connected with the so-called urn models and demonstrate the contribution of the Russian mathematicians to this area of discrete probability theory. Of particular value in the development of the theory of urn models are the works due to the outstanding Russian mathematician A. A. Markov. Unfortunately these works appeared to be disregarded and not cited in the modern publications.

1. The discrete probabilistic problems, a major part of which is constituted by the so-called urn problems, always had a significant place in the probabilistic investigations and in essence originated the theory of probability. These problems form a testing area for general ideas and methods of the probabilistic analysis. In recent years we see the regeneration of interest to discrete mathematics which reflects the new mode of thought on the role of discrete models, from illustrative and academic to fundamental and extremely important in applied mathematical investigations. The growing of interest to discrete probabilistic problems is related to the increasing of general interest to discrete mathematics that in turn is connected with recent development of computer sciences and intensive computerization of all scientific researches.

Historical aspects of a particular branch of science are of independent importance, it can be said that the history of a science has the same significance as the science itself. Therefore one of the rules of scientific work is a profound knowledge of the history of development of the corresponding area of science and of the contribution to this area of the initiators especially from your own country. Our duty to them is to demonstrate in our publications their contribution, to emphasize their priority by citing as a rule the origins.

In the light of the preceding it seems to us that the role of the outstanding Russian mathematician A. A. Markov in the development of the direction in probability theory which is now referred to as the discrete problems of probability theory is underestimated. Andrei Andreevich Markov (14.06.1856–20.07.1922) was one of the brilliant persons in mathematics on the boundary of the XIX and XX centuries. He was a founder of a series of fundamental directions in probability theory, number theory, theory of probabilistic discrete mathematics and contain the information which is interesting to those who work in theoretical and applied areas of discrete mathematics.

The Fourth Petrozavodsk Conference was supported by the Russian Foundation for Basic Research. The conference was also supported by CompuLink, VSP International Science Publishers (The Netherlands), and by the Editorial board of *Discrete Mathematics and Applications*.

The Organizing Committee has the pleasant duty to thank the sponsors of the Fourth Petrozavodsk conference without whose financial support the conference could be not held.

The Organizing Committee greatly appreciates the efforts of the Petrozavodsk collegues who took up the organizing duties in Petrozavodsk.

We are deeply indebted to VSP International Science Publishers who agreed to publish the Proceedings.

We wish also to express our gratitude to all participants of the Conference and invite all of them to take part in the next Petrozavodsk Conference which will be held in 2000.

V. F. Kolchin

Preface

The Fourth Petrozavodsk Conference 'Probabilistic Methods in Discrete Mathematics' was held at 3–7 June 1996 in Petrozavodsk in Russia. The conference was organized by the Steklov Mathematical Institute of the Russian Academy of Sciences, the Department of Mathematics and Data Analysis of the Karelian Scientific Centre of the Russian Academy of Sciences, and the Karelian State University. The Organizing Committee of the Conference included V. Ya. Kozlov (the chairman, Moscow), A. D. Sorokin (vice-chairman, Petrozavodsk), V. F. Kolchin (vice-chairman, Moscow), Yu. V. Prokhorov (Moscow), B. A. Sevastyanov (Moscow), A. S. Fomin (Petrozavodsk), G. I. Ivchenko (Moscow), V. I. Khokhlov (Moscow), Yu. I. Medvedev (Moscow), V. G. Mikhailov (Moscow), Yu. L. Pavlov (Petrozavodsk), V. A. Vatutin (Moscow), A. B. Zhizhchenko (Moscow), A. M. Zubkov (Moscow).

During the last three decades the interest in probabilistic problems of combinatorics has grown steadily both in our country and abroad. The First, Second, and Third conferences were held in Petrozavodsk in 1983, 1989, and 1992.

The Fourth Petrozavodsk Conference received about 60 participants, including about 40 guests of Petrozavodsk, among which were A. Bikelis from Lithuania, M. Drmota and B. Gittenberger from Austria, Sh. Formanov from Uzbekistan, B. Harris from the United States of America, J. Jaworski from Poland.

The themes of the Petrozavodsk conferences cover almost all areas of probabilistic discrete mathematics. As usual, the lectures presented at the Fourth Petrosavodsk conference were devoted to probabilistic problems of combinatorics, statistical problems of discrete mathematics, theory of random graphs, systems of random equations in finite fields, some questions of information security. At this time, a considerable part of the lectures contributed to the conference concerned with discrete random processes.

During the three working days of the Fourth Petrozavodsk Conference, 9 plenary and 35 sectional lectures were given. The following plenary lectures were presented at the conference (the lectures are listed in the order they were read at the conference):

G. I. Ivchenko and Yu. I. Medvedev (Moscow), The contribution of the Russian mathematicians to the study of urn models;

Yu. L. Pavlov (Petrozavodsk), Random forests;

J. Jaworski (Poznan, Poland), Random mappings and Abelian sums;

B. Harris (Madison, the USA), Application of the interval graphs to cluster analysis;

M. Drmota (Vienna, Austria), Branching processes and Brownian excursion, analytic approach;

I. A. Semaev (Moscow), A generalization of the number field sieve;

L. Ya. Savelyev (Novosibirsk), Operator and recursion equations for runs in random sequences;

Sh. K. Formanov (Tashkent, Uzbekistan), Combinatorial limit theorems with square weight function;

A. P. Bikelis (Vilnius, Lithuania), Quasi-lattice probability distributions.

The plenary lectures prepared for publication are presented in the Proceedings in the above mentioned order. All sectional reports are arranged in alphabetic order of author's names.

The papers presented in the Proceedings reflect the current state of the art in

5. We obtain a natural generalization of the urn model considered in the preceding section if we assume that the balls can have $N \geq 2$ colours. Thus, we have an urn where initially there are a_j balls of the jth colour, $j = 1, \dots, N$, the total number of balls in the urn at the initial moment of time is equal to $a = a_1 + \dots + a_N$. We consider the same process where the balls are sequentially drawn at random from the urn in such a way that at each draw the probability of obtaining a specific ball is the same for all balls, after each draw we replace the drawn ball together with c balls of the same colour. Let $\eta_j(n)$ denote the number of the balls of the jth colour obtained during n draws. The distribution of the random vector $\bar{\eta}(n) = (\eta_1(n), \dots, \eta_N(n))$ is of the form: for $\bar{n} = (n_1, \dots, n_N)$ such that $n_1 + \dots + n_N = n$

$$\mathbf{P}\{\bar{\eta}(n) = \bar{n}\} = \frac{n!}{n_1!\dots n_N!a(a+c)\dots(a+(n-1)c)} \prod_{j=1}^{N} a_j(a_j+c)\dots(a_j+(n_j-1)). \quad (3)$$

Hence it follows that for $c \neq 0$

$$\mathbf{P}\{\bar{\eta}(n) = \bar{n}\} = \frac{1}{\binom{\alpha+n-1}{n}} \prod_{j=1}^{N} \binom{\alpha_j + n_j - 1}{n_j} = \frac{1}{\binom{-\alpha}{n}} \prod_{j=1}^{N} \binom{-\alpha_j}{n_j}, \quad (4)$$

where $\alpha_j = a_j/c$, $\alpha = \alpha_1 + \dots + \alpha_N = a/c$.

Representation (3) for $c = 0$ is the polynomial distribution $M(n; a_1/a, \dots, a_N/a)$, and the second representation in (4) for $c = -1$ (and $n \leq a$) is the multidimensional hypergeometric distribution.

We refer to the urn model above with an arbitrary finite $N \geq 2$ as the MP-model and to the distribution given by (3) and (4) as the MP-distribution. Apparently Nisnevich [13] was the first who introduced and investigated this scheme in 1953 (we found no papers published before [13]). He gave distribution (4) and pointed out an interesting property of the MP-model connected with the notion of permutability. We recall shortly this property.

Let X_i be the number of the colour of the ith drawn ball. If $c > 0$, then the sequence $\{X_i, i = 1, 2, \dots\}$ is permutable, i.e., the joint distribution of any finite subset $(X_{i_1}, \dots, X_{i_k})$ of these random variables does not depend on their indices and depends only on k. In particular, these random variables are identically distributed on the set $\{1, \dots, N\}$ and

$$\mathbf{P}\{X_i = j\} = \frac{a_j}{a}, \qquad j = 1, \dots, N.$$

It is well known that any permutable sequences of random variables taking values from the finite set $\{1, \dots, N\}$ can be obtained from a sequence of independent identically distributed random variables with a distribution $\bar{p} = (p_1, \dots, p_N)$, $p_1 + \dots + p_N = 1$ (a polynomially distributed sequence), if we randomize the vector $\bar{p}$ with the use of an appropriate measure. The MP-sequence $\{X_i\}$ is obtained by randomization of a polynomial sequence with the use of the Dirichlet measure $D(\bar{\alpha})$, $\bar{\alpha} = (\alpha_1, \dots, \alpha_N)$, with the density

$$g(\bar{p}, \bar{\alpha}) = \frac{\Gamma(\alpha)}{\Gamma(\alpha_1)\dots\Gamma(\alpha_N)} \prod_{j=1}^{N} p_j^{\alpha_j - 1},$$

where $\alpha = \alpha_1 + \ldots + \alpha_N$, $p_1 + \ldots + p_N = 1$.

Hence it follows that, in particular, MP-distribution (4) is the mixture of the polynomial distribution $M(n;\bar{p})$ with $\mathcal{L}(\bar{p}) = D(\bar{\alpha})$ and, as $n \to \infty$, the ratio $\bar{\eta}(n)/n$ converges with probability one to the random variable $\bar{p}$ but not to $\mathbf{E}\bar{\eta}(n)/n = \bar{a}/a$. Thus, if $c > 0$, the sequence $\{X_i\}$ satisfies neither the law of large numbers (this fact was found by Markov) nor the central limit theorem.

A description of explicit and asymptotic properties of the MP-model and the corresponding bibliography can be found in authors' paper [10] (see also [14]). Here we restrict ourselves to the formulations of the main limit theorems for the MP-model in the case of large samples, i.e., as $n \to \infty$.

THEOREM 1. *Let $n \to \infty$, the parameter $\bar{\alpha} = (\alpha_1, \ldots, \alpha_N)$ be fixed, and let $\bar{n}/n \to \bar{x}$. Then*

$$n^{N-1}\mathbf{P}\{\bar{\eta}(n) = \bar{n}\} \to g(\bar{x}, \alpha).$$

The following theorems describe the scheme of arrays where $\alpha = \alpha(n) \to \infty$ as $n \to \infty$. We set $d = n/\alpha$ and $p_j = \alpha_j/\alpha$, $j = 1, \ldots, N$. Depending on the behaviour of these parameters the following four main types of limiting relations are possible.

THEOREM 2. *Let $n, \alpha \to \infty$ and let there exist $\delta > 0$ such that $p_j \in (\delta, 1-\delta)$ for all $j = 1, \ldots, N$ and $d \le c$, where c is a constant, $n_j = np_j + x_j\sqrt{n}$, $j = 1, \ldots, N$, and $\max_j |x_j| = o(n^{1/6})$. Then*

$$\mathbf{P}\{\eta_j(n) = n_j, j = 1, \ldots, N-1\} = (2\pi n)^{-(N-1)/2}|\Sigma|^{-1/2}\exp\left\{-\frac{1}{2}\bar{x}'\Sigma^{-1}\bar{x}\right\},$$

where

$$\bar{x} = (x_1, \ldots, x_{N-1}), \quad \Sigma = \|\sigma_{ij}\|_1^{N-1}, \quad \sigma_{ij} = (1+d)p_i(\delta_{ij} - p_j),$$

and δ_{ij} is the Kronecker symbol.

THEOREM 3. *Let $n, \alpha \to \infty$ in such a way that $0 < c_1 \le d \le c_2 < \infty$ and let α_j be a positive constant, $j = 1, \ldots, N-1$. Then the random variables $\eta_1(n), \ldots, \eta_{N-1}(n)$ are asymptotically independent and*

$$\mathcal{L}(\eta_j(n)) \to \overline{\mathrm{Bi}}(\alpha_j, d/(1+d)), \qquad j = 1, \ldots, N-1.$$

THEOREM 4. *Let $n, \alpha \to \infty$ in such a way that $d \to 0$ and $nd \to \infty$. If $\alpha_j \to \infty$ in such a way that $d\alpha_j \to \lambda_j > 0$, $j = 1, \ldots, N-1$, then the random variables $\eta_1(n), \ldots, \eta_{N-1}(n)$ are asymptotically independent and*

$$\mathcal{L}(\eta_j(n)) \to \Pi(\lambda_j), \qquad j = 1, \ldots, N-1.$$

THEOREM 5. *Let $n, \alpha \to \infty$ in such way that $d \to \infty$ and $d/\alpha \to 0$. If α_j is a constant and $n_j \to \infty$ in such a way that $n_j/d \to 0$, $j = 1, \ldots, N-1$, then*

$$d^{N-1}\mathbf{P}\{\eta_j(n) = n_j, j = 1, \ldots, N-1\} \to \prod_{j=1}^{N-1} \frac{x_j^{\alpha_j - 1}}{\Gamma(\alpha_j)} e^{-x_j}.$$

We add some remarks. Theorem 2 describes the case of the usual normal approximation where the means and variances of all components of the vector $\bar{\eta}(n)$ are of order n, in this case for $j = 1, \dots, N-1$

$$\mathbf{E}\eta_j(n) = np_j = d\alpha_j,$$

$$\mathbf{D}\eta_j(n) = \frac{n\alpha_j(\alpha - \alpha_j(\alpha + n)}{\alpha^2(\alpha + 1)} = np_j(1 - p_j)(1 + d)(1 + O(\alpha^{-1})).$$

Theorems 3 and 4 describe the case of rare events where the total number of balls in the urn is large but at the initial moment the numbers of balls of colours $1, \dots, N-1$ are small and therefore they rarely appear even in a large sample. Finally, Theorem 5 covers the case where the variances of the random variables $\eta_j(n), j = 1, \dots, N-1$, tend to infinity faster than the mean values.

Note that for $N = 2$, in the case considered by Markov [11], these theorems present all possible limit distributions in the MP-model. However for $N > 2$ different combinations are possible, where a part of the components of the vector $\bar{\eta}(n)$ satisfies the conditions of rare events and the others are asymptotically normal. Examples of the corresponding theorems are given in [10].

Note also that MP-distributions (4) can be represented with the use of the so-called generalized allocation scheme introduced by Kolchin [4] in 1968. This scheme is determined by the conditional distribution of the form

$$\mathcal{L}(\bar{\eta}(n)) = \mathcal{L}\left(\bar{\xi} \mid \sum_{j=1}^{N} \xi_j = n\right),$$

where $\bar{\xi} = (\xi_1, \dots, \xi_N)$ is a vector with independent integer-valued components and the components of the vector $\bar{\eta}(n) = (\eta_1(n), \dots, \eta_N(n))$ are regarded as the contents of N cells after allocating n particles. If in this scheme the random variable ξ_j has the negative binomial distribution $\overline{\mathrm{Bi}}(\alpha_j, \theta), j = 1, \dots, N$, and the parameter θ takes an arbitrary value from the interval $(0, 1)$, then $\bar{\eta}(n)$ has distribution (4).

6. Another generalization of the MP-model is connected with a more complicated rule of changing the urn content. Such a generalization was suggested by Bernshtein [15, 16] in 1940. We present shortly the corresponding results following the style and terminology of these papers.

Initially a box contains a white and b black balls. We carry out n trials consisting in drawing one ball, returning it to the box and adding R new balls to the box according to the following rule: if the ball drawn is white, then we add ν white balls and $\rho = R - \nu$ black balls, if the ball drawn is black, then we add ν_1 white balls and $\rho_1 = R - \nu_1$ black balls. We study the random variable x_n equal to the number of white balls in the box after n trials and the random variable m_n equal to the number of white balls drawn during these n trials.

Analysis of this model is more difficult in comparison with the usual MP-model (the case of $\rho = \nu_1 = 0$). For this model Bernshtein found explicit and asymptotic expressions for the first two moments of the random variables x_n and m_n and proved the asymptotic normality of x_n as $n \to \infty$, $2(\nu - \nu_1)/R \leq 1$, and $\rho\nu_1 > 0$. For example, if $\rho + \nu_1 > 0$, then

$$a_n = \mathbf{E}x_n = \frac{\nu_1}{\nu_1 + \rho}(nR + a + b) + \frac{a}{\nu_1 + \rho} \prod_{j=0}^{n-1} \left(1 + \frac{\delta}{a + b + jR}\right), \tag{5}$$

where

$$A = a\rho - b\nu_1, \quad \delta = \nu - \nu_1 = \rho_1 - \rho, \quad \mathbf{E}m_n = (a_n - a - n\nu_1)/\delta.$$

For comparison we point out that for the MP-model, i.e., for $\nu_1 = \rho = 0$

$$\mathbf{E}x_n = \frac{a}{a+b}(a + b + nR), \qquad \mathbf{E}m_n = \frac{na}{a+b}.$$

If $\nu = \rho_1 = 0$, i.e., if balls of the different colour with respect to the ball drawn are added only, then formula (5) takes the form

$$a_n = \frac{1}{2}(a + b + nR) + \frac{(a-b)(a+b-R)}{2(a+b+(n-1)R)}.$$

Thus, the model considered in this section is studied less than the MP-model. It would be interesting both from theoretical and practical point of view to study the similar generalization of the MP-model with balls of an arbitrary number of colours.

In connection with the urn model considered by Bernshtein it is pertinent to note that much later, in 1949, Friedman [17] suggested an urn model which was a particular case of the model considered by Bernshtein, where $\rho_1 = \nu$, $\nu_1 = \rho$, i.e., each time we add to the urn ν balls of the same colour as the colour of the ball drawn and ρ balls of the opposite colour (see also [8]).

In [17], without citing [15, 16] but by the same method, the author derives a difference-differential equation for the moment generating function of the random variable x_n and finds an explicit form of its solution in the following three particular cases: $\rho = 0$ (the MP-model), $\nu = -1$, $\rho = 1$ (the Ehrenfest urn model), and $\nu = 0$. For these cases some asymptotic results are also obtained.

Thus, we can state that our mathematicians have a priority in the study of this more complicated urn model as well.

REFERENCES

1. P. S. Laplace, *Theorie Analytique des Probabilites*. Paris, 1812.
2. A. A. Markov, On trials joined in a chain by non-observable events. *Izv. Acad. Sci.* (1912) **6**, 551–572 (in Russian).
3. A. A. Markov, On a Laplace's problem. *Izv. Acad. Sci.* (1915) **9**, 87–104 (in Russian).
4. V. F. Kolchin, B. A. Sevastyanov, and V. P. Chistyakov, *Random Allocations*. Wiley, New York, 1978.
5. A. A. Markoff, *Wahrscheinlichkeitsrechnung*. Berlin, 1912.

6. V. A. Ivanov, G. I. Ivchenko, and Yu. I. Medvedev, Discrete problems in probability theory. *J. Soviet Math.* (1985) **31**, No. 2, 3–60.
7. I. N. Kovalenko, A. A. Levitskaya, and M. N. Savchuk, *Selected Problems of Probabilistic Combinatorics*. Naukova Dumka, Kiev, 1986 (in Russian).
8. W. Feller, *An Introduction to Probability Theory and its Applications*, Vol. 1, Wiley, New York, 1968.
9. F. Eggenberger and G. Pólya, Über die Statistik verketteter Vorgänge. *Z. Angew. Math. Mech.* (1923) **3**, 279–289.
10. G. I. Ivchenko and Yu. I. Medvedev, On the Markov–Pólya urn model from 1917 to nowadays. *Survey Appl. and Industrial Math.* (1996) **3**, 484–511 (in Russian).
11. A. A. Markov, On some limiting formulae of probability calculus. *Izv. Acad. Sci.* (1917) **11**, 177–186 (in Russian).
12. A. A. Markov, Extension of the law of large numbers to dependent variables. In: *Selected Papers*. Gostechizdat, Moscow, 1946, pp. 339–361 (in Russian).
13. L. B. Nisnevich. On the Markov urn model. *Soviet Math. Uspekhi* (1953) **8**, 131–134 (in Russian).
14. G. I. Ivchenko and A. V. Ivanov, On the waiting time in the Markov–Pólya urn model. *Discrete Math. Appl.* (1997) **7**, 47–63.
15. S. N. Bernshtein, New applications of almost independent random variables. *Izv. Soviet Acad. Sci., Ser. Math.* (1940) **4**, 137–150 (in Russian).
16. S. N. Bernshtein, A problem on an urn with added balls. *Soviet Acad. Sci. Dokl.* (1940) **28**, 5–7 (in Russian).
17. B. Friedman, A simple urn model. *Commun. Pure and Appl. Math.* (1949) **2**, 59–70.

Probabilistic Methods in Discrete Mathematics, pp. 11–18
V. F. Kolchin et al. (Eds.)

RANDOM FORESTS

YU. L. PAVLOV

Petrozavodsk

Abstract — The history of investigations of random forests consisting of N rooted trees with n non-root vertices is surveyed. For random forests with quite general structure of trees we present the limit distributions of the maximum size of trees, the number of trees of a given size, and the height for various domains of variation of N and n.

The history of investigations of random forests includes the three main stages. On the first stage (1977–1980) the set of forests $F'_{N,n}$ consisting of N rooted trees (roots are labelled) and n non-root labelled vertices with the uniform probability distribution was considered. For such forests the limit distributions of the maximum size of trees and the number of trees of a given size were obtained in [9] and [10]. The technique of obtaining these results was based on the generalized allocation scheme suggested by V. F. Kolchin [5, 8]. The results on the random forests were used for investigation of random mappings [9].

During the second stage (1981–1990), as previously, the set $F'_{N,n}$ with the uniform probability distribution was studied. This stage is characterized by the applications of the methods of the theory of branching processes to investigations of random forests. The basis of these investigations was provided by the discovered by V. F. Kolchin [6, 7] relationship between random rooted trees with n labelled non-root vertices and conditional Galton–Watson branching processes starting with a single particle under the condition that the total progeny of the process is equal to n and the number of offspring of one particle has the Poisson distribution. In [3] this relationship was extended to $F'_{N,n}$ which was the basis of obtaining the limit distributions of the height [12] and the number of vertices in strata of a random forest from $F'_{N,n}$ [13].

Note that in [4] Kalugin discovered the similar relationship for random forests with various constraints imposed on degrees of vertices and Britikov (see, e.g. [1]) considered random forests of non-rooted trees.

The third stage began in 1991, when the relationship between random plane planted trees and Galton–Watson branching processes starting with a single particle with the geometrically distributed number of offspring of one particle was discovered [14]. In [17] this relationship was extended to the set of forests $F''_{N,n}$ consisting of N plane planted trees and $N + n$ non-root vertices. Note that in more detail the relationship between Galton–Watson branching processes with Poisson and geometric distributions of offspring of one particle and trees from $\{F'_{1,n}\}_{n=1}^{\infty}$, $\{F''_{1,n}\}_{n=0}^{\infty}$ is considered in [2]. On the basis of these results the limit distributions of various characteristics of random forests from $F''_{N,n}$ were found in [15]. Thus, the methods of obtaining assertions on the behaviour of some characteristics as well as the corresponding limit distributions turn out to be similar for random forests of various structures. Therefore it seems to be natural to obtain general assertions concerning quantitative characteristics of forests from various classes and for various probability measures (not only uniform) on a set of forests. Such results were obtained in [16].

Let G be the Galton–Watson branching process beginning with N particles for which the number of offspring of a particle has the generating function

$$F(z) = \sum_{k=0}^{\infty} p_k z^k. \tag{1}$$

There is a biunique correspondence δ between the set of realizations of the process G and the set of forests consisting of N plane planted trees. In particular, if the number of offspring of one particle has the geometric distribution, then the process G by virtue of δ induces the uniform distribution on $F''_{N,n}$.

We consider the set of forests $F'_{N,n}$. Any realization $G_{N,n}$ of the process G with $N+n$ particles maps onto a set of forests $\Delta(G_{N,n})$ from $F'_{N,n}$. Each forest from $\Delta(G_{N,n})$ differs from the others by numeration of vertices only. Let $F_{N,n}$ is a set of forests consisting of N rooted trees (roots are labelled) with n non-root vertices which are labelled in some way, hence $F_{N,n} \subseteq F'_{N,n}$. We denote $\Delta_F(G_{N,n})$ the class of forests obtaining from $\Delta(G_{N,n})$ by removing the forests which do not belong to $F_{N,n}$. Let $|\Delta_F(G_{N,n})|$ be the number of forests in $\Delta_F(G_{N,n})$, $|F_{N,n}|$ be the number of forests in $F_{N,n}$, and let ν be the total progeny of the process G. We will consider the classes of forests $F_{N,n}$ and branching processes G related by the equality

$$\mathbf{P}\{G = G_{N,n} \mid \nu = N+n\} = \frac{|\Delta_F(G_{N,n})|}{|F_{N,n}|}. \tag{2}$$

We introduce some constraints on the distribution of offspring of a particle. Let an auxiliary random variable ξ have the distribution determined by generating function (1). We assume that this distribution has the maximum span d and the set of the values of ξ having non-zero probability contains 0 and does not coincide with the set $\{0, 1\}$. Let the number of offspring of a particle in the process G have the distribution

$$p_n(\lambda) = \lambda^k p_k / F(\lambda), \quad k = 0, 1, \dots, \quad 0 < \lambda \leq 1, \tag{3}$$

and $\mathbf{E}\xi = 1$, $\mathbf{D}\xi = B$, $F'''(1) < \infty$. As in the book [16], we consider the class of forests $F_{N,n}$ for which there exist branching processes G possessing property (2) with the number of offspring of a particle distributed according to distribution (3).

The next lemma is used in investigations of various numerical characteristics of the forests from $F_{N,n}$. Let $\varphi = \varphi(G_{N,n})$ be some characteristic of a realization $G_{N,n}$. If a forest $f \in F_{N,n}$ is contained in the set $\Delta_F(G_{N,n})$, then we put

$$\varphi_\Delta = \varphi_\Delta(f) = \varphi(G_{N,n}).$$

Thus, the random variable φ_Δ has one and the same value on any element of the set $\Delta_F(G_{N,n})$. The well-known examples of the random variables determined in a such manner are the maximum size of a tree in the forest, the number of trees of a given size, the height of a random forest, the number of vertices of a given degree, the number of vertices of a given height.

LEMMA 1. *The equality*

$$\mathbf{P}\{\varphi_\Delta = x\} = \mathbf{P}\{\varphi = x \mid \nu = N + n\}$$

holds.

The limit theorems which give the complete description of the behaviour of the maximum size of a tree, the number of trees of a given size and the height of a random forest for various domains of variation of N and n are proved in [16].

We denote by η the maximum size of a tree of a random forest from $F_{N,n}$. The branching process G can be considered as the union of N processes $G^{(1)}, \dots, G^{(N)}$ each of which begins with one particle. Let $\nu^{(i)}$, $i = 1, \dots, N$, stand for the total progenies of the processes $G^{(i)}$, then

$$\nu = \nu^{(1)} + \dots + \nu^{(N)}.$$

Let $\nu_r^{(1)}, \dots, \nu_r^{(N)}$ be auxiliary independent identically distributed random variables such that

$$\mathbf{P}\{\nu_r^{(i)} = k\} = \mathbf{P}\{\nu^{(i)} = k \mid \nu^{(i)} \le r + 1\}, \quad k = 0, 1, 2, \dots$$

We set also

$$\nu_r = \nu_r^{(1)} + \dots + \nu_r^{(N)}, \qquad P_r = \mathbf{P}\{\nu^{(1)} > r + 1\}.$$

LEMMA 2. *For n such that $\mathbf{P}\{\nu = N + n\} > 0$*

$$\mathbf{P}\{\eta \le r\} = (1 - P_r)^N \frac{\mathbf{P}\{\nu_r = N + n\}}{\mathbf{P}\{\nu = N + n\}}.$$

Lemma 2 shows that for obtaining the limit distributions of η it suffices to consider the asymptotic behaviour of the binomial $(1 - P_r)^N$ and the probabilities $\mathbf{P}\{\nu_r = N + n\}$, $\mathbf{P}\{\nu = N + n\}$. The following assertions are proved by the use of Lemma 2. We denote by j the least positive integer such that $p_j > 0$ and by l the least natural number not divided by j for which $p_{j+l} > 0$, if such l does not exist, we put $l = 0$.

THEOREM 1. *Let $N, n \to \infty$ in such a way that n takes values which are divided by d, $n/N \to 0$, and let $\lambda = \lambda(N, n)$ be determined by the relation*

$$\frac{\lambda F'(\lambda)}{F(\lambda)} = \frac{n}{N + n}. \tag{4}$$

Let

$$N\mathbf{P}\{\nu^{(1)} = r + 1\} \to \infty, \qquad N\mathbf{P}\{\nu^{(1)} = r + s + 1\} \to \gamma,$$

where γ is a non-negative constant and natural r and s are divided by d and satisfy one of the conditions

$$r \to \infty, \quad s = d,$$

or r is fixed,

$$r \geq j + l, \quad \mathbf{P}\{\nu^{(1)} = r + s + 1\} > 0, \quad \mathbf{P}\{\nu^{(1)} = r + i + 1\} = 0, \quad 0 < i < s.$$

Then

$$\mathbf{P}\{\eta = r + 1\} \to e^{-\gamma},$$
$$\mathbf{P}\{\eta = r + s + 1\} \to 1 - e^{-\gamma}.$$

THEOREM 2. *Let $N, n \to \infty$ in such a way that n takes values which are divided by d and $n/N \to b$, where b is a positive constant. Let $\lambda = \lambda(N, n)$ be given by (4) and $\alpha = (\lambda_b / F(\lambda_b))^d$, where λ_b is the solution of the equation*

$$\lambda F'(\lambda)/F(\lambda) = b/(b + 1).$$

If $r = r(N, n)$ running through the values which are divided by d is such that

$$\frac{N}{F(\lambda)} \left(\frac{\lambda}{F(\lambda)} \right)^r \frac{d}{r^{3/2}\sqrt{2\pi B}} \to \gamma,$$

where γ is a positive constant, then for any fixed $k = 0, 1, \ldots$

$$\mathbf{P}\{\eta \leq r + kd + 1\} \to \exp\{-\gamma\alpha^{k+1}(1 - \alpha)^{-1}\}.$$

THEOREM 3. *Let $N, n \to \infty$ in such a way that n takes values which are divided by d, $n/N \to \infty$, $n/N^2 \to 0$. Let $\lambda = \lambda(N, n)$ be defined by (4). Then*

$$\mathbf{P}\{\beta\eta - u \leq z\} \to e^{e^{-z}},$$

where

$$\beta = \beta(\lambda) = -\ln(\lambda/F(\lambda))$$

and $u = u(\lambda)$ is chosen so that

$$N\beta^{1/2}u^{-3/2}e^{-u} = \sqrt{2\pi B}.$$

THEOREM 4. *Let $N, n \to \infty$ in such a way that n takes values which are divided by d, $Bn/N^2 \to \gamma$, where γ is a positive constant. Then*

$$\mathbf{P}\{\eta/n \leq z\} \to \gamma^{3/2} \exp\{1/(2\gamma)\} \sum_{k=1}^{\infty} (-1)^k (k!)^{-1} I_k(\gamma z, \gamma),$$

where

$$I_0(u,v) = (v^3 \exp\{1/v\})^{-1/2},$$
$$I_k(u,v) = \int_{X_k(u,v)} \frac{\exp\{-1/(2(v - x_1 - \ldots - x_k))\}\, dx_1 \ldots dx_k}{(2\pi)^{k/2}(x_1 \ldots x_k(v - x_1 - \ldots - x_k))^{3/2}},$$
$$X_k(u,v) = \{x_i \geq u,\ i = 1, \ldots, k,\ x_1 + \ldots + x_k \leq v\}, \qquad k = 1, 2, \ldots$$

THEOREM 5. *Let $n \to \infty$ in such a way that n takes values which are divided by d, $n/N^2 \to \infty$. Then for any fixed positive z*

$$\mathbf{P}\left\{\frac{n-\eta}{N^2} B \leq z\right\} \to \frac{1}{\sqrt{2\pi}} \int_0^z y^{-3/2} \exp\{-1/(2y)\} dy.$$

We denote by μ_r the number of trees with r non-root vertices of a random forest from $F_{N,n}$. Let $\nu_{(r)}^{(1)}, \ldots, \nu_{(r)}^{(N)}$ be auxiliary independent identically distributed random variables such that

$$\mathbf{P}\{\nu_{(r)}^{(i)} = k\} = \mathbf{P}\{\nu^{(i)} = k \mid \nu^{(i)} \neq r+1\}, \quad k = 1, 2, \ldots$$

We also put

$$q_r(\lambda) = \mathbf{P}\{\nu^{(1)} = r+1\},$$
$$\zeta_{N-k}^{(r)} = \nu_{(r)}^{(1)} + \ldots + \nu_{(r)}^{(N-k)}.$$

LEMMA 3. *For any λ, $0 < \lambda \leq 1$, and n such that $\mathbf{P}\{\nu = N + n\} > 0$*

$$\mathbf{P}\{\mu_r = k\} = \binom{N}{k} q_r^k(\lambda)(1 - q_r(\lambda))^{N-k} \frac{\mathbf{P}\{\zeta_{N-k}^{(r)} = N + n - k(r+1)\}}{\mathbf{P}\{\nu = N + n\}}.$$

Let $\lambda = \lambda(N, n)$ be given by (4). The following assertions are proved by the use of Lemma 3.

THEOREM 6. *Let $N, n \to \infty$ in such a way that $N/n \to 0$, $N\lambda^{j+l} \to \infty$ and let n be divided by d. Then for $2 \leq r \neq j$*

$$\mathbf{P}\{\mu_r = k\} = (k!)^{-1}(Nq_r(\lambda))^k \exp\{-Nq_r(\lambda)\}(1 + o(1))$$

uniformly in the integers k such that $(k - Nq_r(\lambda))/\sqrt{Nq_r(\lambda)}$ lies in any finite interval.

Let m and B_λ denote the expectation and the variance of distribution (3). It can be easily seen that $a = \mathbf{E}\nu^{(1)} = 1/(1-m)$, $\sigma^2 = \mathbf{D}\nu^{(i)} = B_\lambda/(1-m)^3$. Let

$$\sigma_{rr}^2 = q_r(\lambda)(1 - q_r(\lambda) - \frac{(a - r - 1)^2}{\sigma^2} q_r(\lambda)).$$

THEOREM 7. *Let $N, n \to \infty$ in such a way that $n/N \geq c > 0$ and let n be divided by d. Then*

$$\mathbf{P}\{\mu_r = k\} = (\sigma_{rr}\sqrt{2\pi N})^{-1} e^{-u_r^2/2}(1 + o(1))$$

uniformly in the integers k such that $(k - Nq_r(\lambda))/(\sigma_{rr}\sqrt{N})$ lies in any finite interval.

We denote by s the least natural number such that $p_{j+s} > 0$, if such s does not exist, we put $s = 0$. Thus, s differs from l in such a way that s can be divided by j.

THEOREM 8. *Let $N, n \to \infty$ in such a way that $n/N \to 0$ and let n be divided by d and*

$$\min(N\lambda^r, N\lambda^{\omega(r)}) \to \infty,$$

where

$$\omega(0) = 2j + l, \quad \omega(1) = 3$$

for $j = 1$,

$$\omega(1) = \max(2j, j + l)$$

for $j > 1$,

$$\omega(r) = j + l$$

for $r \geq 2$, $r \neq j$, and

$$\omega(r) = \max\{\min(2j, j + s), j + l\}$$

for $r = j \geq 2$.

Then the assertion of Theorem 7 is valid.

THEOREM 9. *Let $N, n, r \to \infty$ in such a way that $N\lambda^j \to \infty$. Then*

$$\mathbf{P}\{\mu_r = k\} = (k!)^{-1}(Nq_r(\lambda))^k \exp\{-Nq_r(\lambda)\}(1 + o(1))$$

uniformly in the integers k such that $(k - Nq_r(\lambda))/\sqrt{Nq_r(\lambda)}$ lies in any finite interval.

We denote by τ the height of a random forest from $F_{N,n}$. Let $\mu(t)$ be the number of particles of the tth generations of the process G.

LEMMA 4. *The equality*

$$\mathbf{P}\{\tau < t\} = 1 - \mathbf{P}\{\mu(t) > 0\}\mathbf{P}\{\nu = N + n \mid \mu(t) > 0\}/\mathbf{P}\{\nu = N + n\}$$

holds.

We denote by $\mu^{(i)}(t)$, $i = 1, \ldots, N$, the numbers of particles in the processes $G^{(i)}$ at time t. Let $\nu^{(1)}(t), \ldots, \nu^{(N)}(t)$ be independent identically distributed random variables such that

$$\mathbf{P}\{\nu^{(i)}(t) = k\} = \mathbf{P}\{\nu^{(i)} = k \mid \mu^{(i)}(t) > 0\}, \quad k = 1, 2, \ldots$$

We set also $\zeta_N(t) = \nu^{(1)}(t) + \ldots + \nu^{(N)}(t)$.

LEMMA 5. *The equality*

$$\mathbf{P}\{\tau < t\} = \mathbf{P}\{\mu(t) = 0\}\mathbf{P}\{\zeta_N(t) = N + n\}/\mathbf{P}\{\nu = N + n\}$$

holds.

The following assertions are proved with the use of Lemmas 4 and 5.

THEOREM 10. *Let $N, n \to \infty$ in such a way that $n/N \to 0$, $n^2/N \to \infty$, $N\lambda^{j+l} \to \infty$ and let $t = t(N, n)$ be chosen such that $Nm^t \to \infty$, Nm^{t+1} are bounded. Then*

$$\mathbf{P}\{\tau = t\} = \exp\{-Nm^{t+1}\} + o(1), \qquad \mathbf{P}\{\tau = t + 1\} = 1 - \exp\{-Nm^{t+1}\} + o(1).$$

We denote by $F_\lambda(z)$ the generating function of distribution (3). Let for positive integers r the expression $F_\lambda^{*r}(z)$ mean the rth iteration of the function $F_\lambda(z)$, hence

$$F_\lambda^{*1}(z) = F_\lambda(z), \quad F_\lambda^{*2}(z) = F_\lambda(F_\lambda(z)), \quad F_\lambda^{*3}(z) = F_\lambda(F_\lambda(F_\lambda(z))), \ldots$$

THEOREM 11. *Let $N, n \to \infty$ in such a way that n takes values which are divided by d, $n/N \to b$ and let $t = t(N, n)$ be chosen such that $Nm^t \to \beta$, where b, β are positive constants. Then for any fixed k, $k = 0, 1, \ldots$*

$$\mathbf{P}\{\tau < t + k\} \to \exp\{-K\beta(b/(1 + b))^k\},$$

where

$$K = \lim_{r\to\infty} \frac{1 - F_\lambda^{*r}(0)}{m^r}.$$

THEOREM 12. *Let $N, n \to \infty$ in such a way that n takes values which are divided by d, $n/N \to \infty$, $n/N^2 \to 0$. Then for any fixed x*

$$\mathbf{P}\{\tau \ln(1 + N/n) - \ln(2N^2/(Bn)) \le x\} \to e^{-e^{-x}}.$$

THEOREM 13. *Let $N, n \to \infty$ in such a way that n takes values which are divided by d, $N/\sqrt{Bn} \to z$. Then for any fixed $x > 0$, uniformly with respect to z in any interval of the form $0 < z_0 \le z \le z_1 < \infty$*

$$\mathbf{P}\{(B/n)^{1/2}\tau \le x\} \to e^{z^2/2}(z\sqrt{2\pi})^{-1} \int_{-\infty}^{\infty} \exp\{-iu - zf(x, u)/x\}\, du,$$

where

$$f(x, u) = \frac{x\sqrt{-2iu}(1 + \exp\{-x\sqrt{-2iu}\})}{1 - \exp\{-x\sqrt{-2iu}\}}.$$

THEOREM 14. *Let $n \to \infty$ in such a way that n takes values which are divided by d, $n/N^2 \to \infty$. Then for any fixed $x > 0$*

$$\mathbf{P}\{(B/n)^{1/2}\tau \le x\} \to \sum_{k=-\infty}^{\infty} (1 - k^2x^2)\exp\{-k^2x^2/2\}.$$

Thus, to prove Theorems 1–14, we considered the asymptotic behaviour of the probabilities standing in the right-hand sides of the assertions of Lemmas 2–5. For this purpose we proved local limit theorems in the array scheme for sums of independent random variables including theorems on large deviations. The known sufficient conditions of the local convergence of such sums (see, e.g. [18]) do not cover all the domains of variation of N, n, r. Simple examples of such domains for the sum $\nu = \nu^{(1)} + \ldots + \nu^{(N)}$ are provided by the cases where $d = 1, n = O(N^2/\ln N)$ and $d > 1, n/N \to 0, n^2/N \to \infty$.

The proving of local limit theorems is the main technical difficulty for obtaining Theorems 1–14, and the proofs of these local limit theorems occupy the considerable part of the book [16].

REFERENCES

1. V. E. Britikov, The asymptotics for the number of forests of non-rooted trees. *Math. Notes* (1988) **43**, 672–684 (in Russian).
2. V. A. Vatutin, The distribution of the distance to the root of the minimal subtree containing all vertices of a given height. *Theory Probab. Appl.* (1993) **38**, 273–287 (in Russian).
3. N. B. Kalinina and Yu. L. Pavlov, The distribution of the degrees of vertices of a random forest. In: *Branching processes*. Karelian Branch Soviet Acad. Sci., Petrozavodsk, pp. 10–16 (in Russian).
4. I. B. Kalugin, Branching processes and random mappings of finite sets. *Math. Notes* (1983) **34**, 757–771 (in Russian).
5. V. F. Kolchin, B. A. Sevastyanov, and V. P. Chistyakov, *Random Allocations*. Wiley, New York, 1978.
6. V. F. Kolchin, Branching processes, random trees, and a generalized scheme of arrangements of particles. *Math. Notes* (1977) **21**, 386–394.
7. V. F. Kolchin, The moment of extinction of a branching process and the height of a random tree. *Math. Notes* (1978) **24**, 954–961.
8. V. F. Kolchin, *Random Mappings*. Springer, New York, 1986.
9. Yu. L. Pavlov, Limit distributions of the number of trees of a given size in a random forest. *Soviet Math. Sb.* (1977) **32**, 335–345.
10. Yu. L. Pavlov, The asymptotic distribution of the maximum tree size in a random forest. *Theory Probab. Appl* (1977) **22**, 509–520.
11. Yu. L. Pavlov, A case of limit distribution of the maximum size of a tree in a random forest. *Math. Notes* (1979) **25**, 387–392.
12. Yu. L. Pavlov, Limit distributions of the height of a random forest. *Theory Probab. Appl.* (1983) **28**, 471–480.
13. Yu. L. Pavlov, On the distributions of the number of vertices in strata of a random forest. *Proc. 1st World Congress Bernoulli Society*, Vol.1. VNU Science Press, Utrecht, 1987, 239–241.
14. Yu. L. Pavlov, Some properties of plane planted trees. In: *Discrete Math. and its Applications for Modelling of Complicated Systems. Abstracts.* Irkutsk State Univ., Irkutsk, p. 14 (in Russian).
15. Yu. L. Pavlov, Limit distributions of the height of a random forests consisting of plane rooted trees. *Discrete Math. Appl.* (1994) **4**, 73–88.
16. Yu. L. Pavlov, *Random Forests*. Karelian Sci. Centre of the Russian Acad. Sci., Petrozavodsk, 1996 (in Russian).
17. V. N. Zemlyachenko and Yu. L. Pavlov, Forests of plane planted trees and branching processes. In: *Appl. Math. Informatics* (1992) **1**, Petrozavodsk Univ. Press, Petrozavodsk, pp. 130–135 (in Russian).
18. B. V. Gnedenko and A. N. Kolmogorov, *Limit Distributions for Sums of Independent Random Variables*. Addison–Wesley, Reading, MA, 1954.

Probabilistic Methods in Discrete Mathematics, pp. 19–30
V. F. Kolchin et al. (Eds.)

ASYMPTOTIC PROPERTIES OF RANDOM INTERVAL GRAPHS AND THEIR USE IN CLUSTER ANALYSIS

E. GODEHARDT and B. HARRIS

Heinrich Heine Universität, Düsseldorf
University of Wisconsin, Madison

Abstract — Let $x_1, \dots, x_n$ be realizations of independent identically distributed random variables $X_1, \dots, X_n$ with a known distribution function. The objective is to divide $x_1, \dots, x_n$ into clusters $C_1, \dots, C_p$, where $C_i \cap C_j = \varnothing$, $i \neq j$. We choose a dissimilarity measure δ and a threshold d, and if $\delta(x_i, x_j) < d$, then we classify x_i and x_j into the same cluster, in such a case the interval graph with n vertices corresponding to $x_1, \dots, x_n$ has the edge (i, j).

We consider such characteristics of the random interval graph as the number of edges, the number of complete subgraphs of order m, the number of maximal subgraphs of order m, the number of isolated vertices, and the degrees of vertices.

With only rare exceptions, the exact distributions of these characteristics are complicated. Consequently, much of the paper is devoted to asymptotic analysis of these distributions.

This research was partially supported by the Deutsche Forschungsgemeinschaft–DFG Grants Go 490/4-2 and Go 490/4-3.

1. INTRODUCTION AND SUMMARY

Let $x_1, \dots, x_n$ be vectors in S, a subset of E_k. The objective is to divide these n vectors into classes $C_1, \dots, C_p$, where $C_i \cap C_j = \varnothing$, $i \neq j$, and $C_1 \cup \dots \cup C_p \subset S$. These classes are known as clusters. The experimenter seeks to place similar vectors in the same class and dissimilar vectors into different classes. To accomplish this, a dissimilarity measure δ and a threshold d are defined. In many applications, δ satisfies the mathematical postulates of a metric. In other instances, δ is so defined that it may violate the triangle inequality. Then, if $\delta(x, y) < d$, x and y are considered to be similar and are placed in the same cluster. We will assume that $x_1, \dots, x_n$ are realizations of the random variables $X_1, \dots, X_n$ respectively.

To progress further, we need to introduce the notion of a graph. A graph G_n with n vertices is defined as follows. $G_n = (V, E)$, where V is a set with $|V| = n$, and E is a set of unordered pairs of elements of V. Hence $0 \leq |E| \leq n(n-1)/2$. With no loss of generality, we can assume $V = \{1, 2, \dots, n\}$. The elements of V will be referred to as the vertices of the graph G_n and the elements of E will be called the edges of G_n. There is a natural correspondence between graph theory and cluster analysis. Graphs provide a mean of representing a binary relation and clusters are a collection of subsets of the data such that the elements in a given cluster are related to each other. Hence, the use of graph theoretical methods to study clustering processes is inherently reasonable. For an intensive discussion of this relationship, the reader is referred to the book by Bock [2]. The article by Arabie and Hubert in the book [1] is also relevant in this context. If the graphs are obtained from realizations of random variables, they are realizations of random graphs. There is an extensive literature on random graphs and the following two models for random graphs are frequently employed.

Bernoulli (or binomial) model. A sequence of $n(n-1)/2$ independent and identically distributed Bernoulli trials is conducted. If the ith Bernoulli trial is a success, then the ith pair of elements of V is placed in E. In this way, a random graph is constructed. Note that $|E|$ is a random variable. This model was introduced by E. N. Gilbert [5].

Hypergeometric model. M of $n(n-1)/2$ possible elements of E are chosen at random without replacement (see [4]).

Both of these models are not specifically intended for use with real valued data and in fact would violate the triangle inequality if relationships between real valued observations were given in terms of a metric. Therefore probability models for random interval graphs (or random coincidence graphs) are more suited for studying the homogeneity of real valued data (or vector valued data). In the sequel, we will restrict our attention to random interval graphs.

A fundamental concern in the use of clustering algorithms is that the computational technique finds similarities when only random effects are present. In the present paper, we will use graph theoretic methods to develop tests of the hypothesis that the clusters are chance occurrences against the hypothesis that the clusters are real (causal).

Accordingly, the null hypothesis is formulated as follows. We assume that the random variables $X_1, \ldots, X_n$ are independent, identically distributed by the distribution function $F_X(x)$, where $F_X(x)$ is assumed to be known. Then, with this assumption, the probability distribution functions of various characteristics of an induced interval graph can be calculated and utilized to see if the clustering obtained is computable with these probability distributions. In particular, we obtain asymptotic probability distributions for the number of edges, the number of triangles, and in general the number of complete subgraphs of order m; the number of isolated vertices, the number of maximal complete subgraphs of order m, and the degrees of the vertices.

With only rare exceptions, the exact distributions are complicated and cumbersome to utilize. Consequently, much of the paper is devoted to determining the asymptotic behavior of the probability distributions of these characteristics.

There is an extensive literature on clustering methods. The reader is referred to the books [2, 12, 8, 6] for extensive discussions of these methods. There is also a substantial number of related results in the combinatorial literature and in the statistical literature. W. Eberl and R. Hafner [3] investigate the distribution of the number of edges (which they refer to as coincidences) and in [7], these results are extended to multiple coincidences.

In these papers, they require a symmetry condition, which is not needed for the present methodology (however, in the present work their symmetry condition does hold for edges). Also, for the case of multiple coincidences, Hafner [7] assumes that the graph is connected and can be assumed to be a tree, whereas here complete subgraphs are considered. H. Maehara [11] studied similar questions, but restricted his investigation to the case of the uniform distribution on the circle. In [9, 10] goodness-of-fit statistics based on the number of edges are studied.

2. RANDOM INTERVAL GRAPHS

Let $X_1, \dots, X_n$ be independent, identically distributed random variables with common distribution function $F_X(x)$, where $F_X(x)$ is absolutely continuous with respect to Lebesgue measure. Let $f_X(x)$ denote the corresponding probability density function. We define a family of graphs $G_{d,n} = (V_n, E_{d,n})$ as follows. For $d > 0$ and $X_1 = x_1, \dots, X_n = x_n$, define the vertices of the graph as $V = \{1, 2, \dots, n\}$. The pair $(i, j; i < j)$ is in $E_{d,n}$, the edges of the graph, whenever $|x_i - x_j| < d$. Since $x_1, \dots, x_n$ are realizations of random variables, the various characteristics of the graphs, such as the number of edges are realizations of random variables and events, such as, the graph is connected, are realizations of random events. We now introduce the definitions and notation that will be employed in the sequel.

For $m \geq 1$, $K_{m,d}$ is a complete subgraph of order m, if, for the specified subset $V_m \subset V$, with $|V_m| = m$, all $\binom{n}{m}$ pairs of elements of V_m are in $E_{d,n}$. $K_{1,d}$ is a specified vertex. $K^*_{m,d}$ is a maximal complete subgraph of order m, if the vertices of V_m form a complete subgraph of order m and there is no subgraph of order $m + 1$ with $K^*_{m,d} \subset K_{m+1,d}$, for any subset V_{m+1} of V, with $V_m \subset V_{m+1}$. Thus $K^*_{1,d}$ is an isolated vertex.

To simplify notation, we will denote $G_{d,n}$ by G, V_n by V and $E_{d,n}$ by E, when there is no ambiguity concerning n or d. Similarly, we will drop the subscript X from the probability density function and the cumulative distribution function.

We now proceed to obtain the probability that a specified set of m vertices form $K_{m,d}$ or a $K^*_{m,d}$. With no loss of generality, we can assume that these vertices are $\{1, 2, \dots, m\}$.

THEOREM 1. *The probability that $X_1, \dots, X_m$, or equivalently, the vertices $\{1, 2, \dots, m\}$, form a $K_{m,d}$, $m = 1, 2, \dots, n$, is*

$$\mathbf{P}\{\max_{1 \leq i \leq m} X_i - \min_{1 \leq i \leq m} X_i \leq d\} = m \int_{-\infty}^{\infty} (F(x+d) - F(x))^{m-1} f(x)\, dx. \tag{1}$$

PROOF. Since m specified vertices form a $K_{m,d}$ whenever the greatest distance between any pair does not exceed d, the left hand side of (1) is equivalent to this condition. Assume x_1 is the minimum realized value. Then, given $X_1 = x_1$, we must have $X_2, \dots, X_m \leq x_1 + d$. Since there are m equally likely choices for the minimum, Theorem 1 is established.

Note that if $m = 1$, the right-hand side of (1) is unity.

THEOREM 2. *The probability $P_{m,d}$ that $X_1, \ldots, X_m$ form a $K^*_{m,d}$, $m = 2, \ldots, n$, is*

$$P_{m,d} = m(m-1)$$
$$\times \int_{-\infty}^{\infty} \int_{x_1}^{x_1+d} (F(x_2) - F(x_1))^{m-2}(1 - F(x_1+d) + F(x_2-d))^{n-m} f(x_1) f(x_2)\, dx_2\, dx_1. \quad (2)$$

PROOF. A specified subset of m vertices form $K^*_{m,d}$, when these m vertices form a $K_{m,d}$ and if y is any other vertex, either $y > \min_{1 \le i \le m} x_i + d$ or $y < \max_{1 \le i \le m} x_i - d$. Given $X_1 = x_1$ is the minimum and $X_2 = x_2$ is the maximum of $x_1, \ldots, x_m$ and $x_2 - x_1 < d$, the probability that the m vertices form a $K^*_{m,d}$ is

$$(F(x_2) - F(x_1))^{m-2}(1 - F(x_1+d) + f(x_2-d))^{n-m}.$$

The marginal probability is obtained by integrating over x_1 and x_2. Since there are $m(m-1)$ choices for specifying the minimum and maximum values, the conclusion follows.

THEOREM 3. *The probability $P_{1,d}$ that a specified vertex forms a $K^*_{1,d}$ (i.e., is isolated) is*

$$P_{1,d} = \int_{-\infty}^{\infty} (1 - (F(x+d) - F(x-d)))^{n-1} f(x)\, dx. \quad (3)$$

PROOF. Given $X_1 = x_1$, the integrand is the probability that the remaining observations are further from x_1 than d. Multiplying by $f(x)$ and integrating give the required marginal probability.

A vertex has degree ν, $\nu = 0, 1, \ldots, n-1$, if there are exactly ν edges incident with that vertex. We now establish the probability that a specified vertex is of degree ν.

THEOREM 4. *The probability P_ν that a specified vertex has degree ν, $\nu = 0, 1, \ldots, n-1$, is*

$$P_\nu = \int_{-\infty}^{\infty} \binom{n-1}{\nu} (F(x+d) - F(x-d))^\nu (1 - F(x+d) + F(x-d))^{n-\nu-1} f(x)\, dx. \quad (4)$$

PROOF. Vertex 1 has degree ν whenever ν elements from $x_2, \ldots, x_n$ are within d of x_1 and the remaining $n - \nu - 1$ are further from x_1 than d. There are $\binom{n-1}{\nu}$ choices for the vertices which are adjacent to vertex 1. Note that if $\nu = 0$, this coincides with Theorem 3.

Introducing indicator functions and using the inclusion–exclusion formula, we can exhibit the exact probability distributions for each of the random variables introduced. Also, we can then obtain the factorial moments for each of these. Accordingly, we define the following random variables. Let

$$V_{i_1,\dots,i_m} = \begin{cases} 1 & \text{if the vertices } i_1, \dots, i_m \text{ form a } K_{m,d},\ 1 \le i_1 \le \dots \le i_m \le n, \\ 0 & \text{otherwise,} \end{cases}$$

$$W_{i_1,\dots,i_m} = \begin{cases} 1 & \text{if the vertices } i_1, \dots, i_m \text{ form a } K^*_{m,d},\ 1 \le i_1 \le \dots \le i_m \le n, \\ 0 & \text{otherwise,} \end{cases}$$

$$Y_i(\nu) = \begin{cases} 1 & \text{if the vertex } i \text{ is of degree } \nu,\ 1 \le i \le n, \\ 0 & \text{otherwise.} \end{cases}$$

Let $L_{m,d}$ be the number of distinct subsets of m vertices which form a $K_{m,d}$ and let $L^*_{m,d}$ be the number of distinct subsets of m vertices which form a $K^*_{m,d}$, $m = 1, 2, \dots, n$, $d > 0$. Also, let $Y_{n,d}(\nu)$ be the number of vertices of degree ν, $\nu = 0, 1, \dots, n-1$.

Then

$$L_{m,d} = \sum V_{i_1,\dots,i_m}, \tag{5}$$

and

$$L^*_{m,d} = \sum W_{i_1,\dots,i_m}, \tag{6}$$

the sums running over all distinct subsets J_m of $\{1, 2, \dots, n\}$ with $|J_m| = m$. Similarly,

$$Y_{n,d}(\nu) = \sum_{i=1}^{n} Y_i(\nu), \qquad \nu = 0, 1, \dots, n-1. \tag{7}$$

Specifically:

$$L_{1,d} = |V| = n, \qquad L_{2,d} = |E|,$$

$L_{3,d}$ is the number of triangles in G, $L_{n,d} = 1$ if V_n is a complete graph and 0 otherwise, $L^*_{1,d}$ is the number of isolated vertices of G, $L^*_{n,d} = L_{n,d}$.

3. ASYMPTOTIC PROPERTIES

In this section, under the assumption $d \to 0$ and $n \to \infty$ (in some cases only the assumption $d \to 0$ is needed), we obtain asymptotic estimates for the probability of complete subgraphs of order m, of maximal complete subgraphs of order m and the probability P_ν that a given vertex is of degree ν.

To accomplish this and to subsequently obtain asymptotic probability distributions for the number of complete subgraphs of order m, the number of maximal complete subgraphs of order m and the number of vertices of degree ν, we introduce some preliminary lemmas.

LEMMA 1. *Let k and m be positive integers such that* $0 < m \leq n$, $n = 2, 3, \ldots$ *Then*

$$\frac{n^{mk}}{(m!)^k} e^{-m^2k/(n-m)} \leq \binom{n}{m}^k \leq \frac{n^{mk}}{(m!)^k} \tag{8}$$

and for $k \leq n/m$,

$$\frac{n^{mk}}{(m!)^k} e^{-mk(mk-1)/(n-mk+1)} \leq \frac{n!}{(n-mk)!\,(m!)^k} \leq \frac{n^{mk}}{(m!)^k}. \tag{9}$$

Thus,

$$\binom{n}{m}^k = \frac{n!}{(n-mk)!\,(m!)^k}(1+o(1), \tag{10}$$

for $k = o(\sqrt{n})$, *as* $n \to \infty$.

PROOF. Since $n(n-1)\ldots(n-m+1) \leq n^m$, the right-hand side of (8) is trivially established. To verify the left-hand side of (8), write

$$n(n-1)\ldots(n-m+1) = n^m \prod_{j=0}^{m-1}\left(1-\frac{j}{n}\right) \geq n^m \left(1-\frac{m}{n}\right)^m. \tag{11}$$

Since $\exp\{x/(1+x)\} < 1+x$ for $x \geq -1$, letting $x = -m/n$, the conclusion follows.

The right-hand side of (9) follows upon noting that

$$n(n-1)\ldots(n-mk+1) \leq n^{mk}.$$

To verify the left-hand side of (9), write

$$\begin{aligned}
\frac{n!}{(n-mk)!} &= n^{mk}\prod_{j=0}^{mk-1}\left(1-\frac{j}{n}\right) = n^{mk}\exp\left\{\sum_{j=0}^{mk-1}\log\left(1-\frac{j}{n}\right)\right\} \\
&= n^{mk}\exp\left\{-\sum_{j=0}^{mk-1}\sum_{r=1}^{\infty}\frac{j^r}{rn^r}\right\} \geq n^{mk}\exp\left\{-\sum_{j=0}^{mk-1}\sum_{r=1}^{\infty}\left(\frac{j}{n}\right)^r\right\} \\
&= n^{mk}\exp\left\{-\sum_{j=0}^{mk-1}\frac{j}{n-j}\right\} \geq n^{mk}\exp\left\{-\sum_{j=0}^{mk-1}\frac{j}{n-mk+1}\right\} \\
&= n^{mk}\exp\left\{-\frac{mk(mk-1)}{2(n-mk+1)}\right\},
\end{aligned}$$

establishing the lemma.

LEMMA 2. *Let $F_X(x)$ be absolutely continuous with respect to Lebesgue measure with probability density function $f_X(x)$. Let x be fixed and let d be a positive constant with $F_X(x+d) < 1$ and $F_X(x-d) > 0$. Then, if $f'(x)$ exists and is uniformly bounded on $(x-d, x+d)$,*

$$F_X(x+d) - F_X(x-d) = 2df_X(x) + R(d), \tag{12}$$

where

$$|R(d)| \le d^2 \sup_{x-d<y<x+d} f'_X(y). \tag{13}$$

Similarly, if $f''_X(x)$ exists and is uniformly bounded on $(x-d, x+d)$, then

$$F_X(x+d) - F_X(x-d) = 2df_X(x) + R_1(d), \tag{14}$$

where

$$|R_1(d)| \le \frac{d^3}{3} \sup_{x-d<y<x+d} f''_X(y). \tag{15}$$

In the identical manner,

$$F_X(x+d) - F_X(x) = df_X(x) + R_3(d), \tag{16}$$

where

$$|R_3(d)| \le \frac{d^2}{2} \sup_{x<y<x+d} f'_X(y). \tag{17}$$

PROOF. Expand $F_X(x+d)$ and $F_X(x-d)$ in a Taylor series about x, obtaining

$$F_X(x+d) - F_X(x) = df_X(x) + \frac{d^2}{2!}f'_X(x) + \frac{d^3}{3!}f''_X(x) + \ldots + \frac{d^k}{k!}f_X^{(k-1)}(x) + R_{k,d}(x) \tag{18}$$

and

$$F_X(x-d) = F_X(x) - df_X(x) + \frac{d^2}{2!}f'_X(x) - \ldots \pm \frac{d^k}{k!}f_X^{(k-1)}(x) + R'_{k,d}(x), \tag{19}$$

where

$$R_{k,d}(x) = \frac{d^{k+1}}{(k+1)!}f_X^{(k)}(\theta), \tag{20}$$

for some θ in $(x, x+d)$ and

$$R'_{k,d}(x) = \frac{d^{k+1}}{(k+1)!}f_X^{(k+1)}(\theta_1) \tag{21}$$

for some θ_1 in $(x-d, x)$. The conclusion follows upon collecting terms and bounding the corresponding remainders by taking the supremum over θ and θ_1.

LEMMA 3. *Let $F_X(x)$ be absolutely continuous with respect to Lebesgue measure with probability density function $f_X(x)$. Let x be fixed and let d be a positive constant with $F_X(x+d) < 1$ and $F_X(x-d) > 0$. Let y satisfy $x < y < x+d$. Then, if $f_X'(x)$ exists and is uniformly bounded on $(x, x+d)$,*

$$F_X(y) - F_X(x) = (y-x)f_X(x) + R_3(d), \tag{22}$$

where $R_3(d)$ is given in (17).

Similarly,

$$F_X(y-d) - F_X(x+d) = (y-x-2d)f_X(x) + R_1(d), \tag{23}$$

where $R_1(d)$ is given in (15).

PROOF. The proof of Lemma 3 is completely analogous to that of Lemma 2 and is omitted.

THEOREM 5. *Let $F_X(x)$ be a cumulative distribution function, absolutely continuous with respect to Lebesgue measure and let $f_X(x)$ be the corresponding probability density function. Let $f_X(x)$ be uniformly continuous on every compact set K and let $f_X''(x)$ exist and be uniformly bounded on K. Then the asymptotic probability, as $d \to 0$, that the vertices $\{1, 2, \ldots, m\}$ form a $K_{m,d}$ is*

$$md^{m-1} \int_{-\infty}^{\infty} f_X^m(x)\, dx. \tag{24}$$

The asymptotic probability, as $d \to 0$, $n \to \infty$, so that $nd \to 0$, that the vertices $\{1, 2, \ldots, m\}$ form a $K_{m,d}^*$ is

$$d^{m-1} \int_{-\infty}^{\infty} f_X^m(x)(m + (n-m)f_X(x)((m-1)d - 2m))\, dx. \tag{25}$$

The asymptotic probability, as $d \to 0$, $n \to \infty$, so that $nd \to 0$, that vertex 1 is isolated is

$$1 - 2nd \int_{-\infty}^{\infty} f_X^2(x)\, dx. \tag{26}$$

The asymptotic probability, as $d \to 0$, $n \to \infty$, so that $nd \to 0$, that vertex 1 has degree v is

$$\binom{n-1}{v}(2d)^v \int_{-\infty}^{\infty} (f_X(x))^{v+1}(1 - 2ndf_X(x))\, dx. \tag{27}$$

PROOF. Let $\varepsilon > 0$ be chosen. Then there exists an $M = M(\varepsilon)$ so that $\mathbf{P}\{|X| > M\} < \varepsilon$. Therefore, we can write

$$\begin{aligned}
&\left|\int_{-\infty}^{\infty} (F_X(x+d) - F_X(x))^{m-1} f_X(x)\,dx - d^{m-1}\int_{-\infty}^{\infty} f_X^m(x)dx\right| \\
&\le \left|\int_{|x|>M} (F_X(x+d) - F_X(x))^{m-1} f_X(x)dx - d^{m-1}\int_{|x|>M} f_X^m(x)\,dx\right| \\
&\quad + \left|\int_{|x|\le M} (F_X(x+d) - F_X(x))^{m-1} f_X(x)dx - d^{m-1}\int_{|x|\le M} f_X^m(x)\,dx\right| \\
&= 2^{m-1}\varepsilon + d^{m-1}\varepsilon^m + \left|\int_{|x|\le M} (F_X(x+d) - F_X(x))^{m-1} f_X(x)\,dx - d^{m-1}\int_{|x|\le M} f_X^m(x)\,dx\right| \\
&= 2^{m-1}\varepsilon + d^{m-1}\varepsilon^m + \left|\sum_{i=0}^{k-1}\int_{x_i}^{x_i+d} (df_X(x_i) + R_3(d,x_i))^{m-1} f_X(x)\,dx - d^{m-1}\int_{|x|\le M} f_X^m(x)\,dx\right|,
\end{aligned}$$

where $R_3(d,x)$ is given in (17). Expanding the second term of (28) in a binomial series, we observe that the first term of the series coincides with the last term and hence (28) may be written as

$$2^{m-1}\varepsilon + d^{m-1}\varepsilon^m + \sum_{i=0}^{k}\sum_{j=1}^{m-1}\int_{|x|\le M}\binom{m-1}{j}(df_X(x))^{m-1-j}(R_3(d,x_i))^j f_X(x)\,dx. \quad (28)$$

The second term of (28) can be estimated using

$$|R_3(d,x_i)| \le d^2 Q/2,$$

where Q is the uniform bound of $f_X'(x)$. Since ε is arbitrary, we have established that as $d \to 0$, the probability that the vertices $\{1, 2, \ldots, m\}$ form a $K_{m,d}$ is given by

$$md^{m-1}\int_{-\infty}^{\infty} f_X^m(x)\,dx(1 + O(d)).$$

This estimate can be improved using the alternative form (18) and adding the hypothesis that $f_X''(x)$ is uniformly bounded on every compact set.

The other conclusions of Theorem 5 are established in a similar manner and are not specifically provided here.

We will denote the rth factorial moment of the random variable X by $\mu_{[r]}(X)$ and the mean and variance of the random variable X by $\mu(X)$ and $\sigma^2(X)$ respectively (when there is little chance of confusion, by $\mu_{[r]}$, μ and σ^2 respectively).

4. ASYMPTOTIC POISSON DISTRIBUTIONS

THEOREM 6. *If $n^m d^{m-1} \to \tau > 0$ as $n \to \infty$ and $d \to 0$, then $L_{m,d}$ is asymptotically Poisson distributed with mean*

$$\lambda = \binom{n}{m} m d^{m-1} \int_{-\infty}^{\infty} f_X^m(x)\, dx.$$

PROOF. The kth factorial moment of $L_{m,d}$ is given by

$$\mu_{[k]}(K_{m,d}) = k!\, S_k.$$

Let A_1 be the event that the vertices $\{1, 2, \ldots, m\}$ form a $K_{m,d}$. By symmetry, for any subset of m distinct vertices $\{i_1, \ldots i_m\}$, $1 \le i_1 < \ldots < i_m \le n$, the probability that these vertices form a $K_{m,d}$ is $\mathbf{P}(A_1)$. If $\{i_{j1}, \ldots, i_{jm}\}, j = 1, \ldots, k, k = 1, 2, \ldots$ are k disjoint subsets of m distinct vertices, then the probability that each of these forms a $K_{m,d}$ is $(\mathbf{P}(A_1))^k$, since the vertices are realizations of independent random variables.

Now

$$S_k = \sum \mathbf{P}\left\{\bigcup A_{ij}\right\}, \tag{29}$$

where the sums runs over all distinct choices of k subsets of m vertices. We divide the terms in this sum into two sets, those with all km vertices distinct and those with at least one vertex repeated, that is

$$S_k = S_{k1} + S_{k2},$$

where the events in S_{k1} have all vertices distinct and the events in S_{k2} have at least one of the km vertices repeated in each term. The sum S_k has $\binom{n}{m}^k$ terms, of which $n!/((n-mk)!\,(m!)^k)$ are in S_{k1}. Then, from Lemma 1, for every fixed m and k as $n \to \infty$, we have

$$\binom{n}{m}^k = \frac{n!}{(n-km)!(m!)^k}(1 + O(1/n)) = \frac{n^{mk}}{(m!)^k}(1 + O(1/n)). \tag{30}$$

Then,

$$S_k = \frac{n^{mk}}{(m!)^k}(\mathbf{P}(A_1))^k(1 + O(1/n)) + S_{k2}.$$

It remains to show that $S_{k2} \to 0$ as $n \to \infty$ and $d \to 0$, so that $d^{m-1}n^m \to \tau$. The k-tuples of sets in S_{k2} have non-empty intersections with $km - j$ distinct elements, $j = 1, 2, \ldots, km - k$. From Lemma 1, it is easy to see that the number of selections of elements in S_{k2} with $km - j$ distinct elements is asymptotically given by $c_{jm}n^{km-j}$, for a specified constant $c_{jm} > 0$.

Let $\{i_1, \ldots, i_m\}$ and $\{k_1, \ldots, k_m\}$ be two subsets of m vertices with j elements in common. Then, the probability that both of these form a $K_{m,d}$ is

$$2\int_{-\infty}^{\infty}\int_{x}^{x+2d} (F_X(x+d) - F_X(x))^{m-j-1}(F_X(y) - F_X(y-d))^{m-j-1} \times (F_X(x+d) - F_X(y-d))^j f_X(x) f_X(y)\, dx\, dy = c' d^{2m-j-1}, \tag{31}$$

for some positive constant c'.

This can be readily extended to k subsets of the vertices, where the subsets are partitioned into subcollections according to the number of vertices in common. However, as in (32) above, the resulting integral has magnitude $O(d^{km-j-1})$ and hence each such collection of terms in S_{k2} has probability tending to zero, whenever $k > 1$ and $j < m$. Since there are a finite number of such terms, the conclusion follows.

THEOREM 7. *If $n \to \infty$ and $d \to 0$ so that $n^{v+1}d^v \to \tau_1 > 0$, then $Y_{n,d}(v)$ is asymptotically Poisson distributed with mean*

$$n^{v+1}d^v2^v \int_{-\infty}^{\infty} (f_X(x))^{v+1}dx.$$

PROOF. Let A_i be the event that vertex i has degree v, $i = 1, 2, \dots, n$. We proceed by calculating the factorial moments,using the inclusion–exclusion formula. Let $A_{i_1,\dots,i_k}(v_1, \dots, v_k)$ be the event that vertex i_s has degree v_s, $s = 1, \dots, k$. To evaluate S_k, we need to consider two separate cases. We let $S_k = S_{k1} + S_{k2}$, where S_{k1} contains the terms corresponding to the case where the k-tuples of vertices in those term in S_k are not adjacent. The terms in S_{k2} contain at least one pair of adjacent vertices.

If the vertices $i_1, \dots, i_k$ contain no adjacent pairs, then

$$\mathbf{P}\{A_{i_1}, \dots, A_{i_k}\} = \prod_{s=1}^{k} \mathbf{P}\{A_{i_s}(v_{i_s})\}. \tag{32}$$

If $v_1 = \dots = v_k = v$, then this is a term in S_{k1} and since the vertices are realizations of independent identically distributed random variables, (33) reduces to $(\mathbf{P}\{A_1\})^k$. There are

$$\frac{(n-k)!}{(v!)^k(n-kv)!}\binom{n}{k}$$

such terms. Thus, the magnitude of the terms in S_{k1} is of the form $cn^{kv+k}d^{kv}$. It remains to show that the terms in S_{k2} tend to zero under the hypotheses. Select k vertices. Assume that of the vertices to which they are adjacent, r of the k selected vertices have common adjacencies, but that none of the k selected vertices are adjacent to each other. Let i_j be the number of vertices adjacent to exactly j of the k selected vertices. Then each of these contributes j to the total degree of the k vertices, but should only be counted once in the selection of the vertices to which the k specified vertices are adjacent, and result in an overcount by $j - 1$ in each instance. Hence the contribution of such terms has magnitude

$$c_1 n^{kv+k-\sum_{j=1}^{k} i_j(j-1)} d^{kv-\sum_{j=1}^{k} i_j(j-1)},$$

which tends to zero as $n \to \infty$ and $d \to 0$ with $n^{v+1}d^v \to \tau_1 > 0$. If in addition, of the k selected vertices, r_{ij}, $1 \le i < j \le k$, is the number of adjacencies of the ith vertex to the jth vertex, then it follows that the exponent of n is reduced by twice the number of such adjacencies and the exponent of d is reduced by the number of such adjacencies; hence such terms also tend to zero.

THEOREM 8. *If $n \to \infty$ and $d \to 0$ so that $n^m d^{m-1} \to \tau > 0$, then $L^*_{m,d}$ has an asymptotic Poisson distribution with expected value*

$$\binom{n}{m} m d^{m-1} \int_{-\infty}^{\infty} (f_X(x))^m dx, \qquad m = 2, 3, \ldots$$

If $n \to \infty$ and $d \to 0$ so that $dn/\ln n \to \tau' > 0$, then the number of isolated vertices $K^*_{1,d}$ has an asymptotic Poisson distribution with expected value

$$n \int_{-\infty}^{\infty} e^{-2ndf_X(x)} f_X(x)\, dx.$$

The proofs are completely analogous to those of the two preceding theorems and the details are omitted.

5. CONCLUDING REMARKS

The purpose of this investigation was to obtain some results which would be useful in developing statistical tests to determine if clusters are real or simply chance occurrences. The distributions most likely to be assumed by experimenters are the uniform, exponential and the normal distributions, all of which satisfy all the necessary assumptions. The specific choice of a criterion to use in testing would depend on the alternatives. Some initial ideas on this have been advanced, but substantially more work is needed. This work is continuing and additional results should be reported soon.

REFERENCES

1. P. Arabie and L. J. Hubert, An overview of combinatorial data analysis. In: *Clustering and Classification.* World Scientific, Singapore, 1996.
2. H. H. Bock, *Automatische Klassifikation.* Vandenhoeck and Ruprecht, Goettingen, 1974.
3. W. Eberl and R. Hafner, Die asymptotische Verteilung von Koinzidenzen. *Z. Wahr. verw. Geb.* (1971) **18**, 322–332.
4. P. Erdös and A. Rényi, On the evolution of random graphs. *Publ. Math. Inst. Hungarian Acad. Sci.* (1960) **5**, 17–61.
5. E. N. Gilbert, Random graphs. *Ann. Math. Statist.* (1959) **30**, 1141–1144.
6. E. Godehardt, *Graphs ans Structural Models.* Vieweg, Braunschweig, 1990.
7. R. Hafner, Die asymptotische Verteilung von mehrfachen Koinzidenzen. *Z. Wahr. verw. Geb.* (1972) **24**, 96–108.
8. J. A. Hartigan, *Clustering Algorithms.* Wiley, New York, 1975.
9. S. Rao Jammalamadaka and S. Janson, Limit theorems for a triangular scheme of U-statistics with applications to inter-point distances. *Ann. Probab.* (1986) **14**, 1347–1358.
10. S. Rao Jammalamadaka and X. Zhou, Some goodness-of-fit tests in higher dimensions based on interpoint distances. In: *Proc. Bose Sympos. Probab. Statist. Design Experiments.* Wiley Eastern, New Delhi, 1990.
11. H. Maehara, On the intersection graph of random arcs on a circle. In: *Random Graphs.* Wiley, Chichester, 1990.
12. J. van Ryzin, *Classification and Clustering.* Academic Press, New York, 1977.

Probabilistic Methods in Discrete Mathematics, pp. 31–44
V. F. Kolchin et al. (Eds.)

ON NODES OF GIVEN DEGREE IN RANDOM TREES

M. DRMOTA

Technical University of Vienna

Abstract — Let T be a plane rooted tree with n nodes which is regarded as the family tree of a Galton–Watson branching process conditioned on the total progeny. We discuss the process $L_n^{(d)}(t)$ which is the number of nodes of degree d in layer t. It is shown that the process $n^{-1/2}L_n^{(d)}(n^{1/2}t)$ converges weakly to the Brownian excursion local time. This is done via characteristic functions which are obtained by means of generating functions arising from the combinatorial setup and complex contour integration.

This research was supported by the Austrian Science Foundation, grant P10187-PHY.

1. INTRODUCTION

We consider a class $\mathcal{A}$ of plane rooted trees. For each $T \in \mathcal{A}$, we define the size $|T|$ as the number of nodes T consists of and the weight

$$\omega(T) = \prod_{k\geq 0} \varphi_k^{n_k(T)}, \tag{1.1}$$

where $(\varphi_k, k \geq 0)$ are non-negative numbers and $n_k(T)$ is the number of nodes $v \in T$ with out-degree k. Furthermore, we set

$$a_n = \sum_{T:\ |T|=n} \omega(T).$$

Then the corresponding generating function (GF)

$$a(z) = \sum_{n\geq 0} a_n z^n$$

satisfies the functional equation

$$a(z) = z\varphi(a(z)), \tag{1.2}$$

where

$$\varphi(t) = \sum_{k\geq 0} \varphi_k t^k.$$

According to Meir and Moon [15] we will call such a family of trees simply generated. Now equip the sets $\mathcal{A}_n = \{T \in \mathcal{A} : |T| = n\}$ with the probability distribution induced by the weight function $\omega(T)$. Then we call each tree $T \in \mathcal{A}$ a random tree.

As Kolchin [13, 14] pointed out, there is a natural correspondence between random trees and Galton–Watson branching processes (see also [2, 16–20]). Recently Aldous [2], Takacs [18], and Vatutin [19, 20] suggested to study the simply generated random trees with the use of approximations of various processes related to Galton–Watson branching processes by Brownian processes.

Let X be a branching process with the distribution of offspring ξ determined by

$$\mathbf{P}\{\xi = k\} = \frac{\tau^k \varphi_k}{\varphi(\tau)}, \qquad k = 0, 1, \ldots,$$

where τ is an arbitrary non-negative number within the circle of convergence of $\varphi(t)$. Then the distribution of X conditioned on the total progeny $|X|$ is determined by $\mathbf{P}\{X = T \| X| = n\}$ and it is easily seen that this distribution coincides with that induced by (1.1). Furthermore it is obvious that there occurs no loss of generality if only critical branching processes are considered (compare with [12]). The condition for a branching process to be critical, $\mathbf{E}\xi = 1$, translated into the language of trees is $\tau\varphi'(\tau) = \varphi(\tau)$ and the variance of ξ is given by

$$\sigma^2 = \frac{\tau^2 \varphi''(\tau)}{\varphi(\tau)}. \tag{1.3}$$

Consider a simply generated tree $T \in \mathcal{A}_n$. We denote by $L_T^{(d)}(k)$ the number of nodes of degree d at distance k from the root (where the distance of two nodes v and w is defined as usual by the number of edges of the path connecting v and w). If T is a random tree then $L_T^{(d)}(k)$ becomes a random variable denoted by $L_n^{(d)}(k)$. For non-integer k we define $L_n^{(d)}(k)$ by linear interpolation:

$$L_n^{(d)}(t) = (\lfloor t \rfloor + 1 - t) L_n^{(d)}(\lfloor t \rfloor) + (t - \lfloor t \rfloor) L_n^{(d)}(\lfloor t \rfloor + 1), \quad t \geq 0.$$

We will show that the scaled process

$$l_n^{(d)}(t) = \frac{1}{\sqrt{n}} L_n^{(d)}(t\sqrt{n}), \quad t \geq 0,$$

weakly converges to the Brownian excursion local time as n tends to infinity.

THEOREM 1. *Let $\varphi(t)$ be the GF of a family of random trees. Besides, let $W(t)$ denote the Brownian excursion of duration* 1 *and $l(t)$ its local time, i.e.,*

$$l(t) = \lim_{\varepsilon \to 0} \frac{1}{\varepsilon} \int_0^1 I_{[t,t+\varepsilon]}(W(s))\, ds$$

Assume that $\varphi(t)$ has a positive or infinite radius of convergence R and the greatest common divisor $g = \gcd\{i \mid \varphi_i > 0\} = 1$. Furthermore suppose that the equation

$$t\varphi'(t) = \varphi(t)$$

has a minimal positive solution $\tau < R$ and that σ^2 defined by (1.3) is finite.

Then the process $l_n^{(d)}(t)$ converges weakly to the Brownian excursion local time, exactly that means

$$l_n^{(d)}(t) \xrightarrow{w} \frac{c_d \sigma}{2} l(\sigma t/2)$$

in $C[0, \infty)$, *as* $n \to \infty$, *in which* $c_d = \varphi_{d-1} \tau^{d-1} / \varphi(\tau)$.

REMARK 1. It is well known that the expected value of the number of nodes of degree d in trees of size n is approximately $c_d n$. This property is reflected by the processes $l_n^{(d)}(t)$. The author and B. Gittenberger showed that the process

$$l_n(t) = \frac{1}{\sqrt{n}} L_n(t\sqrt{n})$$

converges to $\sigma l(\sigma t/2)/2$, where $L_n(k)$ denotes the total number of nodes at distance k to the root (in trees of size n).

REMARK 2. The case $g > 1$ can be treated analogously. All limit theorems throughout this paper remain unchanged except that we have to require $n \equiv 1 \mod d$. Thus we may restrict ourselves to $g = 1$.

REMARK 3. It is also possible to consider the step function process

$$\frac{1}{\sqrt{n}} L_n^{(d)}(\lfloor t\sqrt{n} \rfloor).$$

The reason that we decided to work with a linear interpolated process instead of a step function process is that the proof of tightness (Section 4) is essentially shorter for the first one since all trajectories of the process are continuous functions in $C[0, \infty)$. More precisely, there is a similar tightness condition for the space $D[0, \infty)$ in which step functions are allowed (see [5]) and in fact, by using direct (but messy) extensions of the method presented in Section 4, we are also able to prove the same assertion for the step function process.

Since the distribution of $\sup_{t \geq 0} l(t)$ is the same as that of $2 \sup_{0 \leq t \leq 1} W(t)$ (see [4] or [2]), which is indeed the beta distribution, Theorem 1 immediately implies the following property for the maximal width of trees.

COROLLARY 1. *Under the assumption of Theorem 1, we have*

$$\sup_{t \geq 0} l_n^{(d)}(t) \xrightarrow{w} c_d \sigma \sup_{0 \leq t \leq 1} W(t)$$

as $n \to \infty$.

2. PLAN OF THE PROOF

In order to prove Theorem 1, we use a proper variant of Prohorov's theorem [5, Theorem 12.3]. Thus we have to show that the finite dimensional distributions of $l_n^{(d)}(t)$ converges weakly to those of the Brownian excursion local time and that these sequences are tight.

We consider a random tree $T \in \mathcal{A}_n$ and set

$$a_{kmn}^{(d)} = \sum_{T \in \mathcal{A}_n, L_T^{(d)}(k)=m} \omega(T).$$

Thus the distribution of $L_n^{(d)}(k)$ is given by

$$\mathbf{P}\left\{L_n^{(d)}(k) = m \mid T \in \mathcal{A}_n\right\} = \frac{a_{kmn}^{(d)}}{a_n}.$$

In order to obtain this distribution, we use the immediate translation technique of combinatorial constructions into GFs which is widely used in combinatorial enumeration (for a description see e.g. [21]). If we denote vertices by $\circ$, then $\mathcal{A}$ may be described by the symbolic recursion

$$\mathcal{A} = \Psi(\mathcal{A}),$$

where the operator Ψ is defined by

$$\Psi(X) = \{\circ\} \times (\varphi_0 \cup \varphi_1 X \cup \varphi_2 X \times X \cup \ldots)$$

As $\cup$ and $\times$ correspond to sum and product, we immediately get the functional equation (1.2).

Now let us mark all nodes of degree d in layer k and denote the marked nodes by $\bullet$. We call the tree family obtained from $\mathcal{A}$ in that way by $\mathcal{B}^{(d)}$ and its GF by $a_k^{(d)}(z, u)$, i.e.,

$$a_k^{(d)}(z, u) = \sum_{n,m \ge 0} a_{kmn}^{(d)} u^m z^n.$$

Then it is easy to see that

$$\begin{aligned}\mathcal{B}^{(d)} = \Psi^k(\varphi_0\{\circ\} \cup \varphi_1\{\circ\} \times \mathcal{A} \cup \ldots \cup \varphi_{d-2}\{\circ\} \times \mathcal{A}^{d-2} \\ \cup \varphi_{d-1}\{\bullet\} \times \mathcal{A}^{d-1} \cup \varphi_d\{\circ\} \times \mathcal{A}^d \cup \ldots)\end{aligned}$$

holds (compare also with [9]). Due to the correspondence

$$\circ \leftrightarrow z,$$
$$\bullet \leftrightarrow uz.$$

we immediately get

$$\begin{aligned} a_k^{(d)}(z, u) &= y_k(z, z\varphi(a(z)) + (u-1)z\varphi_{d-1}a(z)^{d-1}) \\ &= y_k(z, (u-1)z\varphi_{d-1}a(z)^{d-1} + a(z)) \end{aligned}$$

where

$$y_0(z,u) = u,$$
$$y_{i+1}(z,u) = z\varphi(y_i(z,u)), \quad i \geq 0. \tag{2.1}$$

Thus the distribution of $L_n^{(d)}(k)$ can be obtained by extracting the coefficient of $z^n u^m$ of $a_k^{(d)}(z,u)$ which we denote by $[z^n u^m]a_k^{(d)}(z,u)$. In order to prove Theorem 1, we need only weak limit theorems and thus it suffices to work with characteristic functions. The characteristic function of $L_n^{(d)}(k)/\sqrt{n}$ is

$$\phi_{kn}^{(d)}(t) = \frac{1}{a_n}[z^n]a_k^{(d)}(z, e^{it/\sqrt{n}}) = \frac{1}{a_n}[z^n]y_k(z, z(e^{it/\sqrt{n}} - 1)\varphi_{d-1}a(z)^{d-1} + a(z)).$$

In a similar way we get the characteristic function for the higher dimensional distributions. The characteristic function of

$$\left(\frac{1}{\sqrt{n}}L_n^{(d)}(k_1), \frac{1}{\sqrt{n}}L_n^{(d)}(k_2)\right),$$

for $k_1 < k_2$, is given by

$$\begin{aligned}
&\phi_{k_1k_2n}^{(d)}(t_1,t_2) \\
&\quad = \frac{1}{a_n}[z^n]y_{k_1}(z, z(e^{it_1/\sqrt{n}} - 1)\varphi_{d-1}y_{k_2-k_1-1}(z, z(e^{it_2/\sqrt{n}} - 1)\varphi_{d-1}a(z)^{d-1} + a(z))^{d-1} \\
&\qquad\qquad + y_{k_2-k_1}(z, z(e^{it_2/\sqrt{n}} - 1)\varphi_{d-1}a(z)^{d-1} + a(z))).
\end{aligned}$$

and that of

$$\left(\frac{1}{\sqrt{n}}L_n^{(d)}(k_1), \dots, \frac{1}{\sqrt{n}}L_n^{(d)}(k_p)\right),$$

for $k_1 < k_2 < \dots < k_p$, by

$$\begin{aligned}
\phi_{k_1\dots k_pn}^{(d)}(t_1,\dots,t_p) &= \frac{1}{a_n}[z^n]y_{k_1}\left(z, z\left(e^{it_1/\sqrt{n}} - 1\right)\varphi_{d-1}\right. \\
&\times y_{k_2-k_1-1}\left(z, \dots + y_{k_p-k_{p-1}}\left(z, z\left(e^{it_p/\sqrt{n}} - 1\right)\varphi_{d-1}a(z)^{d-1} + a(z)\right)\dots\right)^{d-1} \\
&+ y_{k_2-k_1}\left(z, \dots + y_{k_p-k_{p-1}}\left(z, z\left(e^{it_p/\sqrt{n}} - 1\right)\varphi_{d-1}a(z)^{d-1} + a(z)\right)\right)\dots\left.\right).
\end{aligned}$$

In order to extract the desired coefficient asymptotically, we use Cauchy's integral formula with a suitably chosen integration contour and approximate the integrand there. Therefore we need a detailed knowledge of the behaviour of the recursion (2.1) (see [7, 8]).

LEMMA 1. *Let z_0 be the point on the circle of convergence of $a(z)$ which lies on the positive real axis. Set $z = z_0(1 + x/n)$ and $\alpha = z\varphi'(a(z))$. Furthermore assume that $|u - a(z)| = O(1/\sqrt{n})$ and $x/n \to 0$ in such a way that $|\arg(-x)| < \pi$ and*

$$|1 - \sqrt{-x/n}| \leq 1 + \frac{C}{\sqrt{n}}$$

are satisfied. Then $y_k(z,u)$ *admits the local representation*

$$y_k(z,u) = a(z) + \frac{2(u-a(z))\alpha^k\sqrt{-x/n}}{\sqrt{-x/n} + \sigma(\tau-u)/(\tau\sqrt{2}) + (\sqrt{-x/n} - \sigma(\tau-u)/(\tau\sqrt{2}))\alpha^k + O(|x|/n)}$$

uniformly for $k = O(\sqrt{n})$.

REMARK 4. As $a(z)$ has only non-negative coefficients and $g = \gcd\{i \mid \varphi_i > 0\} = 1$, the function $a(z)$ has only one singularity on the circle of convergence which is located at z_0. The function $y_k(z,u)$ is closely related to $a(z)$ and hence it has also exactly one singularity near z_0 the nature of which is described by the above result.

By means of this lemma we will be able to derive the characteristic function of the limiting distribution (see Section 3).

THEOREM 2. *Let* $k_i = \kappa_i\sqrt{n}, i = 1, \ldots, k$ *where* $0 < \kappa_1 < \ldots < \kappa_p$. *Then the characteristic function*

$$\phi_{\kappa_1\ldots\kappa_p}(t_1,\ldots,t_p) = \lim_{n\to\infty}\phi_{k_1\ldots k_p n}(t_1,\ldots,t_p)$$

of the limiting distribution of

$$\left(\frac{1}{\sqrt{n}}L_n^{(d)}(k_1), \ldots, \frac{1}{\sqrt{n}}L_n^{(d)}(k_p)\right)$$

satisfies

$$\phi_{\kappa_1\ldots\kappa_p}(t_1,\ldots,t_p) = 1 + \frac{\sigma}{i\sqrt{2\pi}}\int_\gamma f_{\kappa_1,\ldots,\kappa_p}(x, c_d t_1, \ldots, c_d t_p)e^{-x}dx, \tag{2.2}$$

where

$$f_{\kappa_1,\ldots,\kappa_p}(x,t_1,\ldots,t_p) = \Psi_{\kappa_1}(x,t_1,\Psi_{\kappa_2-\kappa_1}(\ldots\Psi_{\kappa_{k-1}-\kappa_{k-2}}(x,t_{k-1},\Psi_{\kappa_p-\kappa_{k-1}}(x,t_p,0))\ldots)$$

with

$$\Psi_\kappa(x,t,y) = \frac{\sqrt{-x}e^{-\kappa\sigma\sqrt{-x/2}}(it+y)}{\sqrt{-x}e^{\kappa\sigma\sqrt{-x/2}} - (it+y)\sigma\sinh(\kappa\sigma\sqrt{-x/2})/\sqrt{2}},$$

and $\gamma = \gamma_1 \cup \gamma_2 \cup \gamma_3$ *defined by*

$$\begin{aligned}\gamma_1 &= \{s \mid |s| = 1 \text{ and } \Re s \le 0\},\\ \gamma_2 &= \{s \mid \Im s = 1 \text{ and } \Re s \ge 0\},\\ \gamma_3 &= \overline{\gamma}_2\end{aligned} \tag{2.3}$$

is a Hankel contour.

REMARK 5. The name Hankel contour, Hankel integral, etc. originates from Hankel's representation of the Gamma function,

$$\frac{1}{2\pi i}\int_{\gamma}(-s)^{-\alpha}e^{-s}\,ds = \frac{1}{\Gamma(\alpha)},$$

and have become usual due to the quite frequent occurrence of integration contours similar to γ in asymptotical problems in combinatorics.

REMARK 6. The author and B. Gittenberger showed that $\phi_{\kappa_1\ldots\kappa_p}(t_1,\ldots,t_p)$ (given in (2.2)) with $\sigma = 2$ and $c_d = 1$ is exactly the finite dimensional characteristic function of the Brownian excursion local time.

REMARK 7. Note that by using the method presented in this section the above theorem can only be established for the step function process $L_n(\lfloor t\sqrt{n}\rfloor)/\sqrt{n}$. However, in Section 4 we prove the inequality

$$\mathbf{P}\left\{\left|L_n(\rho\sqrt{n}) - L_n((\rho+\theta)\sqrt{n})\right| \ge \varepsilon\sqrt{n}\right\} \le C\theta^2/\varepsilon^4$$

for some $C > 0$. This implies

$$\mathbf{P}\left\{\left|L_n^{(d)}(t) - L_n^{(d)}\left(\lfloor\rho\sqrt{n}\rfloor\right)\right| \ge \varepsilon\sqrt{n}\right\} \le \frac{C}{\varepsilon^{\beta}n} \to 0,$$

as $n \to \infty$. Therefore it suffices to prove the theorem for the step function process.

We now turn to the problem of tightness. By Theorem 12.3 in [5] the tightness of

$$l_n^{(d)}(t) = n^{-1/2}L_n^{(d)}(t\sqrt{n}), \quad 0 \le t \le A,$$

follows from the tightness of $L_n^{(d)}(0)$ (which is obviously satisfied) and from an estimate of the form

$$\mathbf{P}\{|L_n^{(d)}(\rho\sqrt{n}) - L_n^{(d)}((\rho+\theta)\sqrt{n})| \ge \varepsilon\sqrt{n}\} \le C\theta^{\alpha}/\varepsilon^{\beta} \tag{2.4}$$

for some $\alpha > 1$, $\beta \ge 0$, and $C > 0$ uniformly for $0 \le \rho \le \rho + \theta \le A$. We will derive (2.4) from the following property.

THEOREM 3. *There exists a constant $C > 0$ such that*

$$\mathbf{E}(L_n^{(d)}(k) - L_n^{(d)}(k+h))^4 \le Ch^2 n \tag{2.5}$$

holds for all non-negative integers n, k, h.

Obviously Theorem 3 proves (2.4) for $\alpha = 2$ and $\beta = 4$ if $\rho\sqrt{n}$ and $\theta\sqrt{n}$ are non-negative integers. However, it is an easy exercise (see [11]) that (in the case of linear interpolation) an estimate of type (2.4) can be extended to arbitrary $\rho, \theta \geq 0$ (probably with a different constant C).

Therefore the tightness follows from Theorem 3. Since

$$\begin{aligned}\mathbf{P}\left\{L_n^{(d)}(k) = m, L_n^{(d)}(k+h) = l\right\} \\ = \frac{1}{a_n}[z^n u^m v^l] y_k(z, z(u-1)\varphi_{d-1} y_{h-1}(z, z(v-1)\varphi_{d-1} a(z)^{d-1} + a(z)) \\ + y_h(z, z(v-1)\varphi_{d-1} a(z)^{d-1} + a(z))),\end{aligned}$$

we obtain

$$\begin{aligned}\mathbf{P}\{L_n^{(d)}(k) - L_n^{(d)}(k+h) = m\} \\ = \frac{1}{a_n}[z^n u^m] y_k(z, z(u-1)\varphi_{d-1} y_h(z, z(u^{-1}-1)\varphi_{d-1} a(z)^{d-1} + a(z)) \\ + y_h(z, z(u^{-1}-1)\varphi_{d-1} a(z)^{d-1} + a(z)))\end{aligned}$$

and consequently,

$$\mathbf{E}(L_n^{(d)}(k) - L_n^{(d)}(k+h))^4 = \frac{1}{a_n}[z^n] H_{kh}(z), \tag{2.6}$$

in which

$$\begin{aligned}H_{kh}(z) = \left(\frac{\partial}{\partial u} + 7\frac{\partial^2}{\partial u^2} + 6\frac{\partial^3}{\partial u^3} + \frac{\partial^4}{\partial u^4}\right) \\ (y_k(z, z(u-1)\varphi_{d-1} y_{h-1}(z, z(u^{-1}-1)\varphi_{d-1} a(z)^{d-1} + a(z)) \\ + y_h(z, z(u^{-1}-1)\varphi_{d-1} a(z)^{d-1} + a(z))))\Big|_{u=1}.\end{aligned}$$

Hence, $H_{kh}(z)$ is more or less explicit (see Section 4) and the transfer lemma by Flajolet and Odlyzko [10] can be applied in order to prove Theorem 3.

3. FINITE DIMENSIONAL CONVERGENCE

We will indicate how to prove Theorem 2 in the one and two-dimensional case. It will be then clear how the general case can be treated.

First, the characteristic function $\phi_{k,n}^{(d)}(t)$ of $L_n^{(d)}(k)/\sqrt{n}$ is given by

$$\begin{aligned}\phi_{k,n}^{(d)}(t) = \frac{1}{a_n}[z^n] y_k(z, z(e^{is/\sqrt{n}} - 1)\varphi_{d-1} a(z)^{d-1} + a(z)) \\ = \frac{1}{a_n}\int_\Gamma y_k\left(z, z(e^{is/\sqrt{n}} - 1)\varphi_{d-1} a(z)^{d-1} + a(z)\right)\frac{dz}{z^{n+1}},\end{aligned}$$

where $\Gamma = \Gamma_1 \cup \Gamma_2 \cup \Gamma_3 \cup \Gamma_4$ is given by

$$\begin{aligned}
\Gamma_1 &= \{z = z_0(1 + x/n) \mid \Re x \le 0,\ |x| = 1\}, \qquad (3.1)\\
\Gamma_2 &= \{z = z_0(1 + x/n) \mid \Im x = 1,\ 0 \le \Re x \le \log^2 n\},\\
\Gamma_3 &= \bar{\Gamma}_2,\\
\Gamma_4 &= \left\{z \mid |z| = z_0 \left|1 + (\log^2 n + i)/n\right|,\ \arg\left(1 + (\log^2 n + i)/n\right) \le |\arg(z)| \le \pi\right\}.
\end{aligned}$$

The integral over Γ_4 is asymptotically negligible. Therefore we concentrate on $\gamma' = \Gamma_1 \cup \Gamma_2 \cup \Gamma_3$ which is the truncated Hankel contour.

It is well known that $a(z)$ has the local expansion

$$a(z) = \tau - \frac{\sqrt{2}\tau}{\sigma}\sqrt{1 - z/z_0} + \dots.$$

Hence

$$\alpha = z\varphi'(a(z)) = 1 - \sqrt{2}\sigma\sqrt{1 - z/z_0} + \dots$$

and we obtain (by substituting $z = z_0(1 + x/n)$)

$$\begin{aligned}
z(e^{is/\sqrt{n}} - 1)\varphi_{d-1}a(z)^{d-1} &= n^{-1/2} z_0 is\varphi_{d-1}\tau^{d-1}(1 + O(\sqrt{|x|/n}))\\
&= n^{-1/2}\tau isc_d(1 + O(\sqrt{|x|/n})),\\
\tau - a(z) &= \frac{\sqrt{2}\tau}{\sigma}\sqrt{-x/n}(1 + O(\sqrt{|x|/n})),\\
\alpha^k &= \exp\left\{\frac{-2k\sigma\sqrt{-x}}{\sqrt{2n}}\right\}(1 + O(k|x|/n)).
\end{aligned}$$

Thus, we can apply Lemma 1 for $z \in \gamma'$ and obtain

$$y_k(z, z(e^{is/\sqrt{n}} - 1)\varphi_{d-1}a(z)^{d-1} + a(z)) = a(z) + R_k,$$

in which

$$R_k = \frac{i\tau sc_d\alpha^k/\sqrt{n} + O(\sqrt{|x|/n}}{A_1 + A_2\alpha^k + O(\sqrt{|x|/n})},$$

where

$$\begin{aligned}
A_1 &= \frac{\sqrt{-x/n} + (\sqrt{-x/n} - is\sigma c_d\sigma/\sqrt{2n})}{2\sqrt{-x/n}},\\
A_2 &= \frac{\sqrt{-x/n} - (\sqrt{-x/n} - is\sigma c_d\sigma/\sqrt{2n})}{2\sqrt{-x/n}}.
\end{aligned}$$

If we set $\kappa = k/n$, then

$$R_k \sim \frac{\tau}{\sqrt{n}}\frac{i\sqrt{-x}sc_d\exp\{-\kappa\sigma/\sqrt{-x/2}\}}{\sqrt{-x}\exp\{\kappa\sigma/\sqrt{-x/2}\} - (i\sigma sc_d/\sqrt{2})\sinh(\kappa\sigma/\sqrt{-x/2})}.$$

This directly leads to

$$[z^n]y_k \sim a_n + \frac{\tau z_0^{-n}}{n^{3/2}} \frac{1}{2\pi i} \int_\gamma \frac{i\sqrt{-x}sc_d \exp\{-\kappa\sigma/\sqrt{-x/2}\}}{\sqrt{-x}\exp\{\kappa\sigma/\sqrt{-x/2}\} - (i\sigma sc_d/\sqrt{2})\sinh(\kappa\sigma/\sqrt{-x/2})} e^{-x}dx,$$

where we used the expansion

$$z^{-n} = z_0^{-n}e^{-x}(1 + O(|x|^2/n)).$$

Since

$$a_n = [z^n]a(z) = \frac{\tau}{\sigma z_0^n \sqrt{2\pi n^3}}(1 + O(1/n)),$$

Theorem 2 follows in the one-dimensional case.

Next let us consider the two-dimensional case. We denote by $\phi^{(d)}_{k,k+h,n}(s,t)$ the characteristic function of the joint distribution of

$$\frac{1}{\sqrt{n}}L_n^{(d)}(k), \quad \frac{1}{\sqrt{n}}L_n^{(d)}(k+h)$$

and set $l = k + h$. Then we have

$$\begin{aligned}\phi^{(d)}_{kln}(s,t) &= \frac{1}{a_n}[z^n]y_k(z, z(e^{is/\sqrt{n}} - 1)\varphi_{d-1}y_{h-1}(z, z(e^{it/\sqrt{n}} - 1)\varphi_{d-1}a(z)^{d-1} + a(z))^{d-1} \\ &\quad + y_h(z, z(e^{it/\sqrt{n}} - 1)\varphi_{d-1}a(z)^{d-1} + a(z))) \\ &= \frac{1}{a_n}\int_\Gamma y_k(z, z(e^{is/\sqrt{n}} - 1)\varphi_{d-1}y_{h-1}(z, z(e^{it/\sqrt{n}} - 1)\varphi_{d-1}a(z)^{d-1} + a(z))^{d-1} \\ &\quad + y_h(z, z(e^{it/\sqrt{n}} - 1)\varphi_{d-1}a(z)^{d-1} + a(z)))\frac{dz}{z^{n+1}},\end{aligned}$$

where Γ is the integration path (3.1). For $z \in \gamma'$ we again use the substitution $z = z_0(1+x/n)$ and setting

$$y_h(z, z(e^{it/\sqrt{n}} - 1)\varphi_{d-1}a(z)^{d-1} + a(z)) = a(z) + R_h$$

we obtain as above

$$R_h \sim \frac{\tau}{\sqrt{n}} \frac{itc_d\sqrt{-x}\exp\{-\eta\sigma/\sqrt{-x/2}\}}{\sqrt{-x}\exp\{\eta\sigma/\sqrt{-x/2}\} - (itc_d\sigma/\sqrt{2})\sinh(\eta\sigma/\sqrt{-x/2})}, \tag{3.2}$$

in which $\eta = h/\sqrt{n}$. Similarly we have

$$y_k(z, z(e^{it/\sqrt{n}} - 1)\varphi_{d-1}(a(z) + R_{h-1})^{d-1} + a(z) + R_h) = a(z) + R_k,$$

where

$$R_k \sim \frac{\tau}{\sqrt{n}} \frac{(isc_d\tau + \sqrt{n}R_h/\tau)\exp\{-\kappa\sigma/\sqrt{-x/2}\}}{\sqrt{-x}\exp\{\kappa\sigma/\sqrt{-x/2}\} - (isc_d\tau + \sqrt{n}R_h/\tau)(\sigma/\sqrt{2})\sinh(\kappa\sigma/\sqrt{-x/2})} \tag{3.3}$$

and $\kappa = k/\sqrt{n}$. Thus Theorem 2 follows in the two-dimensional case. The proof of the general case just requires an iterative use of the preceding arguments.

4. TIGHTNESS

In this section we show that the sequence of random variables

$$l_n^{(d)}(t) = n^{-1/2} L_n^{(d)}(t\sqrt{n}), \qquad t \geq 0,$$

is tight in C[0, ∞).

Recall that

$$a_n = \frac{\tau}{\sigma z_0^n \sqrt{2\pi n^3}} \left(1 + O\left(\frac{1}{n}\right)\right)$$

Therefore, in view of Theorem 3, it suffices to prove

$$[z^n] H_{kh}(z) = O(h^2 n^{-1/2} z_0^{-n}). \tag{4.1}$$

For this purpose we use the following lemma due to Flajolet and Odlyzko from singularity analysis [10].

LEMMA 2. *Let $F(z)$ be analytic in the region*

$$\Delta = \{z\colon |z| < z_0 + \eta,\ |\arg(z - z_0)| > \vartheta,$$

in which z_0 and η are positive real numbers and $0 < \vartheta < \pi/2$. Furthermore suppose that there exists a real number $\beta \notin \{-1, -2, \ldots\}$ such that

$$F(z) = O((1 - z/z_0)^{-\beta}), \qquad z \in \Delta.$$

Then

$$[z^n] F(z) = O(z_0^{-n} n^{\beta - 1}).$$

For example, it follows that if a function $f(z)$ satisfies

$$f(z) = O((1 - \alpha)^{-1})$$

in Δ, where $\alpha = z\varphi'(a(z))$, then

$$[z^n] f(z) = O(z_0^{-n} n^{-1/2}). \tag{4.2}$$

We only have to observe that α has the local expansion

$$\alpha = 1 - \sqrt{2}\sigma\sqrt{1 - z/z_0} + O(1 - z/z_0).$$

We also make use of the fact that

$$\sup_{z \in \Delta} |\alpha| = 1 \tag{4.3}$$

and that

$$\sup_{z \in \Delta} \left| \frac{1 - \alpha^h}{1 - \alpha} \right| = \sup_{z \in \Delta} \left| \sum_{j=0}^{h-1} \alpha^j \right| = h. \tag{4.4}$$

Now we derive a more or less explicit representation for $H_{kh}(z)$. From this and (4.2)–(4.4) estimate (4.1) will follow immediately.

LEMMA 3. *Set* $\alpha = z\varphi'(a(z))$, $\beta = z\varphi''(a(z))$, $\gamma = z\varphi'''(a(z))$, *and* $\delta = z\varphi''''(a(z))$. *Then*

$$\frac{\partial y_k}{\partial u}(z,1) = \left(z\varphi'(a(z))\right)^k = \alpha^k,$$

$$\frac{\partial^2 y_k}{\partial u^2}(z,1) = \frac{\beta}{\alpha}\alpha^k\frac{1-\alpha^k}{1-\alpha},$$

$$\frac{\partial^3 y_k}{\partial u^3}(z,1) = \frac{\gamma}{\alpha}\alpha^k\frac{1-\alpha^{2k}}{1-\alpha^2} + 3\frac{\beta^2}{\alpha}\alpha^k\frac{(1-\alpha^k)(1-\alpha^{k-1})}{(1-\alpha)(1-\alpha^2)},$$

$$\begin{aligned}\frac{\partial^4 y_k}{\partial u^4}(z,1) = {} & \frac{\delta}{\alpha}\alpha^k\frac{1-\alpha^{3k}}{1-\alpha^3} \\ & + \left(2\beta\gamma(2+5\alpha+5\alpha^k+3\alpha^{k+1}) + 3\beta^3/\alpha\right)\alpha^k\frac{(1-\alpha^k)(1-\alpha^{k-1})}{(1-\alpha^2)(1-\alpha^3)} \\ & + 3\beta^3(1+5\alpha)\alpha^k\frac{(1-\alpha^k)(1-\alpha^{k-1})(1-\alpha^{k-2})}{(1-\alpha)(1-\alpha^2)(1-\alpha^3)}.\end{aligned}$$

PROOF. From $y_{k+1}(z,u) = z\varphi(y_k(z,u))$, we directly obtain the recurring relations

$$\frac{\partial y_{k+1}}{\partial u} = z\varphi'(y_k)\frac{\partial y_r}{\partial u},$$

$$\frac{\partial^2 y_{k+1}}{\partial u^2} = z\varphi''(y_k)\left(\frac{\partial y_k}{\partial u}\right)^2 + z\varphi'(y_k)\frac{\partial^2 y_k}{\partial u^2},$$

$$\frac{\partial^3 y_{k+1}}{\partial u^3} = z\varphi'''(y_k)\left(\frac{\partial y_k}{\partial u}\right)^3 + 3z\varphi''(y_k)\frac{\partial y_k}{\partial u}\frac{\partial^2 y_k}{\partial u^2} + z\varphi'(y_k)\frac{\partial^3 y_k}{\partial u^3},$$

$$\begin{aligned}\frac{\partial^4 y_{k+1}}{\partial u^4} = {} & z\varphi''''(y_k)\left(\frac{\partial y_k}{\partial u}\right)^4 + 6z\varphi'''(y_k)\left(\frac{\partial y_k}{\partial u}\right)^2\frac{\partial^2 y_k}{\partial u^2} \\ & + 3z\varphi''(y_k)\left(\frac{\partial^2 y_k}{\partial u^2}\right)^2 + 4z\varphi''(y_k)\frac{\partial y_k}{\partial u}\frac{\partial^3 y_k}{\partial u^3} + z\varphi'(y_k)\frac{\partial^4 y_k}{\partial u^4}.\end{aligned}$$

Since $y_k(z,1) = a(z)$ for all $k \geq 0$, this system of recurring relations has the explicit solution for $u = 1$ which are given in Lemma 3.

Let

$$\begin{aligned}h_{kh}(z,u) = y_k(z, z(u-1)\varphi_{d-1}y_{h-1}(z, z(u^{-1}-1)\varphi_{d-1}a(z)^{d-1} + a(z)) \\ + y_h(z, z(u^{-1}-1)\varphi_{d-1}a(z)^{d-1} + a(z)))\end{aligned}$$

and set

$$\begin{aligned}B = B(z,u) = z(u-1)\varphi_{d-1}y_{h-1}(z, z(u^{-1}-1)\varphi_{d-1}a(z)^{d-1} + a(z)) \\ + y_h(z, z(u^{-1}-1)\varphi_{d-1}a(z)^{d-1} + a(z)).\end{aligned}$$

Since $H_{kh}(z)$ is a linear combination of

$$\frac{\partial h_{kh}(z,1)}{\partial u}, \quad \frac{\partial^2 h_{kh}(z,1)}{\partial u^2}, \quad \frac{\partial^3 h_{kh}(z,1)}{\partial u^3}, \quad \frac{\partial^4 h_{kh}(z,1)}{\partial u^4},$$

we concentrate on these functions. First observe that

$$\begin{aligned}
\frac{\partial h_{kh}(z,u)}{\partial u} &= \frac{\partial y_k}{\partial u}(z,u)\frac{\partial B}{\partial u},\\
\frac{\partial^2 h_{kh}(z,u)}{\partial u^2} &= \frac{\partial^2 y_k}{\partial u^2}(z,u)\left(\frac{\partial B}{\partial u}\right)^2 + \frac{\partial y_k}{\partial u}(z,u)\frac{\partial^2 B}{\partial u^2},\\
\frac{\partial^3 h_{kh}(z,u)}{\partial u^3} &= \frac{\partial^3 y_k}{\partial u^3}(z,u)\left(\frac{\partial B}{\partial u}\right)^3 + 3\frac{\partial^2 y_k}{\partial u^2}(z,u)\frac{\partial B}{\partial u}\frac{\partial^2 B}{\partial u^2} + \frac{\partial y_k}{\partial u}(z,u)\frac{\partial^3 B}{\partial u^3},\\
\frac{\partial^4 h_{kh}(z,u)}{\partial u^4} &= \frac{\partial^4 y_k}{\partial u^4}(z,u)\left(\frac{\partial B}{\partial u}\right)^4 + 6\frac{\partial^3 y_k}{\partial u^3}(z,u)\left(\frac{\partial B}{\partial u}\right)^2\frac{\partial^2 B}{\partial u^2}\\
&\quad + 3\frac{\partial^2 y_k}{\partial u^2}(z,u)\left(\frac{\partial^2 B}{\partial u^2}\right)^2 + 4\frac{\partial^2 y_k}{\partial u^2}(z,u)\frac{\partial B}{\partial u}\frac{\partial^3 B}{\partial u^3} + \frac{\partial y_k}{\partial u}(z,u)\frac{\partial^4 B}{\partial u^4}.
\end{aligned}$$

Furthermore, simple calculations provide

$$\begin{aligned}
\frac{\partial B(z,1)}{\partial u} &= z\varphi_{d-1}a(z)^{d-1}(1-\alpha^h) = O(1),\\
\frac{\partial^2 B(z,1)}{\partial u^2} &= -2z^2\varphi_{d-1}^2 a(z)^{2d-3}\alpha^{h-1} + 2z\varphi_{d-1}a(z)^{d-1}\alpha^h + z^2\varphi_{d-1}^2 a(z)^{2d-2}\beta\alpha^{h-1}\frac{1-\alpha^h}{1-\alpha}\\
&= O(1) + O\left(\frac{1-\alpha^h}{1-\alpha}\right) = O(h),\\
\frac{\partial^3 B(z,1)}{\partial u^3} &= O(1) + O\left(\frac{1-\alpha^h}{1-\alpha}\right) + O\left(\left(\frac{1-\alpha^h}{1-\alpha}\right)^2\right) = O(h^2),\\
\frac{\partial^4 B(z,1)}{\partial u^4} &= O(1) + O\left(\frac{1-\alpha^h}{1-\alpha}\right) + O\left(\left(\frac{1-\alpha^h}{1-\alpha}\right)^2\right) + O\left(\left(\frac{1-\alpha^h}{1-\alpha}\right)^3\right) = O(h^3).
\end{aligned}$$

Hence, we have uniformly for $z \in \Delta$

$$\begin{aligned}
\frac{\partial h_{kh}(z,1)}{\partial u} &= O(1),\\
\frac{\partial^2 h_{kh}(z,1)}{\partial u^2} &= O\left(\frac{(1-\alpha^k)(1-\alpha^h)}{1-\alpha}\right) + O(h) = O(h),\\
\frac{\partial^3 h_{kh}(z,1)}{\partial u^3} &= O\left(\frac{(1-\alpha^h)^3}{(1-\alpha)^2}\right) + O\left(\frac{1-\alpha^h}{1-\alpha}h\right) + O(h^2) = O(h^2),\\
\frac{\partial^4 h_{kh}(z,1)}{\partial u^4} &= O\left(\frac{(1-\alpha^h)^4}{(1-\alpha)^3}\right) + O\left(\frac{(1-\alpha^h)^2}{(1-\alpha)^2}h\right) + O\left(\frac{h^2}{1-\alpha}\right) + O\left(\frac{(1-\alpha^h)^4}{(1-\alpha)^3}\right)\\
&= O\left(\frac{h^2}{1-\alpha}\right),
\end{aligned}$$

which implies

$$[z^n]\frac{\partial h_{kh}(z,1)}{\partial u} = O(n^{-1}) = O(h^2 n^{-1/2}), \qquad [z^n]\frac{\partial^2 h_{kh}(z,1)}{\partial u^2} = O(hn^{-1}) = O(h^2 n^{-1/2}),$$

$$[z^n]\frac{\partial^3 h_{kh}(z,1)}{\partial u^3} = O(h^2 n^{-1}) = O(h^2 n^{-1/2}), \quad [z^n]\frac{\partial^3 h_{kh}(z,1)}{\partial u^3} = O(h^2 n^{-1/2}).$$

Thus, (4.1) follows and we have proved tightness.

REFERENCES

1. D. J. Aldous, The continuum random tree I. *Ann. Prob.* (1991) **19**, 1–28.
2. D. J. Aldous, The continuum random tree II: an overview. In: *Stochastic Analysis*. Cambridge Univ. Press, Cambridge, 1991, 23–70.
3. D. J. Aldous, The continuum random tree III, *Ann. Prob.* (1993) **21**, 248-289.
4. P. Biane and M. Yor, Valeurs principales associees aux temps locaux Browniens. *Bull. Sci. Math.* (1987) **111**, 23–101.
5. P. Billingsley, *Convergence of Probability Measures*. Wiley, New York, 1968.
6. J. W. Cohen and G. Hooghiemstra, Brownian excursion, the *M/M/*1 queue and their occupation times, *Math. Oper. Res.* (1981) **6**, 608–629.
7. M. Drmota, The height distribution of leaves in rooted trees. *Discrete Math. Appl.* (1994) **4**, 45–58.
8. M. Drmota and P. Kirschenhofer, On generalized independent subsets of trees, *Random Structures and Algorithms* (1991) **2**, 187–208.
9. M. Drmota and M. Soria, Marking in combinatorial constructions: generating functions and limiting distributions, *Theoretical Comp. Sci.* (1995) **144**, 67–99.
10. P. Flajolet and A. M. Odlyzko, Singularity analysis of generating functions, *SIAM J. Discrete Math.* (1990) **3**, 216–240.
11. W. Gutjahr and G. Ch. Pflug, The asymptotic contour process of a binary tree is a Brownian excursion, *Stoch. Proc. Appl.* (1992) **41**, 69–89.
12. D. P. Kennedy, The Galton–Watson process conditioned on the total progeny, *J. Appl. Prob.* (1975) **12**, 800–806.
13. V. F. Kolchin, Branching processes, random trees, and a generalized scheme of arrangement of particles. *Math. Notes* (1977) **21**, 386–394.
14. V. F. Kolchin, *Random Mappings*, Optimization Software, New York, 1986.
15. A. Meir and J. W. Moon, On the altitude of nodes in random trees, *Canadian J. Math.* (1978) **30**, 997–1015.
16. Yu. L. Pavlov, Some properties of plane planted trees. In: *Abstr. All-Union Conf. 'Discrete Math. and its Appl.'* Irkutsk, 1991, p. 14 (in Russian).
17. Yu. L. Pavlov, Some properties of planar planted trees. *Discrete Math. Appl.* (1993) **3**, 97–102.
18. L. Takács, Conditional limit theorems for branching processes. *J. Appl. Math. Stoch. Analysis* (1991) **4**, 263–292.
19. V. A. Vatutin, Branching processes with final types of particles and random trees. In: *Abstr. Intern. Conf. on Random Mappings, Partitions and Permutations.* Univ. South California, 1992, pp. 42–43.
20. V. A. Vatutin, Branching processes with final types of nodes and random trees. *Theory Probab. Appl.* (1994) **39**, 699–714 (in Russian).
21. J. S. Vitter and P. Flajolet, Average-case analysis of algorithms and data structures. In: *Handbook of Theoretical Computer Science, Algorithms and Complexity.* North Holland, Amsterdam, 1990, 431–524.

Probabilistic Methods in Discrete Mathematics, pp. 45–63
V. F. Kolchin et al. (Eds.)

A GENERALIZATION OF THE NUMBER FIELD SIEVE

I. A. SEMAEV

Moscow

Abstract — We present a generalization of the number field sieve (NFS) which is currently the most efficient method to factor integers. A prerequisite is the representation of the integer N to be factored as a resultant of polynomials over $\mathbf{Z}$ with small coefficients. We show that for integers of special forms this representation may be found easily. The heuristic running time of the method in its first-order approximation equals

$$L[1/3; c] = e^{c}(\ln N)^{1/3}(\ln \ln N)^{2/3})(1 + o(1)),$$

where $c = 1.923$ or 1.902 as in the NFS. In fact, the factoring is likely to be more efficient. Furthermore, we extend significantly the set of integers factored in time $L[1/3; 1.526]$. Among them there are some algebraic factors of $A^n - 1$, where the composite n has small divisors.

1. Let N be an integer to be factored. We use the standard notation

$$L_N[\alpha; \beta] = L[\alpha; \beta] = e^{\beta}(\ln N)^{\alpha}(\ln \ln N)^{1-\alpha}(1 + o(1)), \qquad N \to \infty.$$

The number field sieve (NFS) is a method for factoring integers. It was suggested by Pollard [1]. Its heuristic running time is estimated as $L[1/3; c]$, where $c = 1.923$ [2]. The modification of it suggested in [3] reduces the exponent to $c = 1.902$. For integers of a special form this exponent is equal to $c = 1.526$ [4]. In the NFS the integer N is represented as a value of some irreducible polynomial

$$f(x) = x^n + a_1x^{n-1} + \ldots + a_n \tag{1}$$

over integers $\mathbf{Z}$. There is a variant of this algorithm with the leading coefficient of $f(x)$ not equal to 1 [5], that is, $f(x)$ has a root $m \in \mathbf{Z}$ modulo N, where $|a_i| \le |m| \le N^{1/n+\varepsilon}$ for small $\varepsilon > 0$. One may suppose that the greatest common divisor of N and Δ_f is equal to 1, where Δ_f is the discriminant of $f(x)$. This polynomial can be found easily. Let $\alpha \in \mathbf{C}$ be a root of $f(x)$. Multiplicative relations provided the factorization are resulted from the congruences

$$cm - d \equiv c\alpha - d \pmod{\mathfrak{N}}, \tag{2}$$

where the absolute values of integers c, d do not exceed some bound L and $\mathfrak{N}$ is an ideal of the ring of integers $\mathfrak{D}$ of the field $\mathbf{Q}(\alpha)$ which is the greatest common divisor of the ideals

$(\alpha - m)\mathfrak{D}$ and $N\mathfrak{D}$. The norm of $\mathfrak{N}$ equals N. Hence it follows that the absolute value of the integer in the left-hand side of (2) is no greater than $(m+1)L$ and the norm of the algebraic number in the right-hand side of (2) is no greater than $m(n+1)L^n$. The running time of the NFS is determined by the probability of the simultaneous B-smoothness, for some bound B, of the integers

$$cm - d, \qquad \mathrm{Norm}(c\alpha - d) = d^n + a_1 c d^{n-1} + \ldots + a_n c^n.$$

An integer is B-smooth if all its prime factors are no greater than B. Obviously, this probability is determined by the values $(m+1)L$ and $m(n+1)L^n$. As $n \to \infty$, the latter value essentially exceeds the first one. For the optimal choice of n, the second value is equal to the cube of the first [2].

Roughly speaking, the main idea of this work is to find the congruences of the form (2), whose left-hand side and the right-hand side are approximately equal. Such a choice raises the probability of the simultaneous smoothness and renders the factorization more efficient. However the heuristic running time of the method remains unchanged in its first-order approximation and is equal to $L[1/3; c]$, where $c = 1.923$ or $c = 1.902$.

A prerequisite to the application of the method is a representation of N or its multiple as a resultant of polynomials $f(x), g(x)$ over $\mathbf{Z}$ whose coefficients are small, that is

$$\mathrm{Res}(f, g) \equiv 0 \pmod{N}. \tag{3}$$

At present an efficient method of finding the coefficients of the polynomials $f(x), g(x)$ in (3) bounded by $N^{1/(n_1+n_2)+\varepsilon}$ for small $\varepsilon > 0$, where n_1 and n_2 are the degrees of the polynomials $f(x)$ and $g(x)$, remains unknown for N of the general form. However for N of some special forms one may get it easily. For example we show that for random integers A, B, C such that $|A|, |B|, |C| \le N^{(1/n)+\varepsilon}$, $\varepsilon > 0$, one can obtain the representation $N = B^n - A^k C^{n-k}$, where the absolute values of the coefficients of the polynomials $f(x)$, $g(x)$ of degree n are no greater than $N^{(1/(2n))+\varepsilon}$, with probability tending to one as $N^{1/n} \to \infty$.

In addition we extend significantly the set of integers which can be factored in time $L[1/3; 1.526]$. In particular, among them there are some algebraic factors of $A^n - 1$, where a composite n has small divisors.

It should be noted that the NFS is a particular case of our method when $\deg f(x)$ or $\deg g(x)$ is equal to one. Besides this introduction, this paper contains three sections. In Section 2, we describe the algorithm and prove a few propositions needed for this description. In Section 3 we prove some identities for resultants. In Section 4 these identities are used for producing congruences (3) for integers of special forms. We present also an example of application of the method.

2. Let

$$\begin{aligned} f(x) &= a_0 x^{n_1} + a_1 x^{n_1 - 1} + \ldots + a_{n_1}, \\ g(x) &= b_0 x^{n_2} + b_1 x^{n_2 - 1} + \ldots + b_{n_2} \end{aligned} \tag{4}$$

be polynomials over **Z**. By the definition [6] of the resultant

$$\mathrm{Res}(f,g) = a_0^{n_2} b_0^{n_1} \prod_{\alpha,\beta}(\alpha - \beta) = a_0^{n_2} \prod_{\alpha} g(\alpha) = (-1)^{n_1 n_2} b_0^{n_1} \prod_{\beta} f(\beta),$$

where α, β run over all the roots of the polynomials $f(x)$, $g(x)$, respectively, with regard to their multiplicities. We apply Euclid's algorithm modulo N to polynomials (4) assuming that (3) is valid. We suppose also that these polynomials are primitive over **Z**. In other words, their coefficients are coprime. We use the following step.

Let us suppose that $n_1 \geq n_2 \geq 1$ and $\gcd(b_0, N) = 1$. Then there exists a polynomial $h(x)$ over **Z** such that $\deg h(x) = n_1 - n_2$ and $\deg g_1(x) = n_2' < n_2$, where

$$g_1(x) \equiv f(x) - h(x)g(x) \pmod{N}. \tag{5}$$

We consider the coefficients of $g_1(x)$ as the least non-negative residues modulo N. For some polynomial $u(x)$ with $\deg u(x) \leq n_1$ we have the equalities

$$\mathrm{Res}(g_1, g) = \mathrm{Res}(f - hg + Nu, g) = (-1)^{n_2' n_2} b_0^{n_2'} \prod_{\beta}(f - hg + Nu)(\beta)$$

$$\equiv (-1)^{(n_2' - n_1) n_2} b_0^{n_2' - n_1} \,\mathrm{Res}(f,g) \pmod{N},$$

where β runs over the set of all roots of $g(x)$ with regard to their multiplicities. Let $\mathrm{Res}(g_1, g) = 0$. We set

$$t(x) = \begin{cases} \gcd(g_1(x), g(x)), & \text{if } g_1(x) \neq 0, \\ g(x), & \text{if } g_1(x) = 0. \end{cases}$$

As $t(x)$ one can take some primitive polynomial over **Z** with the leading coefficient coprime to N. It is clear that $\deg t(x) \geq 1$. It is clear that $t(x)$ is a common factor of $f(x), g(x)$ modulo N.

Let us suppose now that $\mathrm{Res}(f,g) \equiv 0 \pmod{N}$. If $\gcd(b_0^{(0)}, N) = 1$, where $b_0^{(0)} = b_0$, we determine the polynomial $g_1(x)$ by formula (5), otherwise we factor N. We have $\mathrm{Res}(g_1, g) \equiv 0 \pmod{N}$. We can suppose that $g_1(x)$ is primitive. Let the leading coefficient $b_0^{(1)}$ of the polynomial $g_1(x)$ be coprime to N and $\mathrm{Res}(g_1, g) \neq 0$. We apply the step described above to the polynomials $g(x), g_1(x)$, where $\deg g_1(x) < \deg g(x)$. We find a polynomial $g_2(x)$ and so on. We denote the leading coefficient of $g_k(x)$ by $b_0^{(k)}$. Since $\deg g_k(x)$ decreases, we get either $\gcd(b_0^{(k)}, N) \neq 1$ for some $k \geq 0$ and so we factor N, or the polynomials $f(x), g(x)$ have a common factor $t(x) \pmod{N}$ of degree at least one whose leading coefficient is coprime to N.

Let

$$f(x) = a_0 x^n + a_1 x^{n-1} + \ldots + a_n, \qquad a_i \in \mathbf{Z}, \quad i = 0, 1, \ldots, n,$$

be an irreducible polynomial. A root α of $f(x)$ is not an algebraic integer if $a_0 \neq \pm 1$. However $\alpha_1 = \alpha a_0$ is an algebraic integer and it is the root of the polynomial

$$f_1(x) = x^n + a_1 x^{n-1} + a_2 a_0 x^{n-2} + \ldots + a_n a_0^{n-1}.$$

LEMMA 1. *Let* $\gcd(a_0, a_n) = 1$. *Let* $\mathfrak{N}$ *be an ideal of the ring of integers* $\mathfrak{O}$ *of* $\mathbf{Q}(\alpha)$ *which is the greatest common divisor of the ideals* $\alpha_1\mathfrak{O}$ *and* $a_0\mathfrak{O}$. *Then*

$$\mathrm{Norm}(\mathfrak{N}) = |a_0|^{n-1}.$$

PROOF. Let $\mathfrak{Q}$ be a prime ideal of $\mathfrak{O}$ which divides $q\mathfrak{O}$, where q^l is the maximal power of a prime integer q which divides a_0, that is, el is the maximal exponent of $\mathfrak{Q}$ in $a_0\mathfrak{O}$, where e is the ramification index of $\mathfrak{Q}$. Let $m \geq 1$ be the maximal exponent of $\mathfrak{Q}$ in $\alpha_1\mathfrak{O}$. We have the equality

$$a_0(-a_1\alpha_1^{n-2} - \dots - a_0^{n-2}a_n) = \alpha_1^{n-1}(\alpha_1 + a_1).$$

We suppose that $m > el$. Since $\gcd(a_0, a_n) = 1$, the exponent of $\mathfrak{Q}$ corresponding to the left-hand side of the equality is equal to $el(n-1)$. At the same time this exponent corresponding to the right-hand side is no less than $m(n-1)$. We obtain a contradiction. Thus, $m \leq el$.

Let $\alpha_1\mathfrak{O} = \mathfrak{N}\mathfrak{A}$. Then the ideals $\mathfrak{N}$ and $\mathfrak{A}$ are coprime. Otherwise there exists a prime ideal $\mathfrak{Q}$ which divides both $\mathfrak{N}$ and $\mathfrak{A}$. Therefore the exponent of $\mathfrak{Q}$ in $\mathfrak{N}$ is not maximal, that is, $\mathfrak{N}$ is not the greatest common divisor of $a_0\mathfrak{O}$ and $\alpha_1\mathfrak{O}$. Since $\mathrm{Norm}(\alpha_1\mathfrak{O}) = |a_0^{n-1}a_n|$, the norm of $\mathfrak{N}$ divides a_0^{n-1}. Let there exist a prime integer q, which is a factor of a_0, such that q divides $\mathrm{Norm}(\mathfrak{A})$. Thus there exists a prime ideal $\mathfrak{Q}$ which divides both $\mathfrak{A}$ and $a_0\mathfrak{O}$. Since $\gcd(\mathfrak{N}, \mathfrak{A}) = 1$, this contradicts the equality $\gcd(\alpha_1\mathfrak{O}, a_0\mathfrak{O}) = \mathfrak{N}$. Hence $\mathrm{Norm}(\mathfrak{N}) = |a_0|^{n-1}$. The lemma is proved.

Let us outline the algorithm of factorization of N. We suppose that (3) holds for polynomials (4). We suppose also that these polynomials are primitive and irreducible over $\mathbf{Q}$. Otherwise one can hope to factor N with the use of the congruence

$$\mathrm{Res}(f_1, g)\,\mathrm{Res}(f_2, g) \equiv 0 \pmod N$$

if, for example, $f(x) = f_1(x)f_2(x)$. At least if $\deg g(x) = 1$, then the integer N can be factored in this way [5]. Applying Euclid's algorithm modulo N to polynomials (4), we either factor N or obtain a polynomial $t(x)$ of degree at least one, with the leading coefficient coprime to N, which divides both $f(x)$ and $g(x)$ modulo N. Let us consider the latter possibility. Let α and β be roots of $f(x)$ and $g(x)$ respectively. Then $\alpha_1 = \alpha a_0$ and $\beta_1 = \beta b_0$ are roots of the polynomials

$$f_1(x) = x^{n_1} + a_1x^{n_1-1} + a_2a_0x^{n_1-2} + \dots + a_{n_1}a_0^{n_1-1},$$
$$g_1(x) = x^{n_2} + b_1x^{n_2-1} + b_2b_0x^{n_2-2} + \dots + b_{n_2}b_0^{n_2-1}$$

respectively. To simplify the presentation of the algorithm, we suppose that

$$\gcd(a_0, a_{n_1}) = \gcd(b_0, b_{n_2}) = 1.$$

We can also suppose that

$$\gcd(a_0b_0\Delta_{f_1}\Delta_{g_1}, N) = 1,$$

otherwise we factor N. We consider the ring

$$D = (\mathbf{Z}/N)[x]/(t(x)).$$

In other words, the ring D is the quotient ring of $\mathbf{Z}[x]$ modulo $t(x)$ and modulo N at a time. Let $\deg t(x) = l$. It is obvious that $t_1(x) = a_0^l t(x/a_0)$ and $t_2(x) = b_0^l t(x/b_0)$ divide the polynomials $f_1(x)$ and $g_1(x)$ respectively. Let $\mathfrak{O}_1$ be the ring of integers of the field $\mathbf{Q}(\alpha)$. Since $\gcd(\Delta_{f_1}, N) = 1$, for the ideal $\mathfrak{N}_1$ which is the greatest common divisor of the ideals $t_1(\alpha_1)\mathfrak{O}_1$ and $N\mathfrak{O}_1$ we have $\mathrm{Norm}(\mathfrak{N}_1) = N^l$ and $\mathfrak{O}_1/\mathfrak{N}_1 \cong D$. We denote the ring

$$\mathfrak{O}_1 \cup a_0^{-1}\mathfrak{O}_1 \cup a_0^{-2}\mathfrak{O}_1 \cup \ldots$$

by $\mathfrak{O}_1'$. We set $\mathfrak{N}_1' = \mathfrak{N}_1\mathfrak{O}_1'$. Since $\gcd(a_0, N) = 1$, we have the isomorphism

$$\mathfrak{O}_1'/\mathfrak{N}_1' \cong D$$

such that $\alpha \to x \pmod{N, t(x)}$. Similarly, for the ring $\mathfrak{O}_2$ of the integers of $\mathbf{Q}(\beta)$ we have $\mathfrak{O}_2/\mathfrak{N}_2 \cong D$, where $\mathfrak{N}_2$ is the greatest common divisor of the ideals $t_2(\beta_1)\mathfrak{O}_2$ and $N\mathfrak{O}_2$ with $\mathrm{Norm}(\mathfrak{N}_2) = N^l$. Then for

$$\mathfrak{O}_2' = \mathfrak{O}_2 \cup b_0^{-1}\mathfrak{O}_2 \cup b_0^{-2}\mathfrak{O}_2 \cup \ldots$$

and $\mathfrak{N}_2' = \mathfrak{N}_2\mathfrak{O}_2'$ we have the isomorphism

$$\mathfrak{O}_2'/\mathfrak{N}_2' \cong D$$

such that $\beta \to x \pmod{N, t(x)}$.

LEMMA 2. *Let u and v be non-zero elements of D represented as*

$$u = u(x) = u_1x^{l-1} + u_2x^{l-2} + \ldots + u_l,$$
$$v = v(x) = v_1x^{l-1} + v_2x^{l-2} + \ldots + v_l,$$

where u_i, v_i, $i = 1, \ldots, l$, are some residues modulo N. Let $uv = 0$ in D. Then we factor either N or $t(x)$ modulo N in $O(l^2)$ operations modulo N.

PROOF. We apply Euclid's algorithm modulo N to the polynomials $u(x)$ and $t(x)$. Euclid's algorithm modulo N can be realized if the leading coefficients of the sequence of intermediate polynomials are coprime to N. In such a case we find either a non-trivial factor of $t(x)$ modulo N or $u' \in D$ such that $uu' = 1$ in D. Otherwise we factor N. Similarly, by applying Euclid's algorithm modulo N to the polynomials $v(x)$ and $t(x)$, we factor N either $t(x)$ modulo N or find $v' \in D$ such that $vv' = 1$ in D. Thus, either the lemma is valid or we obtain the equalities $uu' = 1$ and $vv' = 1$. Since $uv = 0$, this is impossible. The running time of Euclid's algorithm is $O(l^2)$ operations modulo N. The lemma is proved.

We define in advance some sets of prime ideal $\mathfrak{Q}_{ij} \in \mathfrak{O}_i$ of degree one such that $\mathrm{Norm}(\mathfrak{Q}_{ij}) \leq B_i$ for some bounds B_i, $i = 1, 2$. In addition, we define $O(\ln N)$ prime ideals $\mathfrak{P}_{ik} \in \mathfrak{O}_i$ of degree one such that $\mathrm{Norm}(\mathfrak{P}_{ik}) > B_i$.

It follows from Lemma 1 that for coprime integers c and d, the equality $(c\alpha - d)\mathfrak{O}_1 = \mathfrak{A}\mathfrak{R}_1^{-1}$ holds, where

$$\mathfrak{R}_1 = (a_0\mathfrak{O}_1)(\gcd(a_0\mathfrak{O}_1, \alpha_1\mathfrak{O}_1))^{-1}$$

and $\mathfrak{A}$ is an integral ideal such that

$$\mathrm{Norm}(\mathfrak{A}) = a_0 d^{n_1} + a_1 c d^{n_1 - 1} + \ldots + a_{n_1} c^{n_1}. \tag{6}$$

Similarly, $(c\beta - d)\mathfrak{O}_2 = \mathfrak{B}\mathfrak{R}_2^{-1}$, where

$$\mathfrak{R}_2 = (b_0\mathfrak{O}_2)(\gcd(b_0\mathfrak{O}_2, \beta_1\mathfrak{O}_2))^{-1}$$

and $\mathfrak{B}$ is an integral ideal such that

$$\mathrm{Norm}(\mathfrak{B}) = b_0 d^{n_2} + b_1 c d^{n_2 - 1} + \ldots + b_{n_2} c^{n_2}. \tag{7}$$

1. We sieve through pairs of integers c, d in the range $|c|, |d| \leq L$ for some bound L to find the pairs such that the integers defined by (6) and (7) are B_1 and B_2-smooth respectively.

2. We build a matrix M whose rows are indexed by the pairs c, d found above and whose columns are indexed by the ideals $\mathfrak{R}_i$, $\mathfrak{Q}_{ij}$, $\mathfrak{P}_{ij}$, $i = 1, 2$. The entry of the matrix corresponding to the row indexed by c, d and the column indexed by $\mathfrak{R}_1(\mathfrak{R}_2)$ or $\mathfrak{Q}_{1j}(\mathfrak{Q}_{2j})$ contains 0 if this ideal has an even exponent in $c\alpha - d(c\beta - d)$ and 1 otherwise. If the column corresponds to $\mathfrak{P}_{1i}(\mathfrak{P}_{2i})$ (it is a character columns), the entry contains 0 if $c\alpha - d(c\beta - d)$ is a quadratic residue modulo $\mathfrak{P}_{1i}(\mathfrak{P}_{2i})$ and 1 otherwise. The resulting matrix is roughly square with sides equal to

$$\pi(B_1) + \pi(B_2) + O(\ln N)$$

where $\pi(x)$ is the number of prime integers not exceeding x. The matrix M is sparse with $O(\ln N)$ non-zero entries per row.

3. We find a linear combination of the rows of M which vanishes modulo 2. This linear combination represents a subset S of the pairs c, d. Then

$$\gamma = \prod_S (c\alpha - d) \in \mathfrak{O}_1', \qquad \delta = \prod_S (c\beta - d) \in \mathfrak{O}_2',$$

where $\gamma \pmod{\mathfrak{N}_1'} \equiv \delta \pmod{\mathfrak{N}_2'}$. The ideals $\gamma\mathfrak{O}_1$ and $\delta\mathfrak{O}_2$ are squares of some fractional ideals of the fields $\mathbf{Q}(\alpha)$ and $\mathbf{Q}(\beta)$ respectively. The cancellation of the entries in the character columns gives grounds to expect that the algebraic numbers γ and δ are squares in $\mathbf{Q}(\alpha)$ and $\mathbf{Q}(\beta)$ respectively [2]. By the definition of the rings $\mathfrak{O}_i'$ there are natural l_i, $i = 1, 2$, such that $\gamma_1 = a_0^{2l_1}\gamma$ and $\delta_1 = b_0^{2l_2}\delta$. Since $|S|$ is even, one can take

$l_i = |S|/2$. Indeed, the ideals $\mathfrak{R}_1$ and $\mathfrak{R}_2$ are in $\gamma\mathfrak{D}_1$ and $\delta\mathfrak{D}_2$, respectively, with the same exponent $-|S|$. Furthermore, it is known [5] that

$$\gamma_2 = (f_1'(\alpha_1))^2\gamma_1 \in \mathbf{Z}[\alpha_1],$$
$$\delta_2 = (g_1'(\beta_1))^2\delta_1 \in \mathbf{Z}[\beta_1],$$

where $f_1'(x)$ and $g_1'(x)$ are, respectively, the derivatives of $f_1(x)$ and $g_1(x)$ with respect to x. Since γ and δ are squares in $\mathbf{Q}(\alpha)$ and $\mathbf{Q}(\beta)$, respectively, the numbers γ_2 and δ_2 are squares in the rings $\mathbf{Z}[\alpha_1]$ and $\mathbf{Z}[\beta_1]$.

4. We find $\gamma_3 \in \mathbf{Z}[\alpha_1]$ and $\delta_3 \in \mathbf{Z}[\beta_1]$ such that $\gamma_2 = \gamma_3^2$ and $\delta_2 = \delta_3^2$ by the procedure described in [5]. Then

$$\gamma_4^2 \pmod{\mathfrak{N}_1'} \equiv \delta_4^2 \pmod{\mathfrak{N}_2'},$$

where

$$\gamma_4 = \frac{\gamma_3}{a_0^{l_1} f_1'(\alpha_1)}, \qquad \delta_4 = \frac{\delta_3}{b_0^{l_2} g_1'(\beta_1)}.$$

We denote by $\bar\gamma_4$ and $\bar\delta_4$ the images in D of these elements. Then in D,

$$(\bar\gamma_4 - \bar\delta_4)(\bar\gamma_4 + \bar\delta_4) = 0.$$

5. If $\bar\gamma_4 = \pm\bar\delta_4$, we add a few new rows to the matrix M and subsequently find a new linear combination which involves these new rows and vanishes modulo 2 and so on. Otherwise by Lemma 2 we either factor N or find a divisor $t^{(1)}(x)$ of the polynomial $t(x)$ modulo N of degree lesser than l. In the latter case we construct a new ring D_1 determined by $t^{(1)}(x)$, new ideals $\mathfrak{N}_i' \in \mathfrak{D}_i$. We find the images of γ_4 and δ_4 in D_1 and so on. Since $\deg t(x) > \deg t^{(1)}(x) > \ldots \geq 1$, we factor N in this way.

Let us find the heuristic running time of the algorithm under the assumption that the absolute values of the coefficients of $f(x)$ and $g(x)$ in (3) do not exceed $N^{1/(n_1+n_2)+\varepsilon}$ for small $\varepsilon > 0$. We put

$$n_1 = n_2 = (3\ln N/(8\ln\ln N))^{1/3}, \qquad L = B_1 = B_2 = L[1/3; (8/9)^{1/3}].$$

The standard analysis shows that the running time is equal to $L[1/3; c]$, where $c = (64/9)^{1/3} \approx 1.923$. However, in practice, our algorithm offers some advantage over the NFS. Indeed, to obtain rows of the matrix M, we have to test the smoothness of two integers whose absolute values do not exceed $L[2/3; (8/3)^{1/3}]$, while in the NFS [2] one of the integers is no greater than $L[2/3; 9^{1/3}]$ and the other is no greater than $L[2/3; (1/3)^{1/3}]$. Since $9^{1/3} + (1/3)^{1/3} = 2(8/3)^{1/3}$, it is clear that the probability of the second event is less than the probability of the first one, although the running times of both of the methods are the same in their first-order approximation.

Since $\mathrm{Res}(f + ag, g) = \mathrm{Res}(f, g)$ for $a \in \mathbf{Z}$, the modifications proposed in [3] can be used to reduce the estimate of the running time of our algorithm to $L[1/3; c]$, where $c = 1.902$.

There are integers which can be factored by the NFS in $L[1/3; 1.526]$ operations. For these integers the coefficients of the polynomial (1) do not exceed N^{ε}, where $\varepsilon = o(1/n)$ for $n = (3\ln N/(2\ln\ln N))^{1/3}$ as $N \to \infty$. We extend this set of easily factored integers. We put in (3)

$$n_1 = (3\ln N/(2\ln\ln N))^{1/3}, \quad n_2 = o(n_1), \quad \varepsilon = o(1/n_1)$$

and let the absolute values of the coefficients of $f(x)$ and $g(x)$ do not exceed N^{ε} and $N^{(1/n_1)+\varepsilon}$ respectively. It is obvious that this set is considerably more wide than the set of integers which can be factored in time $L[1/3; 1.526]$ by the NFS and which can be obtained by setting $n_2 = 1$ in (3). However the first-order approximation for the complexity of the factorization of the integers from this extended set is the same. We will show that some algebraic factors of the integers $A^n - 1$, where n has small divisors, are in this set.

3. We prove here several identities for the resultant. We use the following assertion due to Swan [7].

LEMMA 3. *The resultant of two binomials is equal to*

$$\mathrm{Res}(x^j - \xi^j, x^k - \nu^k) = (-1)^j(\nu^m - \xi^m)^d,$$

where $d = \gcd(j,k)$, $m = jk/d$.

We denote by $\Phi_m(x)$ the m-cyclotomic polynomial. By definition

$$\Phi_m(x) = \prod_i (x - \xi_m^i),$$

where the product is taken over i, $1 \le i \le m$, $\gcd(i,m) = 1$, and ξ_m is the mth primitive root of unity. Let $\Phi_m(x,y)$ be the form of degree $\varphi(m)$ associated with $\Phi_m(x)$.

LEMMA 4. *Let for natural integers* l, k, m, n,

$$m_1 = m/\gcd(k/r, m), \qquad n_1 = n/\gcd(l/r, n),$$

where $r = \gcd(l,k)$. *Then for integer* A_0, A_1, B_0, B_1, *where* $A_0, B_0 \neq 0$,

$$\mathrm{Res}(\Phi_m(A_0x^l, A_1), \Phi_n(B_0x^k, B_1)) = \delta\Phi_\nu(A_0^{k/r}B_1^{l/r}, A_1^{k/r}B_0^{l/r})^{r\varphi(m)\varphi(n)/(\varphi(m_1)\varphi(n_1))},$$

where

$$\delta = \begin{cases} (-1)^{l\varphi(m)}, & \text{if } n = 1, \\ 1, & \text{if } n > 1, \end{cases}$$

and

$$\nu = \begin{cases} m_1n_1, & \text{if } \gcd(m_1,n_1) = 1, \\ m_1n_1/2, & \text{if } \gcd(m_1,n_1) = 2,\ m_1n_1/4 \text{ is even}, \\ m_1n_1/4, & \text{if } \gcd(m_1,n_1) = 2,\ m_1n_1/4 \text{ is odd}. \end{cases}$$

PROOF. We represent the polynomials $\Phi_m(A_0x^l, A_1)$ and $\Phi_n(B_0x^k, B_1)$ in the form suitable for the application of Swan's lemma. We have

$$\Phi_m(A_0x^l, A_1) = \prod_j (A_0x^l - \xi_m^j A_1) = A_0^{\varphi(m)} \prod_j (x^l - \xi_m^l(j)),$$

where $\xi^l(j) = \xi_m^j A_1 A_0^{-1}$ and $\gcd(j, m) = 1$. Similarly,

$$\Phi_n(B_0x^k, B_1) = \prod_i (B_0x^k - \xi_n^i B_1) = B_0^{\varphi(n)} \prod_i (x^k - v^k(i)),$$

where $v^k(i) = \xi_n^i B_1 B_0^{-1}$ and $\gcd(i, n) = 1$. Then

$$\begin{aligned} R &= \operatorname{Res}(A_0^{\varphi(m)} \prod_j (x^l - \xi_m^l(j)), B_0^{\varphi(n)} \prod_i (x^k - v^k(i))) \\ &= (A_0^k B_0^l)^{\varphi(m)\varphi(n)} \prod_{i,j} \operatorname{Res}(x^l - \xi^l(j), x^k - v^k(i)). \end{aligned}$$

Hence

$$R = (-1)^{l\varphi(m)\varphi(n)} (A_0^k B_0^l)^{\varphi(m)\varphi(n)} \prod_{i,j} (v^u(i) - \xi^u(j))^r,$$

where $u = lk/r$. We have

$$\begin{aligned} R &= (-1)^{l\varphi(m)\varphi(n)} (A_0^k B_0^l)^{\varphi(m)\varphi(n)} \prod_{i,j} ((\xi_n^i B_1 B_0^{-1})^{l/r} - (\xi_m^j A_1 A_0^{-1})^{k/r})^r \\ &= (-1)^{l\varphi(m)\varphi(n)} \left(\prod_i \xi_n^{li}\right)^{\varphi(m)} \prod_{i,j} (B_1^{l/r} A_0^{k/r} - \xi_m^{kj/r} \xi_n^{-li/r} B_0^{l/r} A_1^{k/r})^r. \end{aligned}$$

Since

$$\prod_i \xi_n^{li} = \begin{cases} 1, & \text{if } n = 2, \\ 1, & \text{if } n = 1 \text{ or } 2/l = 1, \\ (-1)^{\varphi(n)} & \text{otherwise,} \end{cases}$$

where $1 \le i \le n$ and $\gcd(i, n) = 1$, we have

$$R = \delta \prod_{i_1 j_1} (B_1^{l/r} A_0^{k/r} - \xi_{m_1}^{j_1} \xi_{n_1}^{-i_1} B_0^{l/r} A_1^{k/r})^{r\varphi(n)\varphi(m)/(\varphi(n_1)\varphi(m_1))},$$

where the product is taken over i_1, j_1 such that

$$\begin{aligned} &1 \le i_1 \le n_1, \qquad \gcd(i_1, n_1) = 1, \\ &1 \le j_1 \le m_1, \qquad \gcd(j_1, m_1) = 1. \end{aligned}$$

Suppose that $\gcd(n_1, m_1) = 1$. Then $n_1 j_1 - m_1 i_1$ runs over the set of all residues modulo $n_1 m_1$ coprime to the modulus, that is $\xi_{m_1}^{j_1} \xi_{n_1}^{-i_1} = \xi_{n_1 m_1}^{n_1 j_1 - m_1 i_1}$ runs over the set of all primitive $(n_1 m_1)$th roots of unity. Hence

$$R = \delta(\Phi_v(A_0^{k/r} B_1^{l/r}, A_1^{k/r} B_0^{l/r}))^{r\varphi(m)\varphi(n)/(\varphi(m_1)\varphi(n_1))}.$$

Suppose that $\gcd(m_1, n_1) = 2$. Then $m_1 = 2m_1'$ and $n = 2n_1'$. First let m_1' and n_1' be odd. We consider the residues modulo $2m_1'n_1'$ of the form

$$n_1'j_1 - m_1'i_1, \tag{8}$$

where

$$\begin{aligned} 1 \le i_1 \le n_1, \qquad & \gcd(i_1, n_1) = 1, \\ 1 \le j_1 \le m_1, \qquad & \gcd(j_1, m_1) = 1. \end{aligned}$$

Let us note that $2 | n_1'j_1 - m_1'i_1$ and integers (8) modulo $m_1'n_1'$ form the set of all residues coprime to the modulus. Hence $n_1'j_1 - m_1'i_1 = 2r_{ij}$, where r_{ij} runs over the set of all the residues modulo $m_1'n_1'$ coprime to the modulus. Thus

$$\xi_{m_1}^{j_1} \xi_{n_1}^{-i_1} = \xi_{2n_1'm_1'}^{n_1'j_1 - m_1'i_1} = \xi_{n_1'm_1'}^{r_{i_1j_1}}.$$

This proves the lemma if $\gcd(m_1, n_1) = 2$ and $m_1n_1/4$ is odd.

Let us now suppose that $\gcd(m_1, n_1) = 2$ and $m_1n_1/4$ is even. We can assume that m_1' is even and n_1' is odd. In such a case if $1 \le i' \le n_1'$ and $\gcd(i', n_1') = 1$, then either i' or $i' + n_1'$ is coprime to $2n_1'$. Moreover, if $1 \le j' \le m_1'$ and $\gcd(j', m_1') = 1$, then the numbers j' and $j' + m_1'$ are coprime to $2m_1'$. Therefore $n_1'j_1 - m_1'i_1$ is equal either to $n_1'j' - m_1'i'$ or $n_1'j' - m_1'i' + n_1'm_1'$ for i_1 and j_1 such that

$$\begin{aligned} 1 \le i_1 \le n_1, \qquad & \gcd(i_1, n_1) = 1, \\ 1 \le j_1 \le m_1, \qquad & \gcd(j_1, m_1) = 1. \end{aligned}$$

From these considerations it follows that the set of residues $n_1'j_1 - m_1'i_1$ modulo $2m_1'n_1'$ contains all the residues

$$\begin{aligned} n_1'j' - m_1'i', \qquad & n_1'j' - m_1'i' + n_1'm_1', \\ 1 \le i' \le n_1', \qquad & \gcd(i', n_1') = 1, \\ 1 \le j' \le m_1', \qquad & \gcd(j', m_1') = 1. \end{aligned} \tag{9}$$

Residues (9) are different modulo $2m_1'n_1'$ and are coprime to the modulus. Their number is equal to $2\varphi(m_1'n_1') = \varphi(2m_1'n_1')$. Thus residues (9) are all the residues coprime to $2m_1'n_1'$. Therefore

$$\xi_{m_1}^{j_1} \xi_{n_1}^{-i_1} = \xi_{2n_1'm_1'}^{n_1'j_1 - m_1'i_1}$$

runs over the set of all primitive $(2m_1'n_1')$th roots of unity. Thus the lemma is proved in this case as well.

COROLLARY 1. *For natural numbers m, n, s, and l such that $\gcd(m, n) = 1$ and integers $a, b \ne 0$,*

$$\Phi_{mn}(a^s, b) = \delta \operatorname{Res}(\Phi_m(a^{s_0}x^l, b), \Phi_n(x, a^{s_1})),$$

where $s = s_0 + ls_1$, $\delta = (-1)^l$, if $n = 1, m = 1, 2$, and $\delta = 1$ otherwise.

For natural numbers m, n such that $\gcd(m, n) = 1$ *and an integer A,*

$$\Phi_{mn}(A) = \delta \operatorname{Res}(\Phi_m(x), \Phi_n(x, A)),$$

where $\delta = -1$, if $n = 1, m = 1, 2$, and $\delta = 1$ otherwise.

The proof of this corollary is obvious.

REMARK 1. The identity (see [4])

$$a^s - b = f(a^{s_1}),$$

where $f(x) = a^{s_0}x^l - b$, is a particular case of the identity of the first assertion in Corollary 1 for $m = n = 1$.

Let a_i, b_j, $0 \le i, j \le 2$, be arbitrary complex numbers. We assume that

$$\begin{aligned} a_2b_1 - a_1b_2 &= A, \\ a_0b_2 - a_2b_0 &= B, \\ a_1b_0 - a_0b_1 &= C. \end{aligned} \tag{10}$$

LEMMA 5. *The resultant of the polynomials*

$$\begin{aligned} f(x) &= a_0x^n + a_1x^k + a_2, \\ g(x) &= b_0x^n + b_1x^k + b_2, \end{aligned}$$

where $a_0b_0 \ne 0$ and $\gcd(n, k) = 1$, *is equal to*

$$\operatorname{Res}(f, g) = (-1)^{(n+1)(k+1)}(B^n - C^{n-k}A^k).$$

PROOF. First we prove the assertion under the condition $a_2BC \ne 0$. It is easy to see that

$$\begin{aligned} F(x) &= b_0f(x) - a_0g(x) = Cx^k - B, \\ G(x) &= b_2f(x) - a_2g(x) = Bx^n - Ax^k. \end{aligned}$$

Therefore

$$\begin{aligned} \operatorname{Res}(F, G) &= \operatorname{Res}(Cx^k - B, Bx^{n-k} - A) \operatorname{Res}(Cx^k - B, x^k) \\ &= B^k \operatorname{Res}(Cx^k - B, Bx^{n-k} - A) \\ &= B^{2k}C^{n-k} \operatorname{Res}(x^k - BC^{-1}, x^{n-k} - AB^{-1}). \end{aligned}$$

By Swan's lemma, the last resultant is equal to $(-1)^k((AB^{-1})^k - (BC^{-1})^{n-k})$. Therefore

$$\operatorname{Res}(F, G) = (-1)^{k+1}B^k(B^n - C^{n-k}A^k). \tag{11}$$

On the other hand,

$$\mathrm{Res}(F,G)=(-1)^{nk}B^k\prod_\beta F(\beta),$$

where the product is taken over all roots β of the polynomial $G(x)$ with account of their multiplicities. Since $b_2 f(\beta)=a_2 g(\beta)$, we have

$$\mathrm{Res}(F,G)=(-1)^{nk}B^k\prod_\beta(b_0 f(\beta)-(a_0b_2/a_2)f(\beta))=\frac{(-1)^{n(k+1)}B^{n+k}}{a_2^n}\prod_\beta f(\beta)$$
$$=\frac{(-1)^{nk}B^k}{a_2^n}\mathrm{Res}(f,G)=\frac{(-1)^{nk}(B^k a_0^n}{a_2^n}\prod\alpha G(\alpha),$$

where α runs over the set of all roots of the polynomial $f(x)$ with account of their multiplicities. The last expression can be represented in the form

$$\left((-1)^n a_0^n\prod_\alpha g(\alpha)\right)(-1)^{nk}B^k=(-1)^{n(k+1)}B^k\,\mathrm{Res}(f,g).$$

Then from (11) we obtain

$$\mathrm{Res}(f,g)=(-1)^{(n+1)(k+1)}(B^n-C^{n-k}A^k).$$

We proved the assertion under the condition that $C=a_1b_0-a_0b_1$, $B=a_0b_2-a_2b_0$, and a_2 are not equal to zero, that is, we proved the equality as an equality between two polynomials in many variables on some open subset of the affine $(2n+2)$-space over $\mathbf{C}$. Then from the considerations in [8, pp. 48–49], the equality of these polynomials holds on the whole space. Thus, these polynomials are equal and lemma is proved.

LEMMA 6. *For complex numbera$_0$, a_2, b_j, $0\le j\le n$, the resultant of the polynomials*

$$f(x)=a_0x^2+a_2,$$
$$g(x)=b_0x^n+b_1x^{n-1}+\ldots+b_n,$$

where $a_0b_0\neq 0$, is of the form

$$\mathrm{Res}(f,g)=(a_0^m b_n-a_0^{m-1}a_2b_{n-2}+\ldots+(-1)^m a_2^m b_0)^2$$
$$+a_0a_2(a_0^{m-1}b_{n-1}-a_0^{m-2}a_2b_{n-3}+\ldots+(-1)^{m-1}a_2^{m-1}b_1)^2$$

for $n=2m$ and

$$\mathrm{Res}(f,g)=a_0(a_0^m b_n-a_0^{m-1}a_2b_{n-2}+\ldots+(-1)^m a_2^m b_1)^2$$
$$+a_2(a_0^m b_{n-1}-a_0^{m-1}a_2b_{n-3}+\ldots+(-1)^m a_2^m b_1)^2$$

for $n=2m+1$.

PROOF. We prove the lemma by induction. For $n = 1$

$$\operatorname{Res}(a_0x^2 + a_2, b_0x + b_1) = a_0b_1^2 + a_2b_0^2.$$

This equality is valid for any $b_1 \in \mathbf{C}$. Suppose that the assertion is true for $n < n_1$ and prove it for $n = n_1$. First we assume that $b_1 \neq 0$. Let us consider the polynomial

$$\begin{aligned} G(x) &= b_0x^{n-2}f(x) - a_0g(x) \\ &= -a_0b_1x^{n-1} + (a_2b_0 - a_0b_2)x^{n-2} - a_0b_3x^{n-3} - \dots - a_0b_n \end{aligned}$$

of degree $n - 1$. It is obvious that $\operatorname{Res}(f, G) = a_0 \operatorname{Res}(f, g)$. On the other hand, by the induction assumption we have for $n - 1 = 2m$

$$\begin{aligned} \operatorname{Res}(f, G) &= (-a_0^{m+1}b_n + a_0^m a_2 b_{n-2} - \dots + (-1)^{m+1}a_0a_2^m b_1)^2 \\ &\quad + a_0a_2(-a_0^m b_{n-1} + a_0^{m-1}a_2b_{n-3} - \dots + (-1)^{m-1}a_2^{m-1}(a_2b_0 - a_0b_2)^2) \\ &= a_0(a_0(a_0^m b_n - a_0^{m-1}a_2b_{n-2} + \dots + (-1)^m a_2^m b_1)^2 \\ &\quad + a_2(a_0^m b_{n-1} - a_0^{m-1}a_2b_{n-3} + \dots + (-1)^m a_2^m b_1)^2). \end{aligned}$$

Similarly, for $n - 1 = 2m - 1 = 2(m - 1) + 1$,

$$\begin{aligned} \operatorname{Res}(f, G) &= a_0(-a_0^m b_n + a_0^{m-1}a_2b_{n-2} - \dots + (-1)^{m-1}a_2^{m-1}(a_2b_0 - a_0b_2))^2 \\ &\quad + a_2(-a_0^m b_{n-1} + a_0^{m-1}a_2b_{n-3} - \dots + (-1)^m a_0a_2^{m-1}b_1)^2 \\ &= a_0((a_0^m b_n - a_0^{m-1}a_2b_{n-2} + \dots + (-1)^m a_2^m b_0)^2 \\ &\quad + a_0a_2(a_0^{m-1}b_{n-1} - a_0^{m-2}a_2b_{n-3} + \dots + (-1)^{m-1}a_2^{m-1}b_1)^2). \end{aligned}$$

Thus, if $b_1 \neq 0$, then the assertion is true for $n = n_1$. As in the proof of the previous lemma, from [8, pp. 48–49] it follows that for $n = n_1$ the assertion is true for all values of b_1. The lemma is thus proved.

4. In this section we derive representation (3) for integers N of some special form, where the coefficients of polynomials (4) do not exceed $N^{1/(n_1+n_2)+\varepsilon}$ in absolute value for small $\varepsilon > 0$.

4.1. Let $N = B^n - A^kC^{n-k}$, where

$$|A|, |B|, |C| \leq X \tag{12}$$

and $\gcd(n, k) = 1$. We may assume that A, B, C are coprime. In order to use Lemma 5, that is, to represent N as the resultant of polynomials of degree n whose coefficients are small, we have to solve system of equations (10). In another paper we prove the following assertion.

THEOREM 1. *Let A, B, C be random integers satisfying constraint* (12). *Then the probability that system* (10) *has a solution such that*

$$\max_{i,j}\{|a_i|, |b_j|\} \geq l$$

for $X^{1/2} = o(l)$ *tends to* 0 *as* $X \to \infty$.

It follows from this theorem that if $X = N^{1/n+\varepsilon}$ for small $\varepsilon > 0$ in (12), then with probability close to one, such N can be represented as a resultant of polynomials of degree n whose coefficients are no greater than $N^{1/2n+\varepsilon}$ in absolute value. The minimal solutions of system (10) can be found in $O(\ln^3 X)$ binary operations.

4.2. We use Lemma 6 to derive (3) for some N in the case where the polynomial $f(x)$ is of degree 2.

THEOREM 2. *Let* $N = \lambda X^2 + \mu Y^2$, *where* $\gcd(\lambda, \mu) = 1$, $|\mu|, |\lambda| \leq N^{1/(n+2)}$, $n \geq 1$, *and* $|X|, |Y| \leq N^{1/2+\delta}$, $\delta \geq 0$. *Let*

$$\varepsilon = \frac{3}{n(n+2)} + \frac{2\delta}{n}.$$

Then there exist polynomials

$$f(x) = a_0 x^2 + a_2,$$
$$g(x) = b_0 x^n + b_1 x^{n-1} + \ldots + b_n$$

over **Z** *such that*

$$|a_0|, |a_2|, |b_i| \leq \frac{3n+9}{2} N^{1/(n+2)+\varepsilon}, \qquad 0 \leq i \leq n,$$

and

$$\operatorname{Res}(f, g) \equiv 0 \pmod{N}.$$

PROOF. We may assume that $0 < |\mu| < \lambda$. Let $\mu = \mu_1 \mu_2^2$, where μ_1 is a square-free integer. Let $\alpha > 0$ be the maximal divisor of μ_1 such that

$$\alpha\lambda \leq 2N^{1/(n+2)+\varepsilon} = 2B.$$

We suppose that $\lambda_1 = [(B/(\alpha\lambda))^{1/2}] + 1$. Then we set $a_0 = \alpha\lambda\lambda_1^2$, $a_2 = \mu_1/\alpha$. Hence it follows that $\gcd(a_0, a_2) = 1$. Since $\gcd(\alpha\lambda, a_2) = 1$, we have to show that $\gcd(\lambda_1, a_2) = \lambda_2 = 1$. Let $\lambda_2 \geq 2$. Hence $\lambda_1 \geq 2$ and therefore $\alpha\lambda \nleq B$. Then λ_2 divides μ_1 while

$$\alpha\lambda\lambda_2 \leq \alpha\lambda([(B/(\alpha\lambda))^{1/2}] + 1) \leq (\alpha\lambda B)^{1/2} + \alpha\lambda \leq 2B.$$

This contradicts the assumption on α. Thus $\gcd(a_0, a_2) = 1$. Furthermore we have the inequalities

$$B < a_0 < 6B, \qquad |a_2| < a_0.$$

First we consider the case $n = 2m + 1$. We have

$$\alpha\lambda_1^2 N = a_0 X^2 + a_2(\alpha\lambda_1\mu_2 Y)^2,$$

where

$$|\alpha\lambda_1\mu_2 Y| \le |(\alpha\mu_2(B/(\alpha\lambda))^{1/2} + \alpha\mu_2)N^{1/2+\delta}| \le 2N^{\nu}, \qquad \nu = \frac{1}{2} + \frac{1}{n+2} + \frac{\varepsilon}{2} + \delta.$$

We find the coefficients b_i, $0 \le i \le n$, from the equalities

$$a_0^m b_n - a_0^{m-1} a_2 b_{n-2} + \ldots + (-1)^m a_2^m b_1 = X,$$
$$a_0^m b_{n-1} - a_0^{m-1} a_2 b_{n-3} + \ldots + (-1)^m a_2^m b_1 = \alpha\lambda_1\mu_2 Y$$

in the following way. Let b_1 be a residue modulo a_0 such that

$$(-1)^m a_2^m b_1 \equiv X \pmod{a_0},$$

let b_3 be a residue modulo a_0 such that

$$(-1)^{m-1} a_2^{m-1} b_3 \equiv (X - (-1)^m a_2^m b_1)/a_0 \pmod{a_0},$$

and so on. Finally, we obtain

$$b_n = (X + a_0^{m-1} a_2 b_{n-2} + \ldots + (-1)^{m-1} a_2^m b_1)/a_0^m.$$

It is clear that the residues can be chosen such that $|b_i| \le a_0/2$, $i = 1, 3, \ldots, n-2$, and therefore

$$|b_n| \le \frac{|X|}{a_0^m} + \frac{m a_0}{2} \le \frac{3n-1}{2} N^{1/(n+2)+\varepsilon}.$$

Similarly, we find $|b_i| \le a_0/2$, $i = 0, 2, \ldots, n-3$, and

$$|b_{n-1}| \le |\alpha\lambda_1\mu_2 Y|/a_0^m + m a_0/2 \le \frac{3n+1}{2} N^{1/(n+2)+\varepsilon}.$$

Then the assertion for $n = 2m + 1$ follows from Lemma 6.

Let $n = 2m$. We have

$$\lambda\lambda_1^2 N = (\lambda\lambda_1 X)^2 + a_0 a_2 (\mu_2 Y)^2,$$

where

$$|\lambda\lambda_1 X| \le |(\lambda(B/(\alpha\lambda))^{1/2} + \lambda)N^{1/2+\delta}| \le 2N^{\nu_1}, \qquad |\mu_2 Y| \le N^{\nu_2},$$

$$\nu_1 = \frac{1}{2} + \delta + \frac{1}{n+2} + \frac{1}{2}\varepsilon, \qquad \nu_2 = \frac{1}{2} + \delta + \frac{1}{2(n+2)}.$$

We find the coefficients b_i, $0 \le i \le n$, from the equalities

$$a_0^m b_n - a_0^{m-1} a_2 b_{n-2} + \ldots + (-1)^m a_2^m b_1 = \lambda\lambda_1 X,$$
$$a_0^{m-1} b_{n-1} - a_0^{m-2} a_2 b_{n-3} + \ldots + (-1)^{m-1} a_2^{m-1} b_1 = \mu_2 Y.$$

The remaining part of the proof in this case is similar to the corresponding part of the proof above.

4.3. If $\gcd(m, n) = 1$ and $n > 1$, then for the integer $N = \Phi_{mn}(A)$ from the corollary to Lemma 4 we have

$$N = \mathrm{Res}(\Phi_m(x), \Phi_n(x, A)). \tag{13}$$

Let $\varphi(m) = (3 \ln N/(2 \ln \ln N))^{1/3}$ and $\varphi(n) = o(\varphi(m))$ as $N \to \infty$. The coefficients of the polynomials $\Phi_m(x)$ and $\Phi_n(x, A)$ do not exceed $2^{\varphi(m)}$ and $(2A)^{\varphi(n)}$ in absolute value respectively. To obtain one row of the matrix M, we must find two smooth numbers which do not exceed $(\varphi(m) + 1)(2L)^{\varphi(m)}$ and $(\varphi(n) + 1)(2LA)^{\phi(n)}$, respectively, for some bound L. In this case N can be factored using $L[1/3, c]$ operations for $c = 1.526$.

On the other hand, we have

$$\Phi_m(A^n) = \prod_{d|n} \Phi_{md}(A) \equiv 0 \pmod N, \tag{14}$$

where d runs all the divisors of n including 1 and n. Thus A^n is a root of the polynomial $\Phi_m(x)$ modulo N. We can apply the NFS. Here to obtain one row to the matrix, we must find two smooth integers which do not exceed $(\varphi(m) + 1)(2L)^{\varphi(m)}$ and $(L + 1)A^n$ in absolute value, respectively, for some bound L. If n has small divisors, that is, if n is much greater than $\varphi(n)$, then the value A^n is considerably greater than $A^{\varphi(n)}$. Thus, to factor such integers N, (13) should be preferred over (14).

As an illustration of the implementation of this method we factor

$$N = \Phi_{12}(14) = 14^4 - 14^2 + 1 = 38221.$$

We have

$$N = \Phi_{12}(1, 14) = \mathrm{Res}(\Phi_3(x), \Phi_4(14x)),$$

where

$$\Phi_3(x) = x^2 + x + 1,$$
$$\Phi_4(14x) = 14^2x^2 + 1.$$

We take the roots of these polynomials over $\mathbf{Q}$ in the fields $\mathbf{Q}(\alpha)$, $\mathbf{Q}(i)$, respectively, where $\alpha = 1^{1/3}$, $i = \sqrt{-1}$. The rings of integers $\mathbf{Z}[\alpha]$ and $\mathbf{Z}[i]$ of these fields are unique factorization domains. Below we list the primes of these rings of degree 1 of norm no greater than 31 and 29 respectively. For $\mathbf{Z}[i]$ we add the prime algebraic integer 7 of degree 2.

The unities of $\mathbf{Z}[\alpha]$ form a group of order 6 generated by $-\alpha$. The unities of $\mathbf{Z}[i]$ is a group of order 4, it is generated by i. Let β be the root of the polynomial $14^2x^2 + 1$ equal to $\beta = i/14$. It is easy to see that $-14^2 \pmod N$ is a common root of the polynomials x^2+x+1 and 14^2x^2+1 modulo N. Thus we have the homomorphisms from the rings $\mathbf{Z}[\alpha]$ and $\mathbf{Z}[\beta] = \mathbf{Z}[i] \cup \mathbf{Z}[i]/14 \cup \mathbf{Z}[i]/14^2 \cup \ldots$ onto the ring $\mathbf{Z}/N$ defined by $\alpha, \beta \to -14^2 \pmod N$ respectively.

We sieve through the pairs of coprime integers c, d ranging $0 < c \leq 5$, $-23 \leq d \leq 22$ to find the pairs such that $c - d\alpha$ and $c - d\beta = (14c - di/14$ are products of the primes

Table 1.

#	prime $\in Z[i]$	Norm
2	$1-i$	2
3	$2-i$	5
4	$2+i$	5
5	$3-2i$	13
6	$3+2i$	13
7	$4-i$	17
8	$4+i$	17
9	$5-2i$	29
10	$5+2i$	29
11	7	49

Table 2.

#	prime $\in Z[\alpha]$	Norm
12	$1+2\alpha$	3
13	$2-\alpha$	7
14	$1-2\alpha$	7
15	$3-\alpha$	13
16	$1-3\alpha$	13
17	$2-3\alpha$	19
18	$3-2\alpha$	19
19	$5-\alpha$	31
20	$1-5\alpha$	31

given in Tables 1 and 2 respectively. Every such factorization provides a row whose entries are equal to exponents modulo 2 of the primes of $\mathbf{Z}[\alpha]$ and $\mathbf{Z}[i]$ in $c-d\alpha$, $c-d\beta$. In doing so the columns numbered $2, 3, \ldots, 11$ correspond to the primes of $\mathbf{Z}[i]$ and the columns numbered $12, 13, \ldots, 20$ correspond to the primes of $\mathbf{Z}[\alpha]$. The first entry equals 1 if i appears in the factorization of $c-d\beta$ and equals 0 otherwise. No account has been taken in this construction of $\alpha, -1$ in the factorization of $c-d\alpha$ for α is a square in $\mathbf{Z}[\alpha]$ and -1 is a square in $\mathbf{Z}[i]$. Thus we get 17 rows. To obtain further rows, we note that 13 can be factored in $\mathbf{Z}[\alpha]$ and $\mathbf{Z}[i]$ as well:

$$13 = -\alpha^2(3-\alpha)(1-3\alpha), \qquad 13 = (3-2i)(3+2i),$$

the prime integer 7 is prime in $\mathbf{Z}[i]$ and splits in $\mathbf{Z}[\alpha]$:

$$7 = -\alpha^2(2-\alpha)(1-2\alpha).$$

Thus we get 2 rows for 7 appeared in the decomposition of the numbers $c-d\beta$. Finally, since

$$1 \equiv -14^2(14^2-1) \pmod{N}$$

and $14^2-1 = 3\cdot 5\cdot 13$, the congruence

$$-14^2\cdot 5\cdot 13 \equiv 3^{-1} \pmod{N}$$

provides one row for 5 and 13 split in $\mathbf{Z}[i]$ and 3 in $\mathbf{Z}[\alpha]$. Hence we have the matrix 20×20 with the rows numbered by pairs c, d and 3^{-1} for the last row which is given in Table 3.

The rank of this matrix modulo 2 equals 18. Therefore we have two independent combinations of the rows which vanish modulo 2:

$$\begin{matrix} 0&0&1&0&1&0&0&0&0&0&0&0&0&1&1&0&0&0&1 \\ 1&0&0&0&0&0&0&0&1&1&0&1&0&0&1&0&0&0&0&0 \end{matrix}$$

Table 3.

		1	2	3	4	5	6	7	8	9	10	11	12	13	14	15	16	17	18	19	20
0	−1	0	0	0	0	0	0	0	0	0	0	1	0	0	0	0	0	0	0	0	0
1	2	1	1	0	0	0	0	0	0	0	0	1	0	0	1	0	0	0	0	0	0
1	−2	0	1	0	0	0	0	0	0	0	0	1	1	0	0	0	0	0	0	0	0
3	2	1	1	0	0	1	0	1	0	0	0	1	0	0	0	0	0	0	1	0	0
3	−2	0	1	0	0	0	1	0	1	0	0	1	0	1	0	0	0	0	0	0	0
1	5	0	0	0	0	1	0	0	1	0	0	1	0	0	0	0	0	0	0	0	1
5	1	1	0	0	0	0	0	0	0	0	1	1	0	0	0	0	0	0	0	1	0
1	−5	0	0	0	0	0	1	1	0	0	0	1	1	0	1	0	0	0	0	0	0
5	−1	1	0	0	0	0	0	0	0	1	0	1	1	1	0	0	0	0	0	0	0
1	−6	1	1	0	0	0	0	0	0	1	0	1	0	0	0	0	0	0	0	1	0
1	7	0	0	1	0	0	0	0	0	0	0	0	1	0	0	0	0	0	1	0	0
4	7	0	0	0	1	1	0	0	0	0	0	0	1	0	0	0	0	0	0	1	0
2	−7	0	0	0	0	0	0	0	1	0	0	0	1	0	0	0	1	0	0	0	0
3	14	0	1	0	1	0	0	0	0	0	0	0	0	0	0	1	0	1	0	0	0
1	18	0	1	0	1	1	0	0	0	0	0	1	0	1	0	0	0	0	0	0	0
1	22	0	1	1	0	0	0	0	1	0	0	1	1	0	0	0	0	0	0	0	0
1	−23	1	0	0	0	0	0	0	0	0	1	1	1	0	0	0	0	0	0	0	0
13	0	0	0	0	0	1	1	0	0	0	0	0	0	0	0	1	1	0	0	0	0
7	0	0	0	0	0	0	0	0	0	0	0	1	0	1	1	0	0	0	0	0	0
3^{-1}	0	0	0	1	1	1	1	0	0	0	0	0	0	0	0	0	0	0	0	0	0

We consider the first combination. It turns out that the numbers

$$\gamma = (1 + 2\beta)(3 + 2\beta)(1 - 18\beta)(1 - 22\beta)(-14^2 \cdot 5 \cdot 13) \in \mathbf{Q}(i),$$
$$\delta = (1 + 2\alpha)(3 + 2\alpha)(1 - 18\alpha)(1 - 22\alpha)3^{-1} \in \mathbf{Q}(\alpha)$$

have the same image upon the homomorphisms $\mathbf{Z}[\beta], \mathbf{Z}[\alpha] \to \mathbf{Z}/N$. Moreover,

$$1 + 2\beta = (1 - i)(2 + i)^2 7^{-1},$$
$$3 + 2\beta = (1 - i)(3 + 2i)(4 + i)7^{-1},$$
$$1 - 18i = (1 - i)(2 + i)(3 - 2i)7^{-1},$$
$$1 - 22\beta = (1 - i)(2 - i)(4 + i)7^{-1},$$
$$-14^2 \cdot 5 \cdot 13 = (2 - i)(2 + i)(3 - 2i)(3 + 2i)(-14^2).$$

Therefore,

$$\gamma = 2i(1 - i)^2(2 + i)^2(3 + 2i)(3 - 2i)(4 + i)(2 - i)7^{-1} = (260 + 3120\beta)^2.$$

Similarly, since

$$\begin{aligned} 3 + 2\alpha &= -\alpha^2(2 - \alpha), \\ 1 - 18\alpha &= (2 - \alpha)^3, \\ 1 - 22\alpha &= \alpha^2(1 + 2\alpha)(3 - \alpha)^2, \\ 3 &= -(1 + 2\alpha)^2, \end{aligned}$$

we have $\delta = (\alpha^2(2 - \alpha)^2(3 - \alpha))^2 = (27 + 4\alpha)^2$. Consequently,

$$(260 + 3120(-14^2))^2 \equiv (27 + 4(-14^2))^2 \pmod N$$

or $(611260)^2 \equiv (757)^2 \pmod N$. Thus, we find that

$$\gcd(611260 + 757, N) = 37, \qquad \gcd(611260 - 757, N) = 1033,$$

and $N = 37 \cdot 1033$.

REFERENCES

1. J. M. Pollard, Factoring with cubic integers. *Lect. Notes Math.* (1993) **1554**, 4–10.
2. L. M. Adleman, Factoring numbers using singular integers. *Proc. 23rd Annual ACM Sympos. Theory of Computing (STOC)* 1991, pp. 64–71.
3. D. Coppersmith, Modifications to the number field sieve. *J. Cryptology* (1993) **6**, 169–180.
4. A. K. Lenstra, H. W. Lenstra Jr., M. S. Manasse, and J. M. Pollard, The number field sieve. *Proc. 22nd Annual ACM Sympos. Theory of Computing (STOC)* 1990, 564–572.
5. J. P. Buhler, H. W. Lenstra Jr., and C. Pomerance, Factoring integers with the number sieve. *Lect. Notes Math.* (1993) **1554**, 50–94.
6. B. L. van der Waerden, *Algebra* Springer, Berlin, 1971.
7. E. R. Berlekamp, *Algebraic Coding Theory.* McGraw–Hill, New York, 1968.
8. I. R. Shafarevich, *Foundations of Algebraic Geometry*, Vol. 1. Nauka, Moscow, 1988 (in Russian).

Probabilistic Methods in Discrete Mathematics, pp. 65–86
V. F. Kolchin et al. (Eds.)

OPERATOR AND RECURSION EQUATIONS FOR RUNS IN RANDOM SEQUENCES

L. YA. SAVELYEV

Novosibirsk

Abstract — The joint distributions in the explicit form of some run characteristics in finite Markov sequences are considered. Scalar, matrix, and operator equations are derived for calculation of these distributions.
This research was supported by the Russian Foundation for Basic Research, grant 96–01–01496.

1. INTRODUCTION

This paper extends the results of the paper [1]. New characteristics are appended that allow us to replace the previously considered distribution of maximum run lengths by the joint distribution of the maximum run lengths and some other characteristics. This makes it possible to derive recursion equations for the exact joint distribution of the chosen run characteristics in random sequences. For a brief partial review of the results we refer to [2].

1.1. We consider a probability space $(\Omega, \mathcal{F}, \mathbf{P})$, a measurable space $(C, \mathcal{B})$, and a partition $\mathcal{A}$ of the set C into parts $A^\nu \in \mathcal{B}$ with indices ν from a set Γ. Consider a sequence of random variables $\xi(r)\colon \Omega \to C$ with $r \in \mathbf{N} = \{0, 1, 2, \dots\}$. By definition $\xi^{-1}(r, B) = \{\omega\colon \xi(r, \omega) \in B\} \in \mathcal{F}$ for all $r \in \mathbf{N}$ and $B \in \mathcal{B}$. We study characteristics of the ν-runs composed of the ν-events $\xi(r) \in A^\nu$ in the sequence $\xi = \xi(0)\dots\xi(n)$ for random variables $\xi(r)$ with successive numbers $r \le n$. We consider the following basic characteristics:

- the sum $x^\nu = x^\nu(\xi)$ of the ν-run lengths (equal to the total number of ν-events),
- the number $y^\nu = y^\nu(\xi)$ of ν-runs,
- the maximum $z^\nu = z^\nu(\xi)$ of the ν-run lengths,
- the number $v^\nu = v^\nu(\xi)$ of the longest ν-runs,
- the length $w^\nu = w^\nu(\xi)$ of the last ν-run,
- the minimum ${}^*z^\nu = {}^*z^\nu(\xi)$ of the ν-run lengths,

- the number ${}^*v^\nu = {}^*v^\nu(\xi)$ of the shortest ν-runs,

- the length ${}^*w^\nu = {}^*w^\nu(\xi)$ of the first ν-run.

Let also $\mu = \mu(\xi)$ denote the kind of the last event $\xi(n) \in A^\mu$.

The characteristics selected allow us to calculate many other characteristics of ν-runs in the sequence $\xi = \xi(0)...\xi(n)$ such that the mean length x^ν/y^ν, its ratio $x^\nu/(y^\nu z^\nu)$ to the length of the longest ν-run, the range $z^\nu - {}^*z^\nu$, the relative range $1 - {}^*z^\nu/z^\nu$, the compression y^ν/x^ν, the fractions v^ν/y^ν, ${}^*v^\nu/y^\nu$ of the longest and the shortest ν-runs, the sums of the lengths $v^\nu z^\nu$, ${}^*v^{\nu*}z^\nu$ of these runs, the fractions $v^\nu z^\nu/x^\nu$, ${}^*v^{\nu*}z^\nu/x^\nu$ of the events in these runs, and the ratios w^ν/z^ν, ${}^*w^\nu/z^\nu$, ${}^*z^\nu/z^\nu$.

The relative characteristics allow us to compare sequences ξ of different length. Only the ratios with non-zero denominators are considered.

1.2. We denote by $s^\nu = s^\nu(\xi)$ an arbitrary basic characteristic of ν-runs in the sequence ξ. Each of the 8 families $s = s(\xi)$ composed of common selected characteristics $s^\nu = s^\nu(\xi)$ for all $\nu \in \Gamma$ serves as a general ν-run characteristic in ξ. For example, the family $x = (x^\nu)$ describes the sums of the run lengths and the family $y = (y^\nu)$ describes the numbers of runs.

In the set Γ we single out the part $\Gamma(\xi)$ composed of the indices ν of the runs appeared in the sequence ξ:

$$\Gamma(\xi) = \{\nu: y^\nu(\xi) > 0\}.$$

It is clear that $|\Gamma(\xi)| \leq n + 1$. We introduce the general numerical characteristics of runs

$$\wedge s = \min s^\nu, \quad \vee s = \max s^\nu, \quad \Sigma s = \sum_{\nu \in \Gamma(\xi)} s^\nu.$$

It is natural to consider the ratios

$$s^\nu / \wedge s, \quad s^\nu / \vee s, \quad s^\nu / \Sigma s$$

and the general characteristics

$$\vee s - \wedge s, \quad \wedge s / \vee s$$

at the same time.

Provided the set Γ is finite, the minimums, the maximums, and the sums over $\nu \in \Gamma(\xi)$ can be replaced with the minimums, the maximums and the sums over $\nu \in \Gamma$. It follows from the definition that $\min s^\nu(\xi) \neq 0$ if $\nu \in \Gamma(\xi)$, but the equality $\min s^\nu(\xi) = 0$ is possible for $\nu \in \Gamma$.

It depends on special features of the problem which of the characteristics suggested is appropriate to choose for the description of the run structure in the sequence ξ. In 2.4, we introduce the indicators of indices which can help in such a choice.

1.3. The number n of the last variable $\xi(n)$ in the sequence $\xi = \xi(0)\dots\xi(n)$ of length $n+1$ will be used below in the characteristics instead of the argument ξ while deriving recursion equations. Instead of $s^\nu(\xi)$ and $s(\xi)$ we will write $s^\nu(n)$ and $s(n)$.

The families s of the basic characteristics form the vectors

$$\bar{s} = (x, y, z, v, w, {}^*z, {}^*v, {}^*w),$$
$$\bar{s}^\nu = (x^\nu, y^\nu, z^\nu, v^\nu, w^\nu, {}^*z^\nu, {}^*v^\nu, {}^*w^\nu).$$

The arguments ξ or n are written out in these vectors if necessary. The number n is determined by the equality

$$n = \sum_{\nu\in\Gamma} x^\nu(\xi) - 1.$$

Note that $x^\nu(\xi) = 0$ if $\nu \notin \Gamma(\xi)$. Therefore, the summation over Γ means the summation over $\Gamma(\xi)$. We suppose that $\bar{s} = (\bar{s}^\nu)$.

The index $\mu = \mu(\xi)$ of the last event $\xi(n) \in A^\mu$ connects the segment $\xi' = \xi(0)\dots\xi(n')$ with the sequence $\xi = \xi(0)\dots\xi(n)$, where $n' = n - w^\mu(\xi)$ is the number of the last variable $\xi(r)$ which is not contained in the last μ-run of ξ. The sequence ξ' determines the structure vectors $\bar{s}(\xi')$ and $\bar{s}^\nu(\xi')$, which are in correspondence with the vectors $\bar{s}(\xi)$ and $\bar{s}^\nu(\xi')$. In particular, it follows from the definitions that

$$\bar{s}^\lambda(\xi') = \bar{s}^\lambda(\xi)$$

if $\lambda \neq \mu$.

If the sequence $\xi = \xi(0)\dots\xi(n)$ forms a complete μ-run and $w^\mu(n) = n+1$, then by definition $n' = -1$ and $\bar{s}(n') = 0$, $\bar{s}^\nu(n') = 0$ for all $\nu \in \Gamma$.

2. THE FORMULATION OF THE PROBLEM

The problem is stated for arbitrary random sequences, but the results are obtained only for Markov chains. For all necessary definitions, statements, and proofs we refer to [3, 4].

2.1. We assume that the sets C and Γ are finite and that the algebra $\mathcal{B}$ consists of all the subsets of the set C.

The basic characteristics $s^\nu = s^\nu(\xi)$ can be expressed in terms of the indicators $u_n^\nu(r)$ of the events $\xi(r) \in A^\nu$, $0 \le r \le n$, supplemented formally with the values $u_n^\nu(r) = 0$ for integer $r < 0$ and $r > n$:

$$u_n^\nu(r) = \begin{cases} 1, & 0 \le r \le n,\ \xi(r) \in A^\nu, \\ 0, & r < 0, \text{ or } r > n, \text{ or } 0 \le r \le n,\ \xi(r) \notin A^\nu. \end{cases}$$

For integer r, a, b, we set

$$\begin{aligned}
\bar{u}_n^\nu(r) &= 1 - u_n^\nu(r), \\
u_n^\nu(a,b) &= \prod_{a\le r\le b} u_n^\nu(r), \\
u_n^\nu(a,b) &= 0, && a > b, \\
\bar{u}_n^\nu(a,b) &= \prod_{a\le r\le b} \bar{u}_n^\nu(r), \\
\bar{u}_n^\nu(a,b) &= 1, && a > b, \\
s_n^\nu(a,b) &= \bar{u}_n^\nu(a-1)u_n^\nu(a,b)\bar{u}_n^\nu(b+1), \\
x_n^\nu(a,b) &= (b-a+1)s_n^\nu(a,b).
\end{aligned}$$

For $0 \le r \le n$ and $0 \le a \le b \le n$, this equalities determine the indices of the following events: $\xi(r) \notin A^\nu$, $\xi(a) \in A^\nu, \dots, \xi(b) \in A^\nu$, $\xi(a) \notin A^\nu, \dots, \xi(b) \notin A^\nu$, the ν-run on the segment $[a, b]$, and the length $x_n^\nu(a, b)$ of this run (equal to zero if there is no run).

The ν-run characteristics considered are expressed by the following equalities:

$$\begin{aligned}
x^\nu(n) &= \sum_{ab} x_n^\nu(a,b), \\
y^\nu(n) &= \sum_{ab} s_n^\nu(a,b), \\
z^\nu(n) &= \max_{ab} x_n^\nu(a,b), \\
v^\nu(n) &= \sum_{ab} \{1 - \operatorname{sign}(z^\nu(n) - x_n^\nu(a,b))\} s_n^\nu(a,b), \\
w^\nu(n) &= \sum_{ab} x_n^\nu(a,b)\bar{u}_n^\nu(b+1,n), \\
{}^*z^\nu(n) &= \min_{ab}\{x_n^\nu(a,b) \neq 0\}, \\
{}^*v^\nu(n) &= \sum_{ab} \{1 - \operatorname{sign}(x_n^\nu(a,b) - {}^*z_n^\nu)\} s_n^\nu(a,b), \\
{}^*w^\nu(n) &= \sum_{ab} \bar{u}_n^\nu(0,a-1)x_n^\nu(a,b).
\end{aligned} \tag{1}$$

If $x_n^\nu(a,b) = 0$ for all a, b, then by definition ${}^*z^\nu = 0$. We set also $u^\nu(n) = u_n^\nu(n)$. The families u_n^ν, x_n^ν, and s_n^ν have finite support for all ν, n, therefore the right-hand sides of the equalities have a meaning.

Formal expressions of the characteristics considered can be easily algorithmized and can be useful for calculations on a computer. This is why much attention is paid in this paper to formal expressions which have a simple informal meaning.

2.2. The following assertion is valid.

LEMMA 1. *The equalities*

$$x^{\vee}(n) = \sum_{0 \le r \le n} u_n^{\vee}(r),$$
$$y^{\vee}(n) = \sum_{0 \le r \le n} \bar{u}_n^{\vee}(r-1) u_n^{\vee}(r).$$

hold.

PROOF. Let $n = 0$. Then

$$x^{\vee}(0) = x_0^{\vee}(0,0) = s_0^{\vee}(0,0) = \bar{u}_0^{\vee}(0)$$

and

$$y^{\vee}(0) = s_0^{\vee}(0,0) = \bar{u}_0^{\vee}(0-1) u_0^{\vee}(0).$$

Thus for $n = 0$ the equalities of the lemma hold.

Let $m \in \mathbf{N}$ and the equalities hold for $0 \le n \le m$. Then they are valid for $n = m+1$ as well. Indeed, if $u_{m+1}^{\vee}(m+1) = 0$, then $u_{m+1}^{\vee}(r) = u_m^{\vee}(r)$ and $x_{m+1}^{\vee}(a, m+1) = s_{m+1}^{\vee}(a, m+1) = 0$ for all $a \in \mathbf{Z}$ and

$$x^{\vee}(m+1) = x^{\vee}(m) = \sum_{0 \le r \le m} u_m^{\vee}(r) = \sum_{0 \le r \le m+1} u_{m+1}^{\vee}(r),$$
$$y^{\vee}(m+1) = y^{\vee}(m) = \sum_{0 \le r \le m} \bar{u}_m^{\vee}(r-1) u_m^{\vee}(r) = \sum_{0 \le r \le m+1} \bar{u}_{m+1}^{\vee}(r-1) u_{m+1}^{\vee}(r).$$

If $u_{m+1}^{\vee}(m+1) = 1$, then

$$\begin{aligned} u_c^{\vee}(c-1) &= 0, & & \\ u_c^{\vee}(r) &= u_{m+1}^{\vee}(r), & & 0 \le r < c, \\ u_{m+1}^{\vee}(r) &= 1, & & c \le r \le m+1, \\ s_{m+1}^{\vee}(c, m+1) &= 1 & & \end{aligned}$$

for some $c \le m+1$ and

$$\begin{aligned} x^{\vee}(m+1) &= \sum_{0 \le a \le b < c} x_{m+1}^{\vee}(a,b) + ((m+1) - c + 1) \\ &= \sum_{0 \le r < c} u_{m+1}^{\vee}(r) + \sum_{c \le r \le m+1} u_{m+1}^{\vee}(r) = \sum_{0 \le r \le m+1} u_{m+1}^{\vee}(r), \\ y^{\vee}(m+1) &= \sum_{0 \le a \le b < c} s_{m+1}^{\vee}(a,b) + 1 = \sum_{0 \le r < c} \bar{u}_{m+1}^{\vee}(r-1) u_{m+1}^{\vee}(r) \\ &\quad + \sum_{c \le r \le m+1} \bar{u}_{m+1}^{\vee}(r-1) u_{m+1}^{\vee}(r) = \sum_{0 \le r \le m+1} \bar{u}_{m+1}^{\vee}(r-1) u_{m+1}^{\vee}(r). \end{aligned}$$

Thus, we proved by induction that the equalities of the lemma hold for every $n \in \mathbf{N}$.

2.3. Let $\theta \notin C$ and the family of constants $\xi(r)\colon \Omega \to \{\theta\}$, $r = -1, -2, \ldots$ be given. We set $u_n^\nu(r) = x^\nu(n) = y^\nu(n) = z^\nu(n) = v^\nu(n) = w^\nu(n) = {}^*z^\nu(n) = {}^*v^\nu(n) = {}^*w^\nu(n) = 0$ for $n = -1, -2, \ldots$. This complement allows us to consider the random variables $\xi(r)$ and run characteristics of these variables for every $r \in \mathbf{Z}$. It can be easily checked that all the formulas derived for the run characteristics hold for $n < 0$ as well. The supplement of zero components makes it possible not to write out the region of summation without necessity, to summarize over all integers, and not to watch the domains of the variables considered.

Furthermore, the supplement performed permits us to assume that if $\mu \in \Gamma$, then every μ-run $\xi(a) \in A^\mu, \ldots, \xi(b) \in A^\mu$ is preceded by the event $\xi(a-1) \notin A^\mu$.

2.4. In order to have a possibility to choose the characteristics necessary for solution of a particular problem from the set of all characteristics, we introduce the indicators $X = R, S, T, U, V, W, {}^*T, {}^*U, {}^*W$ on Γ and consider the random vectors $u(n)$, $Xs(n)$ composed of the random variables $u^\nu(n), Xs^\nu(n) = X(\nu)s^\nu(n)$, where $X = R, S, T, U, V, W, {}^*T, {}^*U, {}^*W$ and $s = x, y, z, v, w, {}^*z, {}^*v, {}^*w$ respectively. For instance,

$$Rx^\nu(n) = R(\nu)x^\nu(n),$$
$$Sy^\nu(n) = S(\nu)y^\nu(n),$$
$$Tz^\nu(n) = T(\nu)z^\nu(n).$$

If the indicator $X(\nu) = 0$, then the appropriate characteristic $s^\nu(n)$ is excluded, i.e., $Xs^\nu(n) = 0$, and if $X(\nu) = 1$, then $Xs^\nu(n) = s^\nu(n)$. We denote the identity on Γ as Γ as well. In [2], the case $W = \Gamma$ and ${}^*T = {}^*U = {}^*W = 0$ was studied. In this paper, as a rule, we also assume that $W = \Gamma$. In this case, $Ww^\nu(n) = w^\nu(n)$ and the length of the last ν-run in $\xi = \xi(0)\ldots\xi(n)$ remains the same for all $\nu \in \Gamma$. This allows us to derive the necessary recursion equations. If the characteristic w^ν is not necessary for solution of a particular problem, it can be excluded from the joint distribution of characteristics by means of summation over all values of w^ν.

2.5. Let us consider the families of indicators on Γ and their products with the functions $\overline{X} = (R, S, T, \ldots, {}^*W)$ and $\overline{X}(\nu) = (R(\nu), S(\nu), T(\nu), \ldots, {}^*W(\nu))$:

$$\overline{X}\overline{s} = (Rx, Sy, Tz, \ldots, {}^*W{}^*w),$$
$$\overline{X}\overline{s}^\nu = (Rx^\nu, Sy^\nu, Tz^\nu, \ldots, {}^*W{}^*w^\nu),$$
$$\overline{\Gamma} = (\Gamma, \Gamma, \Gamma, \ldots, \Gamma),$$
$$\overline{\Gamma}(\nu) = (1, 1, 1, \ldots, 1).$$

If $\overline{X} = \overline{\Gamma}$, then $\overline{X}\overline{s} = \overline{s}$ and $\overline{X}\overline{s}^\nu = \overline{s}^\nu$.

The run structure of the sequence $\xi = \xi(0)\ldots\xi(n)$ is described by the functions

$$\eta = (\overline{s}, w, u), \quad \eta^\lambda = \left(\overline{s}^\lambda, w^\lambda, u^\lambda\right), \quad X\eta = \left(\overline{X}\overline{s}, w, u\right), \quad X\eta^\lambda = \left(\overline{X}\overline{s}^\lambda, w^\lambda, u^\lambda\right).$$

The arguments ξ or n are written out if necessary. It is due to its exclusive role that the family $w = (w^\nu)$ is written out twice. If $\overline{X} = \overline{\Gamma}$, then $X\eta = \eta$ and $X\eta^\nu = \eta^\nu$. The families of positive integers

$$i = i(\nu), \quad j = j(\nu), \quad k = k(\nu), \quad l = l(\nu),$$
$$m = m(\nu), \quad {}^*k = {}^*k(\nu), \quad {}^*l = {}^*l(\nu), \quad {}^*m = {}^*m(\nu)$$

serve as the values of the random families $x, y, z, v, w, {}^*z, {}^*v, {}^*w$. The vectors

$$\bar{t} = (i, j, k, \dots, {}^*m),$$
$$\bar{t}(\nu) = (i(\nu), j(\nu), k(\nu), \dots, {}^*m(\nu))$$

serve as the values of the random vectors $\bar{s}$, $\bar{s}^\nu$, and the vectors

$$\overline{Xt} = (Ri, Sj, Tk, \dots, {}^*W\,{}^*m), \qquad \overline{Xt}(\lambda) = (Ri(\lambda), Sj(\lambda), Tk(\lambda), \dots, {}^*W\,{}^*m(\lambda))$$

serve as the values of $\overline{X\bar{s}}$ and $\overline{X\bar{s}}^\nu$. Finally, the indicators e^μ of the indices $\mu \in \Gamma$

$$e^\mu(\mu) = 1, \quad e^\mu(\lambda) = 0, \quad \lambda \neq \mu,$$

serve as the values of the random indicator $u = u(\xi)$.

It is convenient to identify the indicator e^μ with the index μ and write

$$\mu(\mu) = 1, \quad \mu(\lambda) = 0, \quad \lambda \neq \mu.$$

In just the same way, it is convenient to identify the indicators $R, S, T, \dots, {}^*W$ with the sets of indices where they are equal to 1. We agree further on, if it is convenient, to identify sets with their indicators.

The triples $t = (\bar{t}, m, \nu)$, $t^\lambda = (\bar{t}(\lambda), m(\lambda), \nu(\lambda))$ serve as the values of the random functions $\eta = (\bar{s}, w, u)$, $\eta^\lambda = (\bar{s}^\lambda, w^\lambda, u^\lambda)$, and the triples $Xt = (\overline{Xt}, m, \nu)$, $Xt^\lambda = (\overline{Xt}(\lambda), m(\lambda), \nu(\lambda))$, as the values of the random functions $X\eta$, $X\eta^\lambda$. In the triples t and t^λ, just as in the random functions considered, the families m and the values $m(\lambda)$ are repeated twice: as a part of $\bar{t}$, $\bar{t}^\lambda$, and separately. In the triples Xt and Xt^λ the components Wm and $Wm(\lambda)$ differ from m and $m(\lambda)$ when $W \neq \Gamma$ and $W(\lambda) = 0$.

The problem is to calculate the probabilities

$$p(t) = \mathbf{P}\{\eta = t\}, \qquad p(Xt) = \mathbf{P}\{X\eta = Xt\},$$

forming the distributions of the random functions η and $X\eta$. In order to solve the problem, we derive a system of recursion equations that will make it possible to compute consecutively the desired probabilities, especially, if the sequence of random variables under consideration is a Markov chain. For homogeneous Markov sequences, we derive also an operator equation. The solution of the operator equation gives us the generating function of the probabilities sought for.

The distribution of $X\eta$ can be obtained from the distribution of η by means of summation over the values of eliminated variables. Therefore the main attention concerns the distribution of η.

3. EQUATIONS FOR THE CHARACTERISTICS

We denote by $u^* = u^*(\xi) = u^*(n)$ the linear functional on $\mathbf{R}^{|\Gamma|}$, which has a coefficient $u = u(\xi) = u(n)$:

$$\varphi u^* = \sum_{\nu\in\Gamma} u^\nu \varphi(\nu) \qquad (\varphi\colon \Gamma \to \mathbf{R}).$$

If $u = e^\mu$, then $\varphi u^* = \varphi(\mu)$. But for $u = e^\mu$, the product φu of the functions φ and u on Γ takes the values $\varphi(\nu)u^\nu = \varphi(\mu)$ for $\nu = \mu$ and $\varphi(\nu)u^\nu = 0$ for $\nu \neq \mu$.

3.1. Let $u(n) = e^\mu$ and $w(n) = m$. Then $w(n)u^*(n) = m(\mu)$ and $n' = n - w(n)u^*(n) = n - m(\mu)$. The structure $\eta' = \eta(\xi')$ of the sequence $\xi' = \xi(0)\ldots\xi(n')$ together with $w = w(n)$ and $u = u(n)$ uniquely defines the structure $\eta = \eta(\xi)$ of the sequence $\xi = \xi(0)\ldots\xi(n)$. For all $\lambda \neq \mu$ the structures of λ-runs in ξ and ξ' coincide:

$$\bar{s}^\lambda(\xi) = \bar{s}^\lambda(\xi').$$

Only the structure of μ-runs is changed:

$$\bar{s}^\mu(\xi) \neq \bar{s}^\mu(\xi').$$

In particular, the sum of μ-run lengths and the number of μ-runs are changed:

$$x^\mu(\xi) = x^\mu(\xi') + m(\mu),$$
$$y^\mu(\xi) = y^\mu(\xi') + 1.$$

We consider the triple (t', Δ, μ) composed of the value t' of the function $\eta' = \eta(\xi')$, an integer $\Delta > 0$, and the index μ. The components of the vector t' are marked by primes. Every triple (t', Δ, μ) can be put in correspondence with a value t of the random function $\eta = \eta(\xi)$ so that $w^\mu = w^\mu(\xi) = \Delta$ and $u = u(\xi) = \mu$. The image $t = D(t', \Delta, \mu)$ of such a map $D\colon (t', \Delta, \mu) \to t$ can be described by a system of equalities for the components of the vector t. In these equalities we use the notation

$$\Delta(\mu) = k'(\mu) - \Delta, \quad {}^*\Delta(\mu) = {}^*k'(\mu) - \Delta$$

and

$$\Delta^\pm(\mu), \quad {}^*\Delta^\pm(\mu)$$

for the positive and the negative parts of $\Delta(\mu)$, ${}^*\Delta(\mu)$.

3.2. The following assertion is valid.

LEMMA 2. *The following equalities hold for the components of the vector $t = D(t', \Delta, \mu)$:*

$$i(\mu) = i'(\mu) + \Delta,\ j(\mu) = j'(\mu) + 1,\ k(\mu) = k'(\mu) \vee \Delta,$$
$${}^*k(\mu) = \operatorname{sign} j'(\mu)({}^*k'(\mu) \wedge \Delta) + (1 - \operatorname{sign} j'(\mu))\Delta,$$
$$l(\mu) = \operatorname{sign} \Delta^-(\mu) + (1 - |\operatorname{sign} \Delta(\mu)|)(l'(\mu) + 1) + \operatorname{sign} \Delta^+(\mu) l'(\mu),$$
$${}^*l(\mu) = \operatorname{sign} {}^*\Delta^+(\mu) + (1 - |\operatorname{sign} {}^*\Delta(\mu)|)({}^*l'(\mu) + 1) + \operatorname{sign} {}^*\Delta^-(\mu) {}^*l'(\mu),$$
$${}^*m(\mu) = \operatorname{sign} j'(\mu) {}^*m'(\mu) + (1 - \operatorname{sign} j'(\mu))\Delta,$$
$$m(\mu) = \Delta, \quad \nu = \mu.$$

Moreover, $\bar{t}(\lambda) = \bar{t}'(\lambda),\ \lambda \neq \mu$.

PROOF. The equality $\eta = t$ means that $\bar{s} = \bar{t}$, $w = m$, and $u = \nu$. The equalities $m(\mu) = \Delta$ and $\nu = \mu$ hold by the definition of the correspondence D. Hence, it is the μ-run of the length Δ which is the last in ξ. The meaning of the remaining equalities is also clear. It is not difficult to prove them using the definitions and equalities (1).

Observe first that for $n' = n - \Delta$,

$$s_n^\nu(a, b) = 0, \qquad n' < b < n,$$
$$s_n^\nu(a, n) = 0, \qquad a \neq n' + 1,$$

for all $\nu \in \Gamma$. For $\nu = \mu$, the first equality follows from

$$\bar{u}^\mu(b + 1) = 0, \quad n' \le b < n,$$

and for $\nu \neq \mu$, from

$$u^\nu(b) = 0, \quad n' < b \le n.$$

For $\nu = \mu$, the second equality follows from $u_n^\mu(n') = u_n^\mu(a, n) = 0$, $a \le n'$, and $\bar{u}_n^\mu(a - 1) = 0$, $n' < a - 1 \le n$, and for $\nu \neq \mu$ from $u^\nu(n) = 0$. Furthermore, it follows from the definitions that $u_n^\nu(r) = u_{n'}^\nu(r)$ if $r \le n'$, therefore

$$s_n^\nu(a, b) = s_{n'}^\nu(a, b), \quad a \le b \le n'.$$

We prove first the equalities of this lemma for $i(\nu)$ and $j(\nu)$. It follows from equalities (1) and the equalities obtained for $s_n^\nu(a, b)$ that

$$y^\nu(n) = \sum_{a \le b < n} s_n^\nu(a, b) + \sum_{a \le b = n} s_n^\nu(a, b)$$
$$= \sum_{a \le b \le n'} s_{n'}^\nu(a, b) + s_n^\nu(n' + 1, n) = y^\nu(n') + s_n^\nu(n' + 1, n).$$

If $\nu = \mu$, then $s_n^\nu(n' + 1, n) = 1$. If $\nu \neq \mu$, then $s_n^\nu(n' + 1, n) = 0$. Therefore

$$y^\mu(n) = y^\mu(n') + 1,$$
$$y^\lambda(n) = y^\lambda(n'), \quad \lambda \neq \mu.$$

In just the same way, replacing $s_n^\nu(a,b)$ with $x_n^\nu(a,b)$, we verify that

$$x^\nu(n) = x^\nu(n') + x^\nu(n'+1,n).$$

If $\nu = \mu$, then

$$x_n^\nu(n'+1,n) = n - (n'+1) + 1 = \Delta.$$

If $\nu \neq \mu$, then

$$x_n^\nu(n'+1,n) = 0.$$

Therefore

$$x^\mu(n) = x^\mu(n') + \Delta,$$
$$x^\lambda(n) = x^\lambda(n'), \quad \lambda \neq \mu.$$

The equalities obtained are equivalent to the equalities of the lemma for $i(\nu)$ and $j(\nu)$.

We prove now the equalities of the lemma for $k(\nu)$ and $^*k(\nu)$. It follows from equalities (1) and the equalities obtained for $s_n^\nu(a,b)$ that

$$\begin{aligned} z^\nu(n) &= (\max_{a\le b<n} x_n^\nu(a,b)) \vee (\max_{a\le b=n} x_n^\nu(a,b)) \\ &= (\max_{a\le b\le n'} x_{n'}^\nu(a,b)) \vee x_n^\nu(n'+1,n) = z^\nu(n') \vee x_n^\nu(n'+1,n). \end{aligned}$$

Therefore

$$z^\mu(n) = z^\mu(n') \vee \Delta,$$
$$z^\lambda(n) = z^\lambda(n'), \quad \lambda \neq \mu.$$

In just the same way, replacing the minimum with the maximum, we verify that

$$^*z^\mu(n) = {}^*z^\mu(n') \wedge \Delta,$$
$$^*z^\lambda(n) = {}^*z^\lambda(n'), \quad \lambda \neq \mu,$$

if $y^\mu(n') > 0$. But if $y^\mu(n') = 0$, then $^*z^\mu(n) = x_n^\mu(n'+1,n) = \Delta$. The equalities obtained are equivalent to the equalities of the lemma for $k(\nu)$ and $^*k(\nu)$.

We prove now the equalities of the lemma for $l(\nu)$ and $^*l(\nu)$. It follows from equalities (1) and the equalities obtained for $s_n^\nu(a,b)$ that

$$\begin{aligned} v^\nu(n) &= \sum_{a\le b<n} (1 - \operatorname{sign}(z^\nu(n) - x_n^\nu(a,b)))s_n^\nu(a,b) \\ &\quad + \sum_{a\le b=n} (1 - \operatorname{sign}(z^\nu(n) - x_n^\nu(a,b)))s_n^\nu(a,b) \\ &= \sum_{a\le b\le n'} (1 - \operatorname{sign}(z^\nu(n) - x_{n'}^\nu(a,b)))s_{n'}^\nu(a,b) \\ &\quad + (1 - \operatorname{sign}(z^\nu(n) - x_n^\nu(n'+1,n)))s_n^\nu(n'+1,n). \end{aligned}$$

Let $\Delta(\mu) < 0$, $z^\mu(n') < \Delta$. Then $\operatorname{sign}\Delta^-(\mu) = |\operatorname{sign}\Delta(\mu)| = 1$, $\operatorname{sign}\Delta^+(\mu) = 0$ and

$$z^\mu(n) = z^\mu(n') \vee \Delta > z^\mu(n') \geq x^\mu_{n'}(a,b),$$
$$\operatorname{sign}(z^\mu(n) - x^\mu_{n'}(a,b)) = 1,$$
$$\operatorname{sign}(z^\mu(n) - x^\mu_n(n'+1,n)) = \operatorname{sign}(z^\mu(n) - \Delta) = 0,$$
$$s^\mu_n(n'+1,n) = 1.$$

Therefore $v^\mu(n) = 1$ and for $\Delta(\mu) < 0$, the equality of the lemma holds for $l(\mu)$.

Let $\Delta(\mu) = 0$, $z^\mu(n') = \Delta$. Then

$$\operatorname{sign}\Delta^-(\mu) = \operatorname{sign}\Delta(\mu) = \operatorname{sign}\Delta^+(\mu) = 0$$

and

$$z^\mu(n) = z^\mu(n') = \Delta,$$
$$z^\mu(n) - x^\mu_{n'}(a,b) = z^\mu(n') - x^\mu_{n'}(a,b),$$
$$\operatorname{sign}(z^\mu(n) - \Delta) = 0.$$

Therefore $v^\mu(n) = v^\mu(n') + 1$ and for $\Delta(\mu) = 0$, the equality of the lemma holds for $l(\mu)$.

Let $\Delta(\mu) > 0$, $z^\mu(n') > \Delta$. Then $\operatorname{sign}\Delta^-(\mu) = 0$, $|\operatorname{sign}\Delta(\mu)| = \operatorname{sign}\Delta^+(\mu) = 1$ and

$$z^\mu(n) = z^\mu(n') > \Delta,$$
$$z^\mu(n) - x^\mu_{n'}(a,b) = z^\mu(n') - x^\mu_{n'}(a,b),$$
$$\operatorname{sign}(z^\mu(n) - \Delta) = 1.$$

Therefore $v^\mu(n) = v^\mu(n')$ and for $\Delta(\mu) > 0$ the equality of the lemma holds for $l(\mu)$.

Thus, for $l(\mu)$, the equality of the lemma always holds true. If $\nu \neq \mu$, it follows from what have been proved that $z^\nu(n) = z^\nu(n')$ and $s^\nu_n(n'+1,n) = 0$. Therefore $v^\nu(n) = v^\nu(n')$ for $\nu \neq \mu$ and the equalities for $l(\nu)$ of this lemma hold for all the indices ν.

The equalities for ${}^*l(\nu)$ can be proved in the same way.

We prove now the equalities of this lemma for ${}^*m(\nu)$. It follows from equalities (1) and the equalities obtained for $s^\nu_n(a,b)$ that

$$\begin{aligned} {}^*w^\nu(n) &= \sum_{a\leq b<n} \bar{u}^\nu_n(0,a-1)x^\nu_n(a,b) + \sum_{a\leq b=n} \bar{u}^\nu_n(0,a-1)x^\nu_n(a,b) \\ &= \sum_{a\leq b\leq n'} \bar{u}^\nu_{n'}(0,a-1)x^\nu_{n'}(a,b) + \bar{u}^\nu_n(0,n')x^\nu_n(n'+1,n). \end{aligned}$$

Let $y^\mu(n') = 0$. Then $x^\mu_{n'}(a,b) = 0$, $\bar{u}^\mu_n(0,n') = 1$, $x^\nu_n(n'+1,n) = \Delta$. Therefore ${}^*w^\mu(n) = \Delta$ if $y^\mu(n') = 0$ and the equality for ${}^*m(\mu)$ holds for $j'(\mu) = 0$.

Let $y^\mu(n') > 0$. Then $s^\mu_{n'}(a,b) = 0$ for some $a \leq b \leq n'$, hence $\bar{u}^\mu_n(0,n') = 0$. Therefore ${}^*w^\mu(n) = {}^*w^\mu(n')$ if $y^\mu(n') > 0$ and the equality for ${}^*m(\mu)$ holds for $j'(\mu) > 0$.

If $\nu \neq \mu$, then $x^\nu_n(n'+1,n) = 0$ and $w^\nu(n) = w^\nu(n')$. Therefore, the equality for ${}^*m(\mu)$ holds for all the indices ν.

The proof of the lemma is complete.

Similar equalities for Ri, Sj, Tk, Ul, Wm, ${}^*T^*k$, ${}^*U^*l$, and ${}^*W^*m$ can be obtained multiplying the equalities of the lemma by the corresponding indices.

3.3. We point out some simple relations between the basic run characteristics in $\xi = \xi(0), \ldots, \xi(n)$:

$$\Sigma x^\nu = n + 1, \quad x^\nu \geq s^\nu, \quad y^\nu \geq v^\nu \vee {}^*v^\nu,$$
$$z^\nu \geq w^\nu \vee {}^*w^\nu, \qquad {}^*z^\nu \leq w^\nu \wedge {}^*w^\nu.$$

To describe more complex relations, it is convenient to introduce the binary enumeration for the values t of the functions $\eta = \eta(\xi)$ describing the run structure in ξ. We assign to $t = (\bar{t}, m, \mu)$ the eight-digit binary number

$$Nt = \text{sign}(|m(\mu) - {}^*m(\mu)|, {}^*l(\mu) - 1, m(\mu) - {}^*k(\mu), i(\mu) - m(\mu),$$
$$l(\mu) - 1, k(\mu) - m(\mu), j(\mu) - 1, n + 1 - i(\mu)).$$

The function sign acts on every component and takes the values 0, 1.

For such an enumeration each complete μ-run, which has $i(\mu) = k(\mu) = m(\mu) = {}^*k(\mu) = {}^*m(\mu) = n + 1$ and $j(\mu) = l(\mu) = {}^*l(\mu) = 1$, has the number 0. Each incomplete unique μ-run, which has $i(\mu) = k(\mu) = m(\mu) = {}^*k(\mu) = {}^*m(\mu) \leq n$ and $j(\mu) = l(\mu) = {}^*l(\mu) = 1$, has the number 1. But if $Nt > 1$, then for $\eta = t$ there are several μ-runs in the sequence ξ. An example with incomplete μ-runs shows that different vectors t can have the same numbers. Indeed, not the individual values are enumerated but specific sets of values.

Let us agree to transfer the run structure value enumeration on the structures itself and write $N\eta = Nt$ if $\eta = t$. By definition

$$N\eta = \text{sign}(|w(\mu) - {}^*w(\mu)|, {}^*v(\mu) - 1, w(\mu) - {}^*z(\mu), x(\mu) - w(\mu),$$
$$v(\mu) - 1, z(\mu) - w(\mu), y(\mu) - 1, \; n + 1 - x(\mu)).$$

It turns out that only 23 numbers from $2^8 = 256$ represent some non-trivial structures η.

3.4. The following assertion is valid.

LEMMA 3. *The following numbers and only they can represent some structures* η: $N\eta =$ 0, 1, 55, 59, 63, 87, 91, 95, 119, 123, 127, 151, 159, 179, 183, 187, 191, 215, 219, 223, 243, 247, 251, 255.

PROOF. We check the numbers $0, 1, \ldots, 255$ consecutively. The numbers $N\eta = 0$ and $N\eta = 1$ represent the structures η of the sequences ξ with a single μ-run. For all the others structures η the relations y^μ, $w^\mu < x^\mu < n + 1$, and $\text{sign}(n + 1 - x^\mu) = \text{sign}(y^\mu - 1) = \text{sign}(x^\mu - w^\mu) = 1$ hold. Therefore $N\eta = b = b(7)b(6)\ldots b(0) \geq 19$ and $b(4) = b(1) = b(0)$. For every triple $c = b(7)b(6)b(5)$, we consider the quadruple of the numbers $b = b(7)b(6)b(5)1b(3)b(2)11$, where $b(3)b(2) = 00, 01, 10, 11$.

Let $c = 000$. Then $b = 19, 23, 27, 31$. There are no structures η with these numbers. Indeed, if $N\eta = b$ and $b(7) = b(6) = b(5) = 0$, then ${}^*w = w^\mu = {}^*z^\mu$, ${}^*v^\mu = 1$, that contradicts $b(1) = 1$ and $y^\mu > 1$.

Let $c = 001$. Then $b = 51, 55, 59, 63$. For this quadruple the equality $N\eta = 51$ is impossible, since otherwise we have ${}^*w^\mu = z^\mu = w^\mu$, $v^\mu = 1$, what contradicts $y^\mu > 1$.

Let $c = 010$. Then $b = 83, 87, 91, 95$. By the same reason the equality $N\eta = 83$ is impossible.

Let $c = 011$. Then $b = 115, 119, 123, 127$. By the same reason the equality $N\eta = 115$ is impossible.

Let $c = 100$. Then $b = 147, 151, 155, 159$. Here the equalities $N\eta = 147, 155$ are impossible, since otherwise we have ${}^*z^\mu = w^\mu = z^\mu$ and ${}^*z^\mu \le {}^*w^\mu \le z^\mu$, ${}^*w^\mu \ne w^\mu$.

Let $c = 101$. Then $b = 179, 183, 187, 191$. There are structures η with these numbers $N\eta = b$.

Let $c = 110$. Then $b = 211, 215, 219, 223$. By the same reason as for $N\eta = 147, 155$ the equalities for $N\eta = 211, 219$ are impossible.

Let $c = 111$. Then $b = 243, 247, 251, 255$. There are structures η with these numbers $N\eta = b$.

Simple examples of binary sequences prove the existence of other structures η with numbers $N\eta$ not marked as impossible.

The number $N\eta$ of the structure η describes some relations between the run characteristics considered. The lemma shows that there are 23 types of such relations. They determine the corresponding types of the structures η. Note that the enumeration introduced is determined by μ-run structure η^μ with the index $\mu = \mu(\xi)$ of the last run in the sequence ξ.

The structures η with numbers $N\eta = 0, 1, 91$ can be called singular, since they describe the sequence ξ with μ-runs of equal lengths. The 20 structures with other numbers are non-singular: they describe sequences ξ which have μ-runs of different lengths. If $N\eta = 0, 1$, then there are only one μ-run, and if $N\eta = 91$ there are several. The non-singular structures are united into 5 groups.

3.5. We divide the set of the 23 possible numbers $N\eta = Nt$ of the structure $\eta = t$ values into 8 parts $M(K)$, where $K = 0, 1, \ldots, 7$:

$$\begin{aligned} &M(0) = \{0\}, && M(1) = \{1\}, && M(2) = \{179, 243\}, \\ &M(3) = \{151, 159\}, && M(4) = \{55{,}63{,}119{,}127{,}183{,}191{,}247{,}255\}, \\ &M(5) = \{87, 95, 215, 223\}, && M(6) = \{59, 123, 187, 251\}, && M(7) = \{91\}. \end{aligned} \tag{2}$$

We denote by $D(K)$ the set of all vectors $t = (\bar{t}, m, v)$ with the number $Nt \in M(K)$. It turns out that for each K, the components of all vectors $t = D(t', \Delta, \mu) \in D(K)$ are identically expressed in terms of the vectors t'.

It follows from Lemma 2 that $v = \mu$, $m(\mu) = \Delta$, and $\bar{t}(\lambda) = \bar{t}'(\lambda)$ for $\lambda \ne \mu$, and the components $i(\mu)$, $j(\mu)$ are expressed in terms of $i'(\mu)$, $j'(\mu)$ identically for all t. Furthermore, $*\, m(\mu) = *\, m'(\mu)$ for $j(\mu) > 1$, since $j(\mu) = 1$ is equivalent to $j'(\mu) = 0$ and $j(\mu) > 1$ to $j'(\mu) > 0$. Therefore ${}^*m(\mu) = \Delta$ for $t \in D(K)$ if $K \le 1$, and ${}^*m(\mu) = {}^*m'(\mu)$ for $t \in D(K)$ if $K > 1$. The cases $K = 0$ and $K = 1$ differ from each other by relations

$i(\mu) = n + 1$ and $i(\mu) \leq n$. The foregoing completely describes the relation between the components $i, j, m, {}^*m$ and $i', j', m', {}^*m'$. It remains to find a relation between $k, l, {}^*k, {}^*l$ and $k', l', {}^*k', {}^*l'$.

3.6. If $t = D(\bar{t}', \Delta, \mu)$ and $t \in D(K)$, then the relation between the components of the vectors $\bar{t}(\mu)$ and $\bar{t}'(\mu)$ are described by the Kth row of Table 1. Everywhere in Table 1 $1 \leq \Delta \leq n$: for $K \neq 0$, the structures $\eta = t \in D(K)$ consist of incomplete runs. Note that $k'(\mu) = \Delta$ for $K = 6, 7$ and ${}^*k'(\mu) = \Delta$ for $K = 5, 7$. In the last row $j'(\mu) = l'(\mu) = {}^*l'(\mu)$, $i'(\mu) = j'(\mu)\Delta$ and ${}^*m'(\mu) = \Delta$. For some numbers Nt, the components $k', l', {}^*k', {}^*l'$ in other rows can be defined more exactly as well (see 3.10).

3.7. The following assertion is valid.

LEMMA 4. *If $t = D(t', \Delta, \mu)$ and $t \in D(K)$, then $\bar{t}(\mu)$ is defined by the Kth row of Table* 1.

PROOF. We check the assertion sequentially for every K. It follows from the results in 3.5 that only the equalities for the components $k, l, {}^*k, {}^*l$ of the vector $t = D(t', \Delta, \mu)$ should be checked.

The cases $K = 0$ and $K = 1$ with a single μ-run were described in 3.3, where we define the binary numbers Nt of the vectors t.

For $K = 2$, the equalities for k and l follow automatically from the definition of $Nt \in M(2)$, and the equalities for ${}^*k$ and ${}^*l$ follow from ${}^*k(\mu) \neq m(\mu) = \Delta$, ${}^*k(\mu) < \Delta$, ${}^*\Delta^-(\mu) < 0$, and Lemma 2. In this case all the minimal μ-runs in ξ belong to ξ'.

For $K = 3$, the situation is dual: the minimal and the maximal μ-runs change places. In this case all the maximal μ-runs in ξ belong to ξ'.

For $K = 4$, it follows from the definition of $Nt \in M(4)$ that $k(\mu) \neq m(\mu)$ and ${}^*k(\mu) \neq m(\mu) = \Delta$, hence $k(\mu) > \Delta$ and ${}^*k(\mu) < \Delta$. Then it follows from Lemma 2 that $k(\mu) = k'(\mu)$ and ${}^*k(\mu) = {}^*k'(\mu)$. Therefore $\Delta^+(\mu) > 0$ and ${}^*\Delta^-(\mu) > 0$, and it follows from Lemma 2 that $l(\mu) = l'(\mu)$ and ${}^*l(\mu) = {}^*l'(\mu)$. In this case all the minimal and maximal μ-runs in ξ belong to ξ'.

For $K = 5$ it follows from the definition of $Nt \in M(5)$ that $k(\mu) \neq m(\mu)$ and ${}^*k(\mu) = m(\mu) = \Delta$; therefore it follows from Lemma 2 that $k(\mu) = k'(\mu) > \Delta$, $\Delta^+(\mu) > 0$, $l(\mu) = l'(\mu)$ and ${}^*k(\mu) = {}^*k'(\mu) = \Delta$, ${}^*\Delta(\mu) = 0$, ${}^*l(\mu) = {}^*l'(\mu) + 1$. In this case all the minimal and maximal μ-runs in ξ except the last one belong to ξ'.

For $K = 6$, the situation is dual: all the minimal and the maximal μ-runs in ξ except the last one belong to ξ'.

For $K = 7$ it follows from the definition of $Nt \in M(7)$ that $k(\mu) = {}^*k(\mu) = {}^*m(\mu) = m(\mu) = \Delta$. Hence it follows, with the use of equalities (1), that $j(\mu) = l(\mu) = {}^*l(\mu)$ and $i(\mu) = j(\mu)\Delta$. In this case all the μ-runs in ξ are of the same length Δ.

Table 1.

K	$i(\mu)$	$j(\mu)$	$k(\mu)$	$l(\mu)$	$m(\mu)$	$^*k(\mu)$	$^*l(\mu)$	$^*m(\mu)$
0	$n+1$	1	$n+1$	1	$n+1$	$n+1$	1	$n+1$
1	Δ	1	Δ	1	Δ	Δ	1	Δ
2	$i'(\mu)+\Delta$	$j'(\mu)+1$	Δ	1	Δ	$^*k'(\mu)$	$^*l'(\mu)$	$^*m'(\mu)$
3	$i'(\mu)+\Delta$	$j'(\mu)+1$	$k'(\mu)$	$l'(\mu)$	Δ	Δ	1	$^*m'(\mu)$
4	$i'(\mu)+\Delta$	$j'(\mu)+1$	$k'(\mu)$	$l'(\mu)$	Δ	$^*k'(\mu)$	$^*l'(\mu)$	$^*m'(\mu)$
5	$i'(\mu)+\Delta$	$j'(\mu)+1$	$k'(\mu)$	$l'(\mu)$	Δ	Δ	$^*l'(\mu)+1$	$^*m'(\mu)$
6	$i'(\mu)+\Delta$	$j'(\mu)+1$	Δ	$l'(\mu)+1$	Δ	$^*k'(\mu)$	$^*l'(\mu)$	$^*m'(\mu)$
7	$i'(\mu)+\Delta$	$j'(\mu)+1$	Δ	$l'(\mu)+1$	Δ	Δ	$^*l'(\mu)+1$	Δ

Table 2.

$i'(\mu)$	$j'(\mu)$	$k'(\mu)$	$l'(\mu)$	$m'(\mu)$	$^*k'(\mu)$	$^*l'(\mu)$	$^*m'(\mu)$
0	0	0	0	0	0	0	0
0	0	0	0	0	0	0	0
$i(\mu)-\Delta$	$j(\mu)-1$	a	b	c	$^*k(\mu)$	$^*l(\mu)$	$^*m(\mu)$
$i(\mu)-\Delta$	$j(\mu)-1$	$k(\mu)$	$l(\mu)$	c	$^*a$	$^*b$	$^*m(\mu)$
$i(\mu)-\Delta$	$j(\mu)-1$	$k(\mu)$	$l(\mu)$	c	$^*k(\mu)$	$^*l(\mu)$	$^*m(\mu)$
$i(\mu)-\Delta$	$j(\mu)-1$	$k(\mu)$	$l(\mu)$	c	$^*k(\mu)=\Delta$	$^*l(\mu)-1$	$^*m(\mu)$
$i(\mu)-\Delta$	$j(\mu)-1$	$k(\mu)=\Delta$	$l(\mu)-1$	c	$^*k(\mu)$	$^*l(\mu)$	$^*m(\mu)$
$i(\mu)-\Delta$	$j(\mu)-1$	$k(\mu)=\Delta$	$l(\mu)-1$	Δ	$^*k(\mu)=\Delta$	$^*l(\mu)-1$	$^*m(\mu)=\Delta$

Table 3.

K	2	3	4						5			6		
Nt	179	151	55	183	63	191	119	247	87	215	95	59	187	123
$l(\mu)=l'(\mu)$		1	1	1			1	1	1	1				
$^*l(\mu)={}^*l'(\mu)$	1		1	1	1	1						1	1	
$^*m(\mu)={}^*m'(\mu)$			Δ		Δ		Δ		Δ		Δ	Δ		Δ

3.8. For $t = D(\bar{t}', \Delta, \mu) \in D(K)$ the feedback between the components of the vectors $\bar{t}(\mu)$ and $\bar{t}'(\mu)$ is described by the Kth row of Table 2.

In this table, $1 \le \Delta \le n$ and a, b, c, $^*a$, $^*b$ are integers which comply with the inequalities $^*k(\mu) \le a < \Delta$, $0 < b < j(\mu)$, $^*k(\mu) \le c \le k(\mu)$, $^*k(\mu) \le {}^*a \le k(\mu)$, and $0 < {}^*b < j(\mu)$. In the last row of the table $j(\mu) = l(\mu) = {}^*l(\mu)$ and $i(\mu) = j(\mu)\Delta$. For some numbers Nt, the components k, l, $^*k$, and $^*l$ in other rows can be defined more exactly as well (see 3.10).

Note that only in the singular cases where $K = 0, 1, 7$ the corresponding rows of the table uniquely define the vector $\bar{t}'(\mu)$. In all the non-singular cases the table gives a set of vectors $\bar{t}'(\mu)$. The values $a = k'(\mu)$, $b = l'(\mu)$, $c = m'(\mu)$, $^*a = {}^*k'(\mu)$, and $^*b = {}^*l'(\mu)$ which determine these sets are constrained by certain relations, therefore they can not independently take arbitrary values from the intervals mentioned. For example, $^*a \le c$ and $c \le a$.

3.9. The following assertion is valid.

LEMMA 5. *The set of vectors $\bar{t}'(\mu)$ such that $D(\bar{t}', \Delta, \mu) = t$ and $t \in D(K)$ is determined by the Kth row of Table 2.*

PROOF. For $K = 0, 1$, we have $i(\mu) = m(\mu)$ and $i'(\mu) = i(\mu) - m(\mu) = 0$. It follows that all the other components of the vector $\bar{t}'(\mu)$ are equal to zero.

For $K = 2$, by Lemma 4, all the necessary equalities for the components of $\bar{t}'(\mu)$ except $k'(\mu)$, $l'(\mu)$, and $m'(\mu)$ follow from Table 1. For $k'(\mu)$ the relations $^*k(\mu) \le {}^*k'(\mu) \le a = k'(\mu) \le k(\mu) = \Delta$ follow from the definitions and Table 1. But since $l(\mu) > 1$ and $l(\mu) = 1$, we have $k'(\mu) \ne k(\mu)$ and $k'(\mu) < \Delta$. For $b = l'(\mu)$ the inequalities $0 < b \le j'(\mu) < j(\mu)$ hold. At last, $^*k(\mu) \le {}^*k'(\mu) \le c = m'(\mu) \le k'(\mu) \le k(\mu)$.

For $K = 3$, the situation is dual: by Lemma 4 the necessary equations for all the components of $\bar{t}'(\mu)$ except $^*k'(\mu)$, $^*l'(\mu)$ and $m'(\mu)$ follow from Table 1. The inequalities for $^*a = {}^*k'(\mu)$, $^*b = {}^*l'(\mu)$, $c = m'(\mu)$ follow from the general inequalities $0 < {}^*l'(\mu) \le j'(\mu) < j(\mu)$ and $^*k(\mu) \le {}^*k'(\mu) \le m'(\mu) \le k'(\mu) \le k(\mu)$.

For $K \ge 4$ by Lemma 4 the necessary equations for all the components of $\bar{t}'(\mu)$ follow from Table 1.

REMARK 1. If $\lambda \ne \mu$, then by Lemma 2, the equalities $\bar{t}'(\lambda) = \bar{t}(\lambda)$ hold. Hence, the vector $\bar{t}'$ is defined completely. The component m' is defined together with the vector. For the index v' in the vector $t' = (\bar{t}', m', v')$ the inequality $v' \ne \mu$ hold. Generally, the index v' is not uniquely defined.

3.10. Table 3 defines more exactly certain component values of the vectors $\bar{t}(\mu)$ and $\bar{t}'(\mu)$ for certain numbers Nt.

The equalities $^*m(\mu) = {}^*m'(\mu) = \Delta$ hold for all the numbers $Nt \le 127$. For $K = 4, 5, 6$, the couples are distinguished in the corresponding groups of numbers Nt such that each couple has exactly one number. For all the numbers Nt which are not indicated in the table the component values of the vectors $\bar{t}(\mu)$ and $\bar{t}'(\mu)$ are not revised: they remain the same as in Tables 1 and 2.

REMARK 2. For the vectors $\bar{X}\bar{t}(\mu)$ and $\bar{X}\bar{t}'(\mu)$ the tables similar to 1, 2, and 3 can be composed. Zero columns corresponding to zero values of the indicators will appear. If some characteristics are excluded the relations between the other characteristics become more complicated and require a special study [1, 2].

4. EQUATIONS FOR PROBABILITIES

In order to obtain a system of recurrence equations which make it possible to consecutively calculate necessary probabilities, we consider the Markov sequence of random variables $\xi(r) : \Omega \to C$ with the initial probabilities $p(\gamma) = \mathbf{P}\{\xi(0) = \gamma\}$ and the transition probabilities $q(r-1, \gamma, r, \delta) = \mathbf{P}\{\xi(r) = \delta/\xi(r-1) = \gamma\}$, where $\mathbf{P}\{\xi(r-1) = \gamma\} \neq 0$, $r \geq 1, \gamma \in C, \delta \in C$.

4.1. We introduce appropriate notation. For $0 \leq n' < n$, except $\xi' = \xi(0) \cdots \xi(n')$, we consider also the supplementary subsequence $\xi'' = \xi(n' + 1) \ldots \xi(n)$ of the sequence $\xi = \xi(0) \ldots \xi(n)$. We denote by $\eta'' = \eta(\xi'')$ the structure of the run ξ'', and by $t'' = (\bar{t}'', m'', \mu'')$ the value of the run structure. If $\eta = \eta(\xi) = t = (\bar{t}, m, \mu)$ and $m(\mu) = \Delta$, $n' = n - \Delta$, then ξ'' is a μ-run with the length Δ and $\mu'' = \mu$, $m''(\mu) = \Delta$, and $\bar{t}''(\lambda) = 0$ where $\lambda \neq \mu$. In this case we will write $t'' = t''(\Delta, \mu)$.

If the sequence ξ is a complete μ-run, then $n' = -1$, $\xi' = \xi(-1) = \theta \notin C, \Delta = n+1$, $\xi'' = \xi$, and $t = t(n + 1, \mu) = t''(n + 1, \mu)$.

For every vector $t = (\bar{t}, m, \mu)$ and every number $\Delta = m(\mu)$ we consider the set

$$D'(t) = \{t' : D(t', \Delta, \mu) = t\}.$$

The correspondence $D' : t \to t'$ is equal to the composition $P'D^{-1}$ of the correspondence $D^{-1} : t \to (t', \Delta, \mu)$ which is inverse to D: $(t', \Delta, \mu) \to t$ and the projector $P' : (t', \Delta, \mu) \to t'$. It follows from Lemma 5 that for all $t \in D(K)$ the set $D'(t)$ is defined uniquely. In order to make use of this fact, we introduce the set $D'(t, K) = D'(t)$ for $t \in D(K)$ and $D'(t, K) = 0$ for $t \notin D(K)$. The correspondences D' and D^{-1} are many-valued.

It is convenient sometimes to reckon θ as an index, and consider the vector $t' = (0, 0, \theta)$ and the set $D'(t, 0) = \{(0, 0, \theta)\}$ for $t = t(n + 1, \mu)$. It allows us often not to distinguish the case $K = 0$.

In order to use the Markov property of the sequence of random variables $\xi(r)$, we consider the couples (η, ζ), (η', ζ') and their values (t, δ), (t', δ'). Here $\zeta = \xi(n)$, $\zeta' = \xi(n')$. In addition to the probabilities

$$p(t) = \mathbf{P}\{\eta = t\}, \qquad p(t') = \mathbf{P}\{\eta' = t'\},$$

we consider also

$$p(t, \delta) = \mathbf{P}\{\eta = t, \zeta = \delta\}, \qquad p(t', \delta') = \mathbf{P}\{\eta' = t', \zeta' = \delta'\}.$$

If $t = (\bar{t}, m, \mu)$ and $\delta \notin A^\mu$, then $p(t, \delta) = 0$.

We consider also the transition probabilities from $\zeta' = \delta'$ to $\zeta = \delta$ with the constraint $\eta'' = t'' = t''(\Delta, \mu)$ for $\Delta = n - n'$

$$q(n', \delta', n, \delta, \mu) = \mathbf{P}\{\eta'' = t'', \zeta = \delta \mid \zeta' = \delta'\}$$

for $\mathbf{P}\{\zeta' = \delta'\} \neq 0$ and $q(n', \delta', n, \delta, \mu) = 0$ for $\mathbf{P}\{\zeta' = \delta'\} = 0$. The constraint $\eta'' = t'' = t''(\Delta, \mu)$ means that the transition from $\zeta' = \delta'$ to $\zeta = \delta$ happens only if $\xi(n' + 1), \ldots, \xi(n) \in A^\mu$. Specifically, if $\delta \notin A^\mu$, then $q(n', \delta', n, \delta, \mu) = 0$.

In the stationary case, the transition probabilities depend only on the difference $\Delta = n - n'$:

$$q(n', \delta', n, \delta, \mu) = q(\delta', \delta, \Delta, \mu).$$

For $n = n' + 1$ and $\Delta = 1$, it is natural to simplify the notation, setting

$$q(n - 1, \gamma, n, \delta) = q(\gamma, \delta), \quad q(n - 1, \gamma, n, \delta, \mu) = q(\gamma, \delta, \mu).$$

4.2. It follows from the definitions that

$$q(-1, \theta, 0, \delta, \mu) = p(\delta), \quad \delta \in A^\mu.$$

If $n > 0$ and $n' < n - 1$, it follows from the Markov property that

$$q(n', \delta', n, \delta, \mu) = \sum_\gamma q(n', \delta', n - 1, \gamma, \mu) q(n - 1, \gamma, n, \delta) \tag{3}$$

for $\delta' \notin A^\mu$ and $\delta \in A^\mu$. The summation is over the index $\gamma \in A^\mu$, since $q(n', \delta', n - 1, \gamma, \mu) = 0$ for $\gamma \notin A^\mu$. Equality (3) allows us to calculate consequently the run probabilities in the end of $\xi = \xi(0) \ldots \xi(n)$ with the final value $\xi(n) = \delta \in A^\mu$.

Equality (3) is equivalent to the equality

$$q(n', \delta', n, \delta, \mu) = \sum_{\gamma(1), \ldots, \gamma(\Delta - 1)} \prod_r q(n' + r - 1, \gamma(r - 1), n' + r, \gamma(r)), \tag{4}$$

where $\gamma(0) = \delta' \notin A^\mu$, $\gamma(\Delta) = \delta \in A^\mu$, $\Delta > 1$, $1 \leq r \leq \Delta$, and the product consists of Δ factors, the summation is over $\gamma(1), \ldots, \gamma(\Delta - 1) \in A^\mu$. In particular, for $n' = -1$, $\delta' = \theta$, $\delta \in A^\mu$, and $n \geq 1$, equality (4) expresses the probability

$$q(-1, \theta, n, \delta, \mu) = p(t(n + 1, \mu), \delta)$$

of a complete μ-run with the last value $\xi(n) = \delta$ in the sequence ξ. In this case, the first factor in the product is $q(-1, \theta, 0, \gamma(1)) = p(\gamma(1))$. For the calculation of the probability of a complete μ-run which is terminated by the value δ, the following equality can be used:

$$p(t(n + 1, \mu), \delta) = \sum_{\gamma \in A^\mu} p(\gamma) \cdot q(0, \gamma, n, \delta, \mu) \tag{5}$$

We agree to designate the μ-runs which are terminated by the value δ as $\mu\delta$-runs.

REMARK 3. Equalities (3–5) can be expressed in the matrix form, if we introduce a suitable vector probabilities. It is often convenient in the case where the values of the sequence are aggregated [1, 5].

4.3. It follows from the definitions that

$$(\eta = t, \zeta = \delta) = \sum_{t',\delta'} (\eta' = t', \zeta' = \delta')(\eta'' = t'', \zeta = \delta) \tag{6}$$

where $t'' = t''(\Delta, \mu)$ and the summation is over $t' \in D'(t)$, $\delta' \notin A^{\mu}$. For $n' = -1$, the event $(\eta' = t', \zeta' = \delta')$ is possible only if $\delta' = \theta$ and $t' = (0, 0, \theta)$.

It is convenient to divide the sets $D'(t)$ into the parts $D'(t, K)$ composed of the vectors t' for $t \in D(K)$. The set $D'(t, K)$ is defined by the kth row of Table 2, where the components of the vector $\bar{t}'(\mu)$ are given for $t = (\bar{t}, m, \mu)$. For $\lambda \neq \mu$, the vector $\bar{t}'(\lambda) = \bar{t}(\lambda)$, thus $\bar{t}'(\mu)$ defines $\bar{t}'$. If $K = 0$, then $\mu = \theta$ and the set $D'(t, K)$ consists of a single element $t' = (0, 0, \theta)$. If $K \neq 0$, all the vectors $t' = (\bar{t}', m', \mu')$ belong to the set $D'(t, K)$ if $\mu' \neq \mu$. Passing from $D'(t)$ to $D'(t, K)$, the sum in formula (6) is essentially simplified.

4.4. The following theorem is valid.

THEOREM 1. *If $t \in D(K)$, then*

$$p(t, \delta) = \sum_{t',\delta'} p(t', \delta') q(n', \delta', n, \delta, \mu), \tag{7}$$

where the summation is over $t' \in D'(t, K)$, $\delta' \notin A^{\mu}$.

PROOF. Equality (7) follows from equality (6), the total probability formula and the Markov property of the sequence of random variables $\xi(r)$.

In order to calculate the transition probabilities $q(n', \delta', n, \delta, \mu)$, equalities (3–5) can be used. In the stationary case, where $q(n', \delta', n, \delta, \mu) = q(\delta', \delta, n - n', \mu)$, the equality of the theorem is essentially simplified.

Using vector probabilities, equality (7) can be derived in the matrix form, just as equalities (3–5).

Equalities (3)–(5), (7) make it possible to calculate consecutively the probabilities $p(t, \delta)$ and obtain the desired probabilities $p(\delta)$, summing over $\delta \in A^{\mu}$. In [1] and in a number of student researches in the Novosibirsk State University there are examples of such calculations.

4.5. If the random variables $\xi(r)$ are independent and identically distributes, the equalities (3–5,7) have especially simple form. In that case, the transition probabilities for one step are equal to the corresponding starting probabilities: $q(\gamma, \delta) = p(\delta)$ for each index γ, δ. The transition probabilities for $\Delta = n - n'$ steps with a constraint $\xi(r) \in A^{\mu}$ $(n' < r \le n)$ are equal to the probability of a μ-run of length Δ terminating in the value δ:

$$q(n', \delta', n\delta, \mu) = a^{\Delta-1}(\mu)p(\delta),$$

where

$$a(\mu) = \mathbf{P}\{\xi(0) \in A^{\mu}\} = \sum_{\delta \in A^{\mu}} p(\delta).$$

COROLLARY 1. *If the random variables $\xi(r)$ are independent and identically distributed, then*

$$p(t) = p(D'(t))a^{\Delta}(\mu), p(D'(t)) = \sum_{t' \in D'(t)} p(t'). \tag{8}$$

PROOF. Using the equality $q(n', \delta', n, \delta, \mu) = a^{\Delta-1}(\mu)p(\delta)$ and summing both sides of equality (7) over $\delta \in A^{\mu}$, we obtain

$$p(t) = \sum_{t', \delta'} p(t', \delta')a^{\Delta}(\mu).$$

It remains to note that

$$\sum_{t', \delta'} p(t', \delta') = \sum_{t'} p(t') = p(D'(t))$$

where the summation is over $t' \in D'(t)$, $\delta' \notin A^{\mu}$.

For $t \in D(K)$, equality (8) is transformed to the equality

$$p(t) = p(D'(t, K))a^{\Delta}(\mu), \; p(D'(t, K)) = \sum_{t' \in D'(t,K)} p(t'). \tag{9}$$

Specifically, if $t = t(n+1, \mu)$, $\Delta = n+1$, and $\Delta'(t) = \{(0, 0, \theta)\}$, then $p(D'(t)) = 1$ and

$$p(t(n+1, \mu)) = a^{n+1}(\mu).$$

In the simple case considered, this equality can be derived from the definitions at once, as well as equalities (8). For the uniformly distributed random variables $\xi(r)$, the problem is to count the number of elements in the set $D'(t)$. This problem is self-important as well. It was studied for certain selected characteristics in [6,7,8].

Equalities (8) let us calculate consecutively the probabilities $p(t)$ which describe the run structures of a sequence of independent identically distributed random variables. Such calculations were carried out, in particular, for Bernoulli random variables and for certain selected characteristics.

5. OPERATOR EQUATIONS

In order to explore the probabilities $p(t, \delta)$, operator equations can be useful which are equivalent to the recursion equations obtained. Examples of the operator equations can be found in [1, 2, 9, 10].

5.1. We introduce the indicator functions $D(t, K) = 1$ for $t \in D(K)$ and $D(t, K) = 0$ for $t \notin D(K)$. Equality (7) for the probabilities $p(t, \delta)$ is equivalent to the equality

$$p(t, \delta) = D(t, 0)p(t(n+1, \mu), \delta) + \sum_{K \neq 0} D(t, K) \sum_{t' \in D'(t,K), \delta' \notin A^{\mu}} p(t', \delta')q(n', \delta', n, \delta, \mu). \tag{10}$$

We consider the space $\mathcal{L} = \mathcal{L}^{\infty}$ of all bounded real-valued functions f of the variables t, δ. It is clear that a function p with the values $p(t, \delta)$ belongs to $\mathcal{L}$. We denote by g the function with the values

$$g(t, \delta) = D(t, 0)p(t(n+1, \mu), \delta).$$

Observe that for every $f \in \mathcal{L}$ and for every t, δ the following inequalities hold:

$$\sum_{K \neq 0} D(t, K) \sum_{t' \in D'(t,K), \delta' \notin A^{\mu}} f(t', \delta')q(n', \delta', n, \delta, \mu) \leq \|f\| \sum_{n' < n, \delta' \notin A^{\mu}} q(n', \delta', n, \delta, \mu)$$

$$\leq \|f\| \sum_{n' < n,\ \delta' \in C} q(n', \delta', n, \delta, \mu).$$

We suppose that

$$\bar{q} = \sup_{n,\delta} \sum_{n' < n,\ \delta' \in C} q(n', \delta', n, \delta, \mu) < \infty \tag{11}$$

and define the operator $Q\colon \mathcal{L} \to \mathcal{L}$ by the equality

$$Qf(t, \delta) = \sum_{K \neq 0} D(t, K) \sum_{t' \in D'(t,K),\ \delta' \notin A^{\mu}} f(t', \delta')q(n', \delta', n, \delta, \mu). \tag{12}$$

It follows from the inequalities obtained and from the condition $\bar{q} < \infty$ that the operator Q is bounded and has the norm $\|Q\| \leq \bar{q}$.

REMARK 4. The general analysis of condition (11) was not performed. For some simple cases this condition can be easily checked. Specifically, the condition holds automatically if a finite sequence is considered and all the transition probabilities with sufficiently large numbers are equal to zero.

5.2. We denote by I the identity operator on $\mathcal{L}$. It follows from equality (10) that the probability p is the solution of the operator equation

$$(I - Q)f = g.$$

The methods of solution of such equations are discussed in [9]. Examples can be found in [10].

REMARK 5. The study of the boundedness and compactness conditions for the operator Q is of interest. The introduction of the indicator functions for the indices makes it possible to distinguish special cases for which more precise norm estimation of the operator Q can be obtained and determine the condition necessary for $\|Q\| < 1$. It is natural to consider other spaces in addition to $\mathcal{L} = \mathcal{L}^{\infty}$ that was done in [9].

REFERENCES

1. L. Ya. Savelyev, Runs in finite Markov chains. *Probabilistic Methods in Discrete Mathematics, Proc. III Intern. Petrozavodsk Conf.* TVP/VSP, Moscow/Utrecht, 1993, pp. 437–450.
2. L. Ya. Savelyev, Operator equations for generating functions. *Soviet Math. Dokl.* (1995) **343**, 740–742 (in Russian).
3. J. G. Kemeny and J. J. Snell, *Finite Markov Chains.* Van Nostrand, Princeton, 1960.
4. A. D. Ventsel, *The Theory of Random Processes.* Nauka, Moscow, 1975 (in Russian).
5. V. A. Ivanov, *Random Sequences of Finite Rank.* MIEM, Moscow, 1992 (in Russian).
6. S. S. Wilks, *Mathematical Statistics.* Wiley, New York, 1962.
7. A. M. Mood, The distribution theory of runs. *Ann. Math. Statist.* (1940) **11**, 367–392.
8. P. Erdős and P. Revesz, On the length of longest head run. In: *Colloq. Math. Soc. J. Bolyai, Topics in Information Theory* **16**. North–Holland, 1975, pp. 219–228.
9. L. Ya. Savelyev, Markov matrix operators. Dep. VINITI, No. 456–B95, 1995 (in Russian).
10. L. Ya. Savelyev. Runs in Markov sequences. *Siberian Math. J.* (1991) **32**, No. 4, 116–132 (in Russian).

Probabilistic Methods in Discrete Mathematics, pp. 87–99
V. F. Kolchin et al. (Eds.)

ON THE LIMIT DISTRIBUTION OF THE HEIGHT OF LEAVES IN A PLANE PLANTED TREE

A. I. ABRAMOV

Vienna University

Abstract — We consider the height of the sth leaf of a random plane planted tree. Earlier for the height of the sth leaf of a random tree from a class of the so-called simply generated trees, Drmota, by a fine analysis of the corresponding generating functions, proved a local limit theorem on the convergence of the distribution of this random variable, under an appropriate normalization, to the Maxwell distribution. In this paper we obtain, by rather simple arguments, results which are close to Drmota's results and partially generalize them.

Let the uniform distribution be given on the set $\mathcal{F}_{n,R}$ of all plane planted trees with n non-root vertices and R end vertices (leaves). We consider the height h_s of the sth leaf numbered in the natural order, in the clockwise direction. We prove that if $s, n, R \to \infty$ in such a way that $s/n \to \alpha$, $0 < \alpha < 1/2$, $R = n/2 + v\sqrt{n}$, $|v| < K$, K is a constant, $\sigma = 2\sqrt{\alpha(1-2\alpha)}$, then the distribution of $h_s/(\sigma\sqrt{n})$ converges weakly to the Maxwell distribution with the density

$$\sqrt{\frac{2}{\pi}}x^2e^{-x^2/2}.$$

Thus, under the conditions given above the limit distribution of the height of the sth leaf does not depend on the fact that the random tree is chosen from the class of trees with a fixed number of leaves provided this number is not far from the mean value of the number of leaves of a random tree from the class of all plane planted trees. In other words, under the conditions given above the height of the sth leaf is asymptotically independent of the number of leaves of the random tree.

1. INTRODUCTION

The study of characteristics of random trees and forests from different classes is of considerable interest in probabilistic combinatorics. The first results on the random rooted trees with labelled vertices concerned the degrees of vertices [19, 20, 23] and the height of such trees [24, 25] were obtained by direct combinatorial methods, by methods based on generating functions, and by the method of moments.

An essential progress in the study of random trees began in 1977 when a connection between random trees and branching processes was discovered [15, 16, 17]. This connection led to a considerable simplification of proofs of the known results and gave a possibility to investigate some new characteristics of random trees and forests [12, 15, 16, 17, 21, 27].

Along with this direction of investigations the methods of study of random trees based on the analysis of generating functions were refined [5, 6, 8, 10, 13, 26].

The next important advancement in the study of random trees and forests in the recent years was related to the possibility to code plane planted trees in such a way that a random tree corresponds to a random walk, which under passing to limit with changing the scale, converges to a Brownian motion with corresponding restrictions [2, 3, 4, 9, 11, 27].

In this paper we investigate the height of the sth leaf of a random plane planted tree. In the paper [5] for the height of the sth leaf of a random tree from a class of the so-called simply generated trees (introduced in [18]) Drmota, by a fine analysis of generating functions, proves a local limit theorem on the convergence of the distribution of this random variable, under an appropriate normalization, to the Maxwell distribution. In this paper we obtain, by rather simple arguments, results which are close to the results of the paper [5] and partially generalize them.

We consider the set $\mathcal{F}_{n,R}$ of all plane planted trees with n non-root vertices and R end vertices (leaves). The leaves of the trees from $\mathcal{F}_{n,R}$ are enumerated in the natural order, in the clockwise direction. We assume h_s to be equal to the height of the leaf labelled s if s does not exceed R, and $h_s = 0$ otherwise. Let the uniform distribution be given on $\mathcal{F}_{n,R}$, then h_s becomes a random variable. In this paper we prove the following assertion.

THEOREM 1. *If $s, n, R \to \infty$ in such a way that $s/n \to \alpha$, $0 < \alpha < 1/2$, $R = n/2 + v\sqrt{n}$, $|v| < K$, K is a constant, $\sigma = 2\sqrt{\alpha(1-2\alpha)}$, then the distribution of $h_s/(\sigma\sqrt{n})$ converges weakly to the Maxwell distribution with the density*

$$\sqrt{\frac{2}{\pi}}x^2 e^{-x^2/2}.$$

Thus, under the conditions of the theorem the limit distribution of the height of the sth leaf does not depend on the fact that the random tree is chosen from the class of trees with a fixed number of leaves provided this number is not far from the mean value of the number of leaves of a random tree from the class of all plane planted trees. It is known [5] (see also [1] devoted to this particular case) that in the class $\mathcal{F}_n$ of all plane planted trees with n vertices the height of the sth leaf has the same Maxwell distribution as $n, s \to \infty$ and $s/n \to \alpha$, $0 < \alpha < 1/2$. Thus, under the conditions of the theorem the height of the sth leaf is asymptotically independent of the number of leaves of the random tree.

This result seems to be natural although it is clear that such independence is generally not valid for many other characteristics of a random tree. One of the simplest examples is provided by the random variables equal to degrees of vertices of a random tree. If μ_r is the number of vertices with outdegree equal to r and R is fixed, then under a proper centering an normalizing, the random variables $\mu_0, \mu_1, \ldots, \mu_R$ have in limit as $n \to \infty$ a joint normal distribution with non-zero correlations.

2. AUXILIARY RESULTS

We consider the set $W(l, m)$ of all forests consisting of m labelled plane planted trees and l non-root vertices. Let the uniform distribution be given on $W(l, m)$, and let $\mu_r^{(t)}(l, m)$ be the random variable which is equal to the number of vertices of height t and outdegree r in a forest from $W(l, m)$. We consider the homogeneous branching process G_λ beginning with m particles with geometrically distributed offspring ξ of one particle:

$$p_k = \mathbf{P}\{\xi = k\} = \lambda^k(1-\lambda), \qquad k = 0, 1, \ldots \tag{1}$$

Let ν denote the total progeny of the process G_λ, and let $\mu^{(r)}(t)$ be the number of particles in the tth generation of the process G_λ that have exactly r direct descendants. We introduce the matrices $\|\mu_r^{(t)}(l, m)\|$, $\|\mu^{(r)}(t)\|$, $t, r = 0, 1, \ldots, l-m$, and a matrix M of the same size constituted by non-negative integers. The relations between the forests from $W(l, m)$ and the branching process G_λ is given by the following lemma proved in [21, 22].

LEMMA 1. *The equality*

$$\mathbf{P}\{\|\mu_r^{(t)}(l, m)\| = M\} = \mathbf{P}\{\|\mu^{(r)}(t)\| = M \mid \nu = l\}$$

is true.

The process G_λ is related to the generalized scheme of allocating $l - m$ particles to l cells with the joint distribution of contents $\eta_1, \ldots, \eta_l$ of the cells of the form

$$\mathbf{P}\{\eta_1 = k_1, \ldots, \eta_l = k_l\} = \mathbf{P}\{\xi_1 = k_1, \ldots, \xi_l = k_l \mid \xi_1 + \ldots + \xi_l = l - m\},$$

where $\xi_1, \ldots, \xi_l$ are independent identically distributed by law (1) random variables.

Let

$$\mu_r = \sum_{t=0}^{\infty} \mu^{(r)}(t)$$

and let $\mu_r(l - m, n)$ be the number of cells containing exactly r particles in the allocation scheme described above.

LEMMA 2. *The equality*

$$\mathbf{P}\{\mu_r = k_r, r = 0, 1, \ldots, l-m \mid \nu = l\} = \mathbf{P}\{\mu_r(l-m, l) = k_r,\ r = 0, 1, \ldots, l-m\}$$

is true.

PROOF. Since

$$\mathbf{P}\{\mu_r = k_r, r = 0, 1, \ldots, l-m \mid \nu = l\} = \frac{\mathbf{P}\{\mu_r = k_r,\ r = 0, 1, \ldots, l-m;\ \nu = l\}}{\mathbf{P}\{\nu = l\}},$$

to obtain the assertion of the lemma, it is sufficient to use the following formulae (see [17]):

$$\mathbf{P}\{\mu_r(l-m,l)=k_r,\ r=0,1,\dots,l-m\}=\frac{l!\,p_0^{k_0}\cdots p_{l-m}^{k_{l-m}}}{k_0!\dots k_{l-m}!\,\mathbf{P}\{\xi_1+\dots+\xi_l=l-m\}},$$

$$\mathbf{P}\{\mu_r=k_r,\ r=0,1,\dots,l-m;\ \nu=l\}=\frac{ml!\,p_0^{k_0}\cdots p_{l-m}^{k_{l-m}}}{lk_0!\dots k_{l-m}!},$$

and

$$\mathbf{P}\{\nu=l\}=\frac{m}{l}\mathbf{P}\{\xi_1+\dots+\xi_l=l-m\}.$$

The following lemma gives the distribution of the number $\mu_r(n,N)$ of the cells containing exactly r particles in the generalized scheme of allocating l particles to N cells (see [17, 19]).

Let $\xi_1,\dots,\xi_l$ be independent identically distributed by law (1) random variables, and let $\xi_1^{(r)},\dots,\xi_l^{(r)}$ be independent identically distributed random variables such that

$$\mathbf{P}\{\xi_1^{(r)}=k\}=\mathbf{P}\{\xi_1=k\mid\xi_1\neq r\},\qquad k=0,1\dots$$

We set

$$\zeta_N=\xi_1+\dots+\xi_N,$$
$$\zeta_N=\xi_1^{(r)}+\dots+\xi_N^{(r)}.$$

LEMMA 3. *The equality*

$$\mathbf{P}\{\mu_r(l,N)=k\}=\binom{N}{k}p_r^k(1-p_r)^{n-k}\frac{\mathbf{P}\{\zeta_{N-k}^{(r)}=l-kr\}}{\mathbf{P}\{\zeta_N^{(r)}=l\}}$$

is true.

The proof of Lemma 3 can be found in [17].

Let $\alpha_0(l,m)$ be the number of trees in $W(l,m)$ consisting of the roots only.

LEMMA 4. *If $l,m\to\infty$ in such a way that $m^2/l\to\beta^2$, β is a constant, $\beta>0$, then*

$$\mathbf{P}\{\alpha_0(l,m)=k\}=\frac{2}{\sqrt{2\pi m}}e^{-w^2/2}(1+o(1))$$

uniformly in the integers k such that $w=(k-m/2)/\sqrt{m}$ lies in any finite interval.

The proof of Lemma 4 can be found in [22].

Let $\xi_1,\dots,\xi_n$ be independent identically distributed by law (1) random variables, and let $\lambda=1/2$. Then

$$p_k=\mathbf{P}\{\xi_1=k\}=\frac{1}{2^{k+1}},\qquad k=0,1,\dots \tag{2}$$

Let $u=(k-n)/\sqrt{2n}$ and $S_n=\xi_1+\dots+\xi_n$.

THEOREM 2. *If* $n \to \infty$ *and* $|u|^3/\sqrt{n} \to 0$, *then*

$$\mathbf{P}\{S_n = k\} = \binom{n+k-1}{n-1}\frac{1}{2^{n+k}} = \frac{1}{2\sqrt{\pi n}}e^{-u^2/2}\left(1 + O\left(\frac{1+|u|^3}{\sqrt{n}}\right)\right).$$

PROOF. The proof of the first equality can be found in [17]. To prove the second equality, we use Stirling's formula. Then

$$\begin{aligned}\mathbf{P}\{S_n = k\} &= \binom{n+k-1}{n-1}\frac{1}{2^{n+k}} = \frac{(n+k-1)!}{(n-1)!k!}\frac{1}{2^{n+k}} \\ &= \frac{(n+k-1)^{n+k-1}\sqrt{2\pi(n+k-1)}e^{-(n+k-1)}(1+O(1/(n+k)))}{(n-1)^{n-1}\sqrt{2\pi(n-1)}e^{-(n-1)}(1+O(1/n)))k^k\sqrt{2\pi k}e^{-k}(1+O(1/k)))}\frac{1}{2^{n+k}} \\ &= \frac{y_1}{2\sqrt{\pi n}y_2y_3}\left(1 + O\left(\frac{1}{n}\right)\right),\end{aligned}$$

where

$$\begin{aligned}y_1 &= \left(1 + \frac{u\sqrt{2n}-1}{2n}\right)^{2n+u\sqrt{2n}-1}, \\ y_2 &= \left(1 - \frac{1}{n}\right)^{n-1}, \\ y_3 &= \left(1 + \frac{u\sqrt{2n}}{n}\right)^{n+u\sqrt{2n}}\end{aligned}$$

Taking the logarithms of y_1, y_2 and y_3, we obtain

$$\begin{aligned}\ln y_1 &= u\sqrt{2n} - 1 + \frac{u^2}{2} + O\left(\frac{|u|^3+1}{\sqrt{n}}\right), \\ \ln y_2 &= -1 + O\left(\frac{1}{n}\right), \\ \ln y_3 &= u\sqrt{2n} - 1 + u^2 + O\left(\frac{|u|^3}{\sqrt{n}}\right).\end{aligned}$$

Therefore

$$\begin{aligned}\mathbf{P}\{S_n = k\} &= \frac{1}{2\sqrt{\pi n}}\left(1 + O\left(\frac{1}{n}\right)\right)\exp\left\{-u^2 + O\left(\frac{|u|^3+1}{\sqrt{n}}\right)\right\} \\ &= \frac{1}{2\sqrt{\pi n}}e^{-u^2/2}\left(1 + O\left(\frac{|u|^3+1}{\sqrt{n}}\right)\right).\end{aligned}$$

The following assertion is an immediate corollary of the theorem.

COROLLARY 1. *If $n \to \infty$, then*

$$\mathbf{P}\{S_n = k\} = \binom{n+k-1}{n-1}\frac{1}{2^{n+k}} = \frac{1}{2\sqrt{\pi n}}e^{-u^2/2}(1+o(1))$$

uniformly in the integers k such that $|u| \leq n^{1/8}$.

Since the distribution of S_n has one maximum, it is monotone on the left-hand and the right-hand sides of the point of this maximum. Then Corollary 1 implies the following assertion.

COROLLARY 2. *If $n \to \infty$, $k = n + u\sqrt{2n}$, $|u| \geq \sqrt{2}n^{1/8}$, then there exists a constant C such that*

$$\mathbf{P}\{S_n = k\} \leq Ce^{-n^{1/4}}.$$

3. THE PROOF OF THEOREM

For a tree $T \in \mathcal{F}_{n,R}$ the path joining the sth leaf with the root of T divides this tree into two forests, the left forest and the right one. Let $h_s = m + 1$, let n_1 be the number of non-root vertices in the left forest, and let n_2 be the number of non-root vertices in the right forest. Let $W(l, m)$ be the set of forests consisting of m labelled plane planted trees and l non-root vertices. Then the tree T can be represented as the union of two forests from $W(n_1, m)$ and $W(n_2, m)$ planted on the vertices of the path joining the sth leaf with the root. Let $N_m(l)$ be the number of forests in $W(l, m)$ and let $N_m(l, s)$ be the number of forests in $W(l, m)$ with exactly s leaves which are not the root of the corresponding tree (such a leaf will be referred to as a real leaf). Let $F_{n,R}$ be the number of trees in $\mathcal{F}_{n,R}$.

The above reasoning implies the following assertion.

LEMMA 5. *The equality*

$$\mathbf{P}\{h_s = m + 1\} = \sum_{n_1+n_2=n-m-1} \frac{N_m(n_1, s)\,N_m(n_2, R-s)}{F_{n,R}} \tag{3}$$

is true.

Let $\mathcal{F}_n$ be the set of all plane planted trees with n non-root vertices, and let F_n be the number of trees in $\mathcal{F}_n$. If P_R is the probability that a tree $T \in \mathcal{F}_n$ has exactly R leaves, and $P_{l,m,s}$ is the probability that a forest $W \in W(l, m)$ has exactly s real leaves, then from (3) we obtain

$$\begin{aligned}\mathbf{P}\{h_s = m + 1\} &= \sum_{n_1+n_2=n-m-1} \frac{N_m(n_1)N_m(n_2)}{F_n}\frac{N_m(n_1, s)}{N_m(n_1)}\frac{N_m(n_2, R-s)}{N_m(n_2)}\frac{F_n}{F_{n,R}} \\ &= \sum_{n_1+n_2=n-m-1} \frac{N_m(n_1)N_m(n_2)}{F_n}\frac{P_{n_1,m,s}P_{n_2,m,R-s}}{P_R}. \end{aligned} \tag{4}$$

Let $n \to \infty$, $m = \beta\sqrt{n}(1 + o(1))$, $s = \alpha n(1 + o(1))$, $R = n/2 + v\sqrt{n}$, where $|v| \le K$, K is a constant. We partition the sum from (4) into two sums. Let $\Sigma_1(m)$ denote the sum over $n_1 = 2s + u\sqrt{n}$, where $|u| \le C$, C is a constant, and let $\Sigma_2(m)$ be the sum over the remaining n_1. We put $\lambda = 1/2$ in distribution (1).

We begin with estimating $\Sigma_1(m)$. We first estimate the factor $A = N_m(n_1)N_m(n_2)/F_n$.

LEMMA 6. *If $n, m \to \infty$ in a such a way that $m = \beta\sqrt{n}(1 + o(1))$, then*

$$A = \frac{\beta^2}{4(2\alpha(1-2\alpha))^{3/2}\sqrt{\pi n}} \exp\left\{-\frac{\beta^2}{8\alpha(1-2\alpha)}\right\}(1 + o(1)) \tag{5}$$

uniformly in the integers n_1, n_2 such that $|u| \le C$, where C is a constant.

PROOF. Since (see [5, 14, 22])

$$N_m(n_1) = \frac{m}{n_1 + m}\binom{2n_1 + m - 1}{n_1},$$

$$N_m(n_2) = \frac{m}{n_2 + m}\binom{2n_2 + m - 1}{n_2},$$

$$F_n = \frac{1}{n}\binom{2n-2}{n-1},$$

using Corollary 1, we obtain

$$N_m(n_1) = \frac{\beta 2^{2n_1+m-1}}{(2\alpha)^{3/2} n\sqrt{\pi}} e^{-\beta^2/(8\alpha)}(1 + o(1)), \tag{6}$$

$$N_m(n_2) = \frac{\beta 2^{2n_2+m-1}}{(1-2\alpha)^{3/2} n\sqrt{\pi}} e^{-\beta^2/(4(1-2\alpha))}(1 + o(1)), \tag{7}$$

$$F_n = \frac{2^{2n-2}}{n\sqrt{n\pi}}\left(1 + O\left(\frac{1}{\sqrt{n}}\right)\right). \tag{8}$$

Taking (6)–(8) and the relation $n_1 + n_2 = n - m - 1$ into account, we obtain the assertion of the lemma.

The following lemma gives an estimate of P_R.

LEMMA 7. *If $n \to \infty$, $R = n/2 + v\sqrt{n} + o(\sqrt{n})$, where v is a fixed constant, then*

$$P_R = \frac{2}{\sqrt{\pi n}} e^{-4v^2}(1 + o(1)). \tag{9}$$

PROOF. Applying Lemmas 1–3, we obtain

$$P_R = \mathbf{P}\{\mu_0(n, n+1) = R\} = \frac{1}{2^{n+1}}\binom{n+1}{R}\frac{\mathbf{P}\{\xi_1^{(0)} + \ldots + \xi_{n+1-R}^{(0)} = n\}}{\mathbf{P}\{\xi_1 + \ldots + \xi_{n+1} = n\}}, \tag{10}$$

where $\mu_0(n, N)$ is the number of empty cells in the generalized scheme of allocating n particles to N cells, $\xi_1, \ldots, \xi_{n+1}$ are independent identically distributed by law (2.2) random variables, and $\xi_1^{(0)}, \ldots, \xi_{n-R+1}^{(0)}$ are independent identically distributed random variables such that

$$\mathbf{P}\{\xi_1^{(0)} = k\} = \mathbf{P}\{\xi_1 = k \mid \xi_1 \neq 0\}, \qquad k = 0, 1, \ldots$$

To estimate the factors in (10), we use Corollary 1. Then

$$\frac{1}{2^{n+1}} \binom{n+1}{R} = \frac{2}{\sqrt{2\pi n}} e^{-2v^2}(1 + o(1)),$$

and

$$\mathbf{P}\{\xi_1 + \ldots + \xi_{n+1} = n\} = \frac{1}{2\sqrt{\pi n}}(1 + o(1)).$$

It is clear that

$$\mathbf{P}\{\xi_1^{(0)} + \ldots + \xi_N^{(0)} = l\} = \mathbf{P}\{\xi_1 + \ldots + \xi_N = l - N\}.$$

Then, using Corollary 1, we obtain

$$\begin{aligned}\mathbf{P}\{\xi_1^{(0)} + \ldots + \xi_{n-R+1}^{(0)} = n\} &= \mathbf{P}\{\xi_1 + \ldots + \xi_{n-R+1} = R - 1\} \\ &= \frac{1}{\sqrt{2\pi n}} e^{-2v^2}(1 + o(1)).\end{aligned}$$

Taking these relations into account, we obtain the assertion of the lemma.

Let us estimate the probabilities $P_{n_1,m,s}$ and $P_{n_2,m,R-s}$. We consider a random forest from $W(n_1, m)$. A tree consisting of only the root is called an empty tree. Let $\alpha_0(n_1, m)$ be the number of empty trees in a random forest from $W(n_1, m)$. If $\alpha_0(n_1, m) = k$, then $S_{m-k} = \xi_1^{(0)} + \ldots + \xi_{m-k}^{(0)}$ is the number of vertices of the next generation, and these vertices are the roots of the forest consisting of S_{m-k} roots and $n_1 - S_{m-k}$ non-root vertices. Then it is easily seen that

$$P_{n_1,m,s} = \sum_k \mathbf{P}\{\alpha_0(n_1, m) = k\} \sum_l \mathbf{P}\{S_{m-k} = \xi_1^{(0)} + \ldots + \xi_{m-k}^{(0)} = l\} P_{n_1-l,l,s}^{(0)}, \qquad (11)$$

where $P_{n_1-l,l,s}^{(0)}$ is the probability that a random forest from $W(n_1 - l, l)$ has exactly s leaves.

Let $l = m + z\sqrt{m}$, $k = m/2 + w\sqrt{m}$, $l_0 = m - Z\sqrt{m}$, $l_1 = m + Z\sqrt{m}$, $Z = m^{1/8}$, and let W be a constant. Then we represent $P_{n_1,m,s}$ as a sum of the four parts

$$P_{n_1,m,s} = S_1 + S_2 + S_3 + S_4, \qquad (12)$$

where

$$S_1 = \sum_{k:\ |w|\le W} \mathbf{P}\{\alpha_0(n_1,m)=k\} \sum_{l_0\le l\le l_1} \mathbf{P}\{S_{m-k}=\xi_1^{(0)}+\ldots+\xi_{m-k}^{(0)}=l\}P^{(0)}_{n_1-l,l,s},$$

$$S_2 = \sum_{k:\ |w|> W} \mathbf{P}\{\alpha_0(n_1,m)=k\} \sum_{l} \mathbf{P}\{S_{m-k}=\xi_1^{(0)}+\ldots+\xi_{m-k}^{(0)}=l\}P^{(0)}_{n_1-l,l,s},$$

$$S_3 = \sum_{k:\ |w|\le W} \mathbf{P}\{\alpha_0(n_1,m)=k\} \sum_{l<l_0} \mathbf{P}\{S_{m-k}=\xi_1^{(0)}+\ldots+\xi_{m-k}^{(0)}=l\}P^{(0)}_{n_1-l,l,s},$$

$$S_4 = \sum_{k:\ |w|\le W} \mathbf{P}\{\alpha_0(n_1,m)=k\} \sum_{l>l_1} \mathbf{P}\{S_{m-k}=\xi_1^{(0)}+\ldots+\xi_{m-k}^{(0)}=l\}P^{(0)}_{n_1-l,l,s}.$$

We begin with estimating $P^{(0)}_{n_1-l,l,s}$ for the interval of summation in S_1, S_2, S_3, and S_4. Applying Lemmas 1–3, we obtain

$$P^{(0)}_{n_1-l,l,s} = \mathbf{P}\{\mu_0(n_1-l,n_1)=s\} = \frac{1}{2^{n_1}}\binom{n_1}{s}\frac{\mathbf{P}\{\xi_1^{(0)}+\ldots+\xi_{n_1-s}^{(0)}=s\}}{\mathbf{P}\{\xi_1+\ldots+\xi_{n_1}=n_1-l\}}.$$

Since

$$\mathbf{P}\{\xi_1^{(0)}+\ldots+\xi_N^{(0)}=p\} = \mathbf{P}\{\xi_1+\ldots+\xi_N=p-N\}, \tag{13}$$

using Corollary 1 and taking into account that $n_1 = 2s + u\sqrt{n}$, we get

$$P^{(0)}_{n_1-l,l,s} = \frac{1}{\sqrt{\pi\alpha n}}e^{-u^2/(4\alpha)}\frac{\mathbf{P}\{\xi_1+\ldots+\xi_{n_1-s}=s-l\}}{\mathbf{P}\{\xi_1+\ldots+\xi_{n_1}=n_1-l\}}. \tag{14}$$

If $l < l_1$, then, using Theorem 2, we obtain

$$P^{(0)}_{n_1-l,l,s} = \frac{1}{\sqrt{\pi\alpha n}}e^{-u^2/(4\alpha)}\frac{e^{-(u\sqrt{n}+l)^2/(4\pi n)}/(2\sqrt{\pi\alpha n})}{e^{-l^2/(8\alpha n)}/(2\sqrt{2\pi\alpha n})}(1+o(1)) \le \frac{c}{\sqrt{n}}. \tag{15}$$

If $l_0 \le l \le l_1$, then

$$P^{(0)}_{n_1-l,l,s} = \frac{2}{\sqrt{2\pi\alpha n}}\exp\left\{-\frac{u^2+(u+\beta)^2}{4\alpha}+\frac{\beta^2}{8\alpha}\right\}(1+o(1)). \tag{16}$$

For $l > l_1$ we use the trivial estimate

$$P^{(0)}_{n_1-l,l,s} \le 1.$$

Let us study S_1. Using (13) and Theorem 2, we obtain

$$\begin{aligned}\mathbf{P}\{S_{m-k}=\xi_1^{(0)}+\ldots+\xi_{m-k}^{(0)}=l\} &= \mathbf{P}\{\xi_1+\ldots+\xi_{m-k}=l-m+k\}\\ &= \frac{1}{\sqrt{2\pi m}}e^{-(z+2w)^2/2}(1+o(1)).\end{aligned}$$

Using Lemma 4 and (16), we obtain

$$S_1 = B \sum_{|w|\le W} \frac{2}{\sqrt{2\pi m}} e^{-w^2/2} \sum_{|z|\le Z} \frac{1}{\sqrt{2\pi m}} e^{-(z+2w)^2/2},$$

where

$$B = \frac{2}{\sqrt{2\pi\alpha n}} \exp\left\{-\frac{u^2+(u+\beta)^2}{4\alpha} + \frac{\beta^2}{8\alpha}\right\}(1+o(1)).$$

Replacing the sums by the integrals and taking into account that the constant W can be chosen as large as desired and $Z \to \infty$, we obtain

$$S_1 = B\left(\int_{-\infty}^{\infty} \frac{1}{\sqrt{2\pi}} e^{-w^2/2} dw - \left(\int_{-\infty}^{-W} \frac{1}{\sqrt{2\pi}} e^{-w^2/2} dw + \int_{W}^{\infty} \frac{1}{\sqrt{2\pi}} e^{-w^2/2} dw\right)\right)$$
$$= B(1-\varepsilon_n),$$

where ε_n can be made as small as desired by the choice of sufficiently large W and n.

Let us estimate

$$S_2 = \sum_{k:\ |w|>W} \mathbf{P}\{\alpha_0(n_1,m)=k\} \sum_{l} \mathbf{P}\{S_{m-k} = \xi_1^{(0)} + \ldots + \xi_{m-k}^{(0)} = l\} P_{n_1-l,l,s}^{(0)}$$
$$= \sum_{k:\ |w|>W} \mathbf{P}\{\alpha_0(n_1,m)=k\} \left(\sum_{l\le l_1} \mathbf{P}\{S_{m-k} = \xi_1^{(0)} + \ldots + \xi_{m-k}^{(0)} = l\} P_{n_1-l,l,s}^{(0)}\right.$$
$$\left. + \sum_{l>l_1} \mathbf{P}\{S_{m-k} = \xi_1^{(0)} + \ldots + \xi_{m-k}^{(0)} = l\} P_{n_1-l,l,s}^{(0)}\right).$$

For the sum over $l \le l_1$ we use (15) and the fact that

$$\sum_{l\le l_1} \mathbf{P}\{S_{m-k} = l\} \le 1.$$

For the sum over $l > l_1$ we estimate the probability $\mathbf{P}\{S_{m-k} = l\}$ using Corollary 2 and the trivial estimate

$$P_{n_1-l,l,s}^{(0)} \le 1.$$

Then we obtain

$$S_2 \le \sum_{k:\ |w|>W} \mathbf{P}\{\alpha_0(n_1,m)=k\} \left(\frac{c}{\sqrt{n}} + \sum_{l>l_1} c_1 e^{-c_2 m^{1/4}}\right)$$
$$\le \sum_{k:\ |w|>W} \mathbf{P}\{\alpha_0(n_1,m)=k\} \left(\frac{c}{\sqrt{n}} + \frac{1}{n}\right) \le \frac{c}{\sqrt{n}} \varepsilon_n',$$

where ε_n' can be made as small as desired by the choice of sufficiently large W and n.

We obtain the estimates

$$S_3 \le \frac{1}{n}, \qquad S_4 \le \frac{1}{n}$$

using Corollary 2 for the estimation of $\mathbf{P}\{S_{m-k} = \xi_1^{(0)} + \ldots + \xi_{m-k}^{(0)} = l\}$ and the trivial inequality $P_{n_1-l,l,s}^{(0)} \le 1$ in both of the cases.

Thus, we obtain

$$P_{n_1,m,s} = \frac{2}{\sqrt{2\pi\alpha n}} \exp\left\{-\frac{u^2 + (u+\beta)^2}{4\alpha} + \frac{\beta^2}{8\alpha}\right\}(1 + o(1)). \tag{17}$$

In the same way we deduce the estimate

$$P_{n_2,m,s} = \frac{2}{\sqrt{\pi(1-2\alpha)n}} \exp\left\{-\frac{(u+2v)^2 + (u+2v+\beta)^2}{2(1-2\alpha)} + \frac{\beta^2}{4(1-2\alpha)}\right\}(1 + o(1)). \tag{18}$$

Using Lemma 6 and (10), (17) and (18), we obtain

$$\Sigma_1(m) = \frac{\beta^2}{2(2\alpha(1-2\alpha))^{3/2}\sqrt{\pi n}} \exp\left\{-\frac{\beta^2}{8\alpha(1-2\alpha)}\right\}$$
$$\times \sum_{n_1=2\alpha n+u\sqrt{n}:\ |u|\le C} \frac{1}{\sqrt{2\pi\alpha(1-2\alpha)n}} \exp\left\{-\frac{(u+\beta/2+4\alpha v)^2}{2\alpha(1-2\alpha)}\right\}(1 + o(1)).$$

Replacing the sum over n_1 by the integral over u and taking into account that $du = 1/\sqrt{n}$, the constant C can be chosen as large as desired, and finally changing the variables according to the formula

$$t = \frac{u + \beta/2 + 4\alpha v}{\sqrt{\alpha(1-2\alpha)}},$$

we see that

$$\sum_{n_1=2\alpha n+u\sqrt{n}:\ |u|\le C} \frac{1}{\sqrt{2\pi\alpha(1-2\alpha)n}} \exp\left\{-\frac{(u+\beta/2+4\alpha v)^2}{2\alpha(1-2\alpha)}\right\}(1 + o(1))$$
$$= \frac{1}{\sqrt{2\pi}} \int_{-\infty}^{\infty} e^{-t^2/2} dt(1 + o(1)) = 1 + o(1),$$

and

$$\mathbf{P}\{h_s = m+1\} = \frac{\beta^2}{2(2\alpha(1-2\alpha))^{3/2}\sqrt{\pi n}} \exp\left\{-\frac{\beta^2}{8\alpha(1-2\alpha)}\right\}(1 + o(1)) + \Sigma_2(m).$$

Setting $\sigma = 2\sqrt{\alpha(1-2\alpha)}$ and introducing the new variable $x = \beta/\sigma$, we obtain

$$\mathbf{P}\{h_s = m\} = \frac{1}{\sqrt{n}} p(x)(1 + o(1)) + \Sigma_2(m), \tag{19}$$

where

$$p(x) = \sqrt{\frac{2}{\pi}} x^2 e^{-x^2/2}.$$

Let us prove that $\mathbf{P}\{h_s \leq z\sigma\sqrt{n}\}$ tends to the Maxwell distribution function at z as $n \to \infty$. To this end it is sufficient to prove that as $n \to \infty$

$$\sum_{m=1}^{\infty} \Sigma_2(m) \to 0.$$

Summing both sides of the equality

$$\mathbf{P}\{h_s = m\} = \Sigma_1(m) + \Sigma_2(m)$$

over m, we obtain

$$\sum_{m=1}^{\infty} \Sigma_1(m) + \sum_{m=1}^{\infty} \Sigma_2(m) = 1.$$

Since $\Sigma_1(m)$ and $\Sigma_2(m)$ are non-negative, for any positive b and B, $b < B$, the relation

$$\sum_{m=1}^{\infty} \Sigma_2(m) = 1 - \sum_{m=1}^{\infty} \Sigma_1(m) \leq 1 - \sum_{b\sqrt{n} \leq m \leq B\sqrt{n}} \Sigma_1(m) \tag{20}$$

holds. It can easily be seen that as $n \to \infty$ relation (19) holds uniformly in integers m such that x lies in any finite interval of the form $b \leq x \leq B$. Therefore

$$\sum_{b\sqrt{n} \leq m \leq B\sqrt{n}} \Sigma_1(m) = \int_b^B p(x)\, dx\, (1 + o(1)). \tag{21}$$

Since $p(x)$ is the density of a distribution concentrated on the positive semi-axis, for any $\varepsilon > 0$ one can choose b and B such that

$$\int_b^B p(x)\, dx > 1 - \frac{\varepsilon}{2}.$$

Then (21) implies that for such b and B and sufficiently large n

$$\sum_{b\sqrt{n} \leq m \leq B\sqrt{n}} \Sigma_1(m) > 1 - \varepsilon. \tag{22}$$

Finally, from (20) and (22) it follows that for any fixed $\varepsilon > 0$ and sufficiently large n

$$\sum_{m=1}^{\infty} \Sigma_2(m) \leq \varepsilon.$$

Hence the assertion of the theorem follows.

REFERENCES

1. A. Abramov, The limit distribution of leaves of plane planted trees. In: *Abstracts of the All-Russian Conference on Stochastic Methods, Yoshkar-Ola, 1995*. TVP, Moscow, 1995, pp. 10–11 (in Russian).
2. D. J. Aldous, The continuum random tree. I. *Ann. Probab.* (1991) **19**, 1–28.
3. D. J. Aldous, The continuum random tree. II. An overview. In: *Stochastic Analysis*. Cambridge Univ. Press, Cambridge, 1991, pp. 23–70.
4. D. J. Aldous, The continuum random tree. III. *Ann. Prob.* (1993) **21**, 248–289.
5. M. Drmota, The height distribution of leaves in rooted trees. *Discrete Math. Appl.* (1994) **4**, 45–58.
6. Ph. Flajolet and A. Odlyzko, The average height of binary trees and other simple trees. *J. Comput. System Sci.* (1982) **25**, 171–213.
7. Ph. Flajolet and A. Odlyzko, Singularity analysis of generating functions. *SIAM J. Discrete Math.* (1990) **3**, 216–240.
8. W. Gutjahr and G. Ch. Pflug, The asymptotic distribution of leaf heights in binary trees. *Graphs and Combinatorics* (1992) **8**, 243–251.
9. W. Gutjahr and G. Ch. Pflug, The asymptotic contour process of a binary tree is a Brownian excursion. *Stoch. Processes and Appl.* (1992) **41**, 69–89.
10. W. Gutjahr and G. Ch. Pflug, The limiting common distribution of two leaf heights in a random binary tree. *Theor. Inf. Appl.* (1992) **26**, 1–18.
11. M. Drmota, On nodes of given degree in random trees. *These Proceedings*, pp. 31–44.
12. I. B. Kalugin, Branching processes and random mapping of finite sets. *Math. Notes* (1983) **34**, 757–771 (in Russian).
13. R. Kemp, The average height of planted plane trees with *M* leaves. *J. Comb. Theory, Ser.B* (1983) **34**, 191–208.
14. D. E. Knuth, *The Art of Computer Programming, Vol.* 1. Addison–Wesley, Reading, MA, 1969.
15. V. F. Kolchin, Branching processes, random trees and the generalized allocation scheme. *Math. Notes* (1977) **21**, 691–705 (in Russian).
16. V. F. Kolchin, The moment of extinction of branching processes and the height of random tree. *Math. Notes* (1978) **24**, 859–870 (in Russian)
17. V. F. Kolchin, *Random Mappings*. Springer, New York, 1986.
18. A. Meir and J. W. Moon, On the altitude of nodes in random trees. *Canad. J. Math.* (1978) **30**, 997–1015.
19. J. W. Moon, A problem on random trees. *J. Comb. Theory* (1971) **2**, 201–205.
20. J. W. Moon, On the maximum degree in a random tree. *Michigan Math. J.* (1968) **15**, 429–432.
21. Yu. L. Pavlov, Limit distributions of the height of a random forest of plane rooted trees. *Discrete Math. Appl.* (1994) **4**, 73–88.
22. Yu. L. Pavlov, *Random Forests*. Karelian Sci. Centre of the Russian Acad. Sci., Petrozavodsk, 1996 (in Russian).
23. A. Renyi, Some remarks on the theory of trees. *Publ. Math. Inst. Hungar. Acad. Sci.* (1954) **4**, No.7, 3–85.
24. A. Renyi and G. Szekeres, On the height of trees. *J. Austral. Math. Soc.* (1967) **7**, 497–507.
25. V. E. Stepanov, On the distribution of the number of leaves in strata of a random tree. *Theory Probab. Appl.* (1969) **14**, 64–77. (in Russian)
26. L. Takacs, Conditional limit theorems for branching processes. *J. Appl. Math. Stoch. Analysis* (1991), **4**, 263–292.
27. V. A. Vatutin, Branching processes with final types of nodes and random trees. *Theory Probab. Appl.* (1994) **39**, 699-714 (in Russian).

Probabilistic Methods in Discrete Mathematics, pp. 101–102
V. F. Kolchin et al. (Eds.)

ON THE DISTRIBUTION OF THE WEIGHTS OF THE RANDOM REED–MULLER CODEWORDS

A. S. AMBROSIMOV

Moscow

Abstract — Let $\mathcal{R}(r, n)$ be the binary Reed–Muller code of order r and length 2^n, $1 \le r \le n$. It is proved that if the uniform measure is given on $\mathcal{R}(r, n)$, then the limit distribution of the weight of a random codeword with corresponding normalization is normal as $n \to \infty$, $r = r(n) \to \infty$.

For $1 \le r \le n$ the binary Reed–Muller code $\mathcal{R}(r, n)$ of order r and length 2^n is the set of all vectors f which are defined by all Boolean functions $f(x_1, \ldots, x_n)$ of degree at most r. We consider the random function $f\colon (\mathrm{GF}(2))^n \to \mathrm{GF}(2)$, $\deg f \le r$, and present it in the algebraic normal form

$$f(x) = a_0 + \sum_{m=1}^{r} \sum_{1 \le i_1 < \ldots < i_m \le n} a_{i_1 \ldots i_m} x_{i_1}^{(x)} \ldots x_{i_m}^{(x)},$$

where $x = (x_1^{(x)}, \ldots, x_n^{(x)})$. This representation in the form of Zhegalkin's polynomial contains exactly $1 + \binom{n}{1} + \ldots + \binom{n}{r}$ independent uniformly distributed on GF(2) coefficients. Let $\xi_{r,n} = \|f(x)\|$ denote the number of arguments x such that $f(x) = 1$.

THEOREM 1. *For any r the moments of orders $1, \ldots, r+1$ of the random variable $\left(\xi_{r,n} - 2^{n-1}\right)/2^{n/2-1}$ converge, as $n \to \infty$, to the corresponding moments of the standard normal distribution.*

If $n \to \infty$ and $r = r(n) \to \infty$, $r \le n$, then the distribution of the random variable $\left(\xi_{r,n} - 2^{n-1}\right)/2^{n/2-1}$ converges weakly to the standard normal distribution.

PROOF. We represent the random variable $\xi_{r,n}$ as the sum of indicators

$$\xi_{r,n} = \sum_{x \in (\mathrm{GF}(2))^n} f(x).$$

We use the method [1] which was suggested by S. N. Bernshtein and now is usually referred to as the Bernshtein method. We consider the random variable

$$\xi'_{r,n} = \sum_{x \in (\mathrm{GF}(2))^n} f'(x)$$

which is the sum of 2^n independent indicators such that $f'(x)$ and $f(x)$ have the same distribution. It is clear that $\xi_{r,n}$ is the sum of pairwise independent random variables. Therefore

$$\mathbf{D}\xi_{r,n} = \mathbf{D}\xi'_{r,n} = 2^{n-2}.$$

The moments of the random variables $\left(\xi'_{r,n} - 2^{n-1}\right)/2^{n/2-1}$ converge to the moments of the standard normal distribution as $n \to \infty$. This fact follows, for example, from Zaremba's result [3]. The main idea is to establish the coincidence of the moments (up to the order $r+1$) of the random variables $(\xi_{r,n} - 2^{n-1})/2^{n/2-1}$, $(\xi'_{r,n} - 2^{n-1})/2^{n/2-1}$. We prove this fact for $r+2 \le n$. To do this, it is sufficient to demonstrate that any $r+1$ indicators in the sum $\xi_{r,n}$ are independent random variables. Let $x_1, \dots, x_{r+1} \in (\mathrm{GF}(2))^n$, $x_i \ne x_j$, $i \ne j$. Without loss of generality we suppose that $\|x_1\| \le \dots \le \|x_{r+1}\|$, where $\|x\|$ is the number of ones in the vector x.

Let us prove that the random variables $f(x_1), \dots, f(x_{r+1})$ are independent for any fixed vectors $x_1, \dots, x_{r+1}$. Each random variable $f(x_j)$ is the sum in GF(2) with index j containing the coefficients $a_0, a_{i_1}, \dots, a_{i_1 \dots i_m}$, $1 \le i_1 < \dots < i_m \le n$, $m = 1, \dots, r$. The sum with index j contains a coefficient which does not belong to any sum with the index $m < j$. Indeed, $f(x_2)$ contains a coefficient $a_{i_1^1}$ which does not belong to the sum $f(x_1)$ since otherwise $x_1 = x_2$. Similarly, the sum $f(x_3)$ contains an element $a_{i_2^2}$ which does not belong to $f(x_2)$ and an element $a_{i_1^2}$ which does not belong to $f(x_1)$. If $i_2^2 = i_1^2$, then the element $a_{i_2^2}$ has the desired property, otherwise the coefficient $a_{i_1^2, i_2^2}$ is in the sum $f(x_3)$, but by construction it is not in $f(x_1)$ and $f(x_2)$.

We consider the sum $f(x_j)$. For any $m \le j-1$ the sum contains an element $a_{i_m^{j-1}}$ which does not belong to $f(x_m)$ because otherwise $x_j = x_m$. Let $\{l_1, \dots, l_s\}$, $l_1 < \dots < l_s$, $1 \le s \le j-1$, be the set of all different indices among $i_1^{j-1}, \dots, i_{j-1}^{j-1}$. Then by construction the coefficient $a_{l_1, \dots, l_s}$ has the desired property. Therefore the random variable $f(x_j)$ and the random vector $(f(x_1), \dots, f(x_{j-1}))$ are independent. Hence it follows from the conditional probability formula that the random variables $f(x_1), \dots, f(x_{r+1})$ are independent. The theorem is proved.

Note that the distribution of the codeword weights of $\mathcal{R}(2, n)$ is described in [2], and for any fixed $r \ge 3$ even the asymptotic behaviour of this random variables was not investigated.

REFERENCES

1. M. Loève, *Probability Theory.* Van Nostrand, Princeton, 1955.
2. F. MacWilliams and N. Sloane, *The Theory of Error-Correcting Codes.* North-Holland, Amsterdam, 1977.
3. S. K. Zaremba, Note on the central limit theorem. *Math. Z.* (1958) **69**, 295–298.

Probabilistic Methods in Discrete Mathematics, pp. 103–109
V. F. Kolchin et al. (Eds.)

ON A METHOD OF PROVING LIMIT THEOREMS FOR BRANCHING PROCESSES WITH IMMIGRATION

M. H. ASADULLIN

Voskresenskoe, Moscow region

Abstract — Previously, S. V. Nagaev and the author obtained a result concerning a scheme of summation of a random number of independent random variables and suggested a new method of proving limit theorems for branching processes with immigration. This method provides a possibility of obtaining asymptotic results on branching processes with immigration from the corresponding results for the branching processes without immigration. In this paper we consider in detail the supercritical and subcritical cases which were not yet considered.

1. INTRODUCTION

Let $X_\tau(t)$, $0 \le t \le \tau$, be a family of non-negative integer-valued stochastic processes depending on a real parameter τ. We assume that for any fixed τ the trajectories of $X_\tau(t)$ are right continuous non-decreasing in t and bounded for any finite t. Under these assumptions we can write $X_\tau(t)$ as

$$X_\tau(t) = \int_0^t \zeta_\tau(u)\, dY_\tau(u),$$

where $Y_\tau(t)$, $0 \le t \le \tau$, $Y_\tau(0) = X_\tau(0)$, is a counting stochastic process such that the jump moments of realizations of $X_\tau(t)$ and $Y_\tau(t)$ coincide and $\zeta_\tau(t)$ is an integer-valued non-negative stochastic process equal to $X_\tau(t) - X_\tau(t-)$ at the jump points of $X_\tau(t)$ and 0 at all other points.

In addition, let for any fixed τ, a sequence of independent non-negative stochastic processes $\mu_\tau^j(t,u)$, $t \ge u$, with $\mu_\tau^j(u,u) = 1$ be given. We assume that for any fixed τ the stochastic processes $X_\tau(t)$, $\mu_\tau^j(t,u_i)$, $j \ge 1$, $0 \le u_1 < u_2 < \dots$, are mutually independent and for any j the distribution of $\mu_\tau^j(t,u)$ coincides with that of a non-negative stochastic process $\mu_\tau(t-u)$ with $\mu_\tau(0) = 1$.

We set

$$Z_\tau(t) = \int_0^t \sum_{j=1}^{\zeta_\tau(u)} \mu\tau^j(t,u)\, dY_\tau(u) = \int_0^t \sum_{j=1}^{\zeta_\tau(u)} \mu_\tau^j(t-u)\, dY_\tau(u). \qquad (1)$$

The counting stochastic process $X_\tau(t)$ may be regarded as the number of particles entering the population during the interval of time $[0,t]$. Each of the $\zeta_\tau(u)$ particles, arriving at a moment u, initiates a branching process $\mu_\tau^j(t,u)$, $t \geq u, j \geq 1$. Finally, we assume that for any fixed τ and j and any set $\{u_i\}$ the stochastic processes $X_\tau(t)$, $\mu_\tau^j(t,u_i)$, $j \geq 1$, $0 \leq u_1 < u_2 < \ldots$, are mutually independent and in addition for any j the distributions of $\mu_\tau^j(t,u)$ coincide with the distribution of $\mu_\tau(t-u)$, where $\mu_\tau(t)$ is a branching process with $\mu_\tau(0) = 1$.

From this point of view $Z_\tau(t)$ is the total number of particles in the branching processes with immigration $X_\tau(t)$ at the moment t.

A highly important place in the theory of branching processes with immigration is occupied by limit theorems for $Z_\tau(t)$ as $t \to \infty$ (see, for example, references in [1 - 2]). The first results in this direction were proved for the processes where a process with independent increments was taken as the immigration process. Later more general immigration processes were considered (see [3, 6–8]).

S. V. Nagaev and the author obtained in [3] a result concerning the summation of a random number of independent random variables (of the form (1)) and developed a general method of proving limit theorems for branching processes with immigration. This method enables to reduce the study of the asymptotic behaviour of $Z_\tau(t)$ to the known results for branching processes without immigration. In [3] we considered the case where the conditional distributions of branching processes converge to the exponential law. It is known that this is true, for example, for the critical (see [1]) or close to critical (the case of transient phenomena) branching processes (see, for example, [1, 4, 5]).

In this paper we consider the supercritical and subcritical cases.

2. SUPERCRITICAL CASE

First we consider the supercritical case. We assume that $t \leq \tau$ and set

$$Q_\tau(t) = \mathbf{P}\{\mu_\tau(t) > 0\}.$$

We say that the family $\mu_\tau(t)$ of branching processes extinct if

$$\lim_{t,\tau\to\infty} Q_\tau(t) = 0.$$

For the family $\mu_\tau(t)$ of branching processes depending on the parameter τ we assume that the following condition is satisfied:
there is a family of monotone increasing and continuous in t functions $A_\tau(t)$, $0 \leq t \leq \tau$, such that $A_\tau(0) < \infty$, $A_\tau(t) \to \infty$ as $t \to \infty$ and $\tau \to \infty$ and for any $x \geq 0$

$$\lim_{t,\tau\to\infty} \mathbf{P}\{\mu_\tau(t)/A_\tau(t) \leq x \mid \mu_\tau(t) > 0\} = F_1(x) \tag{2}$$

(here and below $F_1(x)$ is a distribution function such that $F_1(0+) = 0$).

This condition means that the limit theorems are valid for the branching processes $\mu_\tau(t)$. For the supercritical case, these theorems are well known (see [1, 2]).

In addition, we assume that

$$\lim_{x\to\infty} \sup_{\tau} \sup_{0\le t\le q} \mathbf{P}\{\mu_\tau(t) > x\} = 0 \tag{3}$$

for some finite q.

Let $u = A_\tau(t)/A_\tau$, where $A_\tau = A_\tau(\tau)$. For $u \in (0,1]$ we set $\psi(u,\tau) = A_\tau^{-1}(uA_\tau) \equiv t$, where $A_\tau^{-1}(\cdot)$ is the function inverse to $A_\tau(\cdot)$. It is clear that $\lim_{\tau\to\infty} A_\tau(0)/A_\tau = 0$. By definition $\psi(0,\tau) = 0$.

For the case where the summands in (1) do not vanish, that corresponds to the supercritical case for the branching processes with immigration, the following assertion is true.

THEOREM 1. *Let condition* (2) *be valid for a sequence of families of stochastic processes* $\{\mu_\tau^j(t)\}_{j\ge 1}$, *there exist*

$$\lim_{t,\tau\to\infty} Q_\tau(t) = Q > 0,$$

and let there exist a family of stochastic processes $X_\tau(t)$ *coinciding with the stochastic processes* $\{\mu_\tau^j(t)\}_{j\ge1}$ *such that for a sequence* C_τ *the following condition* (A1) *is valid: as* $\tau \to \infty$,

$$\text{(A1)} \qquad C_\tau \int_v^1 uQ_\tau(\psi(u,\tau))\, dX_\tau(\tau - \psi(u,\tau)) \to T_1(v)$$

in the sense of convergence of finite dimensional distributions, where $T_1(v)$ *is a stochastic process with non-describing trajectories defined on the half-interval* $(0,1]$.

Let, in addition, the following condition (B1) *be valid: for any fixed* $s \ge 0$

$$\text{(B1)} \qquad \lim_{\tau\to\infty,\varepsilon\to0} C_\tau \int_0^{\tau_\varepsilon} \left|\mathbf{E}e^{-s\mu_\tau(t)/A_\tau} - 1\right| dX_\tau(\tau - t) = 0,$$

where the limit is taken in the sense of convergence in probability and $\tau_\varepsilon = \psi(\varepsilon,\tau)$.

Then as $\tau \to \infty$,

$$C_\tau Z_\tau(\tau)/A_\tau \to W_1$$

in distribution and

$$\mathbf{E}e^{-sW_1} = \mathbf{E}\exp\left\{\int_{0+}^1 \ln(1 - Q + Qh_1(su))\, dT_1(u)\right\},$$

where $h_1(s)$ *is the Laplace transform of the distribution function* $F_1(x)$.

REMARK 1. Note that the integral in Theorem 1 may, generally speaking, be improper, i.e.,

$$\int_{0+}^1 \ln(1 - Q + Qh_1(su))\, dT_1(u)$$

is regarded as the limit in probability, as positive $\varepsilon \to 0$, of the integral

$$\int_\varepsilon^1 \ln(1 - Q + Qh_1(su))\, dT_1(u).$$

REMARK 2. It is not difficult to see that if the mathematical expectations of the stochastic processes $\mu_\tau^j(t), j \geq 1$, exist, then condition (B1) follows from the condition: as $\tau \to \infty$ and $\varepsilon \to 0$,

$$C_\tau \int_0^{\tau_\varepsilon} (\mathbf{E}\mu_\tau(t))/A_\tau \, dX_\tau(\tau - t) \to 0$$

in probability.

PROOF. Let $u = A_\tau(t)/A_\tau$ and, therefore, $t = \psi(u, \tau)$. Substituting u for t, we easily see that the distribution of the random variable $C_\tau Z_\tau(\tau)/A_\tau$ coincides with the distribution of $I_\tau^1(\varepsilon) + I_\tau^2(\varepsilon)$, where

$$I_\tau^1(\varepsilon) = \frac{C_\tau}{A_\tau} \int_0^{\tau_\varepsilon} \sum_{j=1}^{\zeta_\tau(\tau-t)} \mu_\tau^j(t) \, dY_\tau(\tau - t) = \frac{C_\tau}{A_\tau} \int_0^{\tau_\varepsilon} \sum_{j=1}^{\zeta_\tau(\tau-\psi(u,\tau))} \mu_\tau^j(\psi(u, \tau)) \, dY_\tau(\tau - \psi(u, \tau)),$$

$$I_\tau^2(\varepsilon) = \frac{C_\tau}{A_\tau} \int_{\tau_\varepsilon}^{\tau} \sum_{j=1}^{\zeta_\tau(\tau-t)} \mu_\tau^j(t) \, dY_\tau(\tau - t) = \frac{C_\tau}{A_\tau} \int_{\tau_\varepsilon}^{\tau} \sum_{j=1}^{\zeta_\tau(\tau-\psi(u,\tau))} \mu_\tau^j(\psi(u, \tau)) \, dY_\tau(\tau - \psi(u, \tau)). \tag{4}$$

First we consider the distribution of the random variable $I_\tau^2(\varepsilon)$. The Laplace transform of $I_\tau^2(\varepsilon)$ is

$$\mathbf{E}\exp\{-sI_\tau^2(\varepsilon)\} = \mathbf{E} \prod_{i=1}^{Y_\tau(\tau-\psi(\varepsilon,\tau))} (\mathbf{E}\exp\{-s\mu_\tau(\psi(u_i, \tau))/A_\tau\})^{C_\tau\zeta_\tau(\tau-\psi(u_i,\tau))}$$

$$= \mathbf{E}\exp\left\{C_\tau \int_\varepsilon^1 \ln \mathbf{E}\exp\{-s\mu_\tau(\psi(u, \tau))/A_\tau\} \, dX_\tau(\tau - \psi(u, \tau))\right\}.$$

It is clear that

$$\mathbf{E}\exp\{-s\mu_\tau(t)/A_\tau\} = 1 - Q_\tau(t) + Q_\tau(t)\,\mathbf{E}\{\exp\{-s\mu_\tau(t)/A_\tau\} \mid \mu_\tau(t) > 0\}.$$

From (2) it follows that for any finite L

$$\lim_{t\to\infty} \sup_{s\leq L} |\mathbf{E}\{\exp\{-s\mu_\tau(t)/A_\tau(t)\} \mid \mu_\tau(t) > 0\} - h_1(s)| = 0.$$

Therefore, for any fixed $s \geq 0$

$$\lim_{r,\tau\to\infty} \sup_{r<t<\tau} |\mathbf{E}\{\exp\{-s\mu_\tau(t)/A_\tau \mid \mu_\tau(t) > 0\}\} - h_1(sA_\tau(t)/A_\tau)| = 0.$$

Since $\tau_\varepsilon \to \infty$ as $\tau \to \infty$ for any fixed $\varepsilon > 0$ and $A_\tau(t)/A_\tau = u$, this relation is equivalent to the fact that for any fixed $s \geq 0$

$$\lim_{\tau\to\infty} \ln \mathbf{E}\exp\{-s\mu_\tau(\psi(u, \tau))/A_\tau\} = \ln(1 - Q + Qh_1(su)).$$

This property along with condition (A1) allows us to use Theorem 1 of [3] for the random variable $I_\tau^2(\varepsilon)$ with $B_\tau(u) = (uQ_\tau(\psi(u, \tau)))^{-1}$ (see the definition of $B_\tau(u)$ in [3]).

Hence it follows that if $\tau \to \infty$, then for any fixed $\varepsilon > 0$ the distribution of $I_\tau^2(\varepsilon)$ converges to the distribution of W_1 and

$$\mathbf{E}\exp\{-sW_1(\varepsilon)\} = \mathbf{E}\exp\left\{\int_\varepsilon^1 \ln(1 - Q + Qh_1(su))\,dT_1(u)\right\}.$$

On the other hand, $W_1(\varepsilon) \to W_1$ in probability as $\varepsilon \to 0$, where the random variable W_1 has the Laplace transform

$$\mathbf{E}e^{-sW_1} = \mathbf{E}\exp\left\{\int_{0+}^1 \ln(1 - Q + Qh_1(su))\,dT_1(u)\right\}.$$

Therefore, if $\tau \to \infty$, $\varepsilon \to 0$, $\varepsilon > \varepsilon(\tau)$, then

$$I_\tau^2(\varepsilon) \to W_1 \tag{5}$$

in distribution.

We turn now to the estimation of $I_\tau^1(\varepsilon)$. It is not difficult to check that

$$\mathbf{E}\exp\{-sI_\tau^1\}(\varepsilon)\} = \mathbf{E}\exp\left\{C_\tau \int_0^{\tau_\varepsilon} \ln \mathbf{E}\exp\{-s\mu_\tau(t)/A_\tau\}\,dX_\tau(\tau - t)\right\}. \tag{6}$$

Observe that for any $\delta > 0$

$$\begin{aligned}|\mathbf{E}\exp\{-s\mu_\tau(t)/A_\tau\} - 1| &\le Q_\tau(t)|\mathbf{E}\{\exp\{-s\mu_\tau(t)/A_\tau\} \mid \mu_\tau(t) > 0\} - 1| \\ &< \delta s + \mathbf{P}\{\mu_\tau(t) > \delta A_\tau \mid \mu_\tau(t) > 0\}.\end{aligned} \tag{7}$$

It follows from (2) and the inequality $A_\tau/A_\tau(t) > \varepsilon^{-1}$, which is valid for $t < \tau_\varepsilon$, that for any δ, ε, and $t < \tau_\varepsilon$

$$\begin{aligned}\mathbf{P}\{\mu_\tau(t) > \delta A_\tau \mid \mu_\tau(t) > 0\} &= \mathbf{P}\{\mu_\tau(t)/A_\tau(t) > \delta A_\tau/A_\tau(t) \mid \mu_\tau(t) > 0\} \\ &< 2(1 - F_1(\delta/\varepsilon))\end{aligned} \tag{8}$$

for sufficiently large t.

Since $F_1(0+) = 0$, it follows from (8) that for any $\delta > 0$ there exist ε and t_0 such that

$$\mathbf{P}\{\mu_\tau(t) > \delta A_\tau \mid \mu_\tau(t) > 0\} < \frac{1}{4} \tag{9}$$

for $t_0 < t < \tau_\varepsilon$ and sufficiently large τ.

On the other hand, condition (3) yields

$$\mathbf{P}\{\mu_\tau(t) > \delta A_\tau \mid \mu_\tau(t) > 0\} < \frac{1}{4} \tag{10}$$

for $0 \le t \le t_0$ and sufficiently large τ.

Taking $\delta < (4s)^{-1}$ in (7) and recalling (9)–(10), we see that for any fixed s there exist ε and τ_ε such that

$$|\mathbf{E}\exp\{-s\mu_\tau(t)/A_\tau\} - 1| < \frac{1}{2} \tag{11}$$

for $0 \le t \le t_0$ and $\tau > \tau_0$.

It is clear that $|\ln x| < 2|x-1|$ for $|x-1| < 1/2$. This fact combined with (11) and condition (B1) implies that as $\tau \to \infty$, $\varepsilon \to 0$

$$C_\tau \int_0^{\tau_\varepsilon} \ln \mathbf{E}\exp\{-s\mu_\tau(t)/A_\tau\}\, dX_\tau(\tau - t) \to 0$$

in probability. Taking this result and (6) into account, we obtain

$$\lim_{\tau\to\infty,\varepsilon\to 0} \mathbf{E}\exp\{-sI_\tau^1(\varepsilon)\} = 1.$$

Thus, as $\tau \to \infty$ and $\varepsilon \to 0$, the random variable $I_\tau^1(\varepsilon) \to 0$ in probability. Taking into account (4) and (5), we see that the proof of Theorem 1 is completed.

3. SUBCRITICAL CASE

In this section we assume that for the family $\mu_\tau(t)$ of branching processes depending on the parameter τ the following condition is satisfied:
for any $x \ge 0$

$$\lim_{t\to\infty,\tau\to\infty} \mathbf{P}\{\mu_\tau(t) \le x \mid \mu_\tau(t) > 0\} = F_2(x), \tag{12}$$

where $F_2(x)$ is the distribution function such that $F_2(0+) = 0$.

This condition means that the limit theorems are valid for the branching processes $\mu_\tau(t)$. For subcritical case these theorems are well known (see [1, 2]).

We set $\varphi(u,\tau) = u\tau$, $u \in [0,1]$. For subcritical branching processes with immigration, the following assertion is valid.

THEOREM 2. *Let $\{\mu_\tau^j(t)\}_{j\ge1}$ be a sequence of families of random processes satisfying conditions* (12) *and*

$$\lim_{t,\tau\to\infty} Q_\tau(t) = 0,$$

and let $X_\tau(t)$ be a family of stochastic processes such that for a sequence C_τ the following condition (A2) *is valid: as $\tau \to \infty$ and $\varepsilon \to 0$,*

$$\text{(A2)} \qquad C_\tau \int_\varepsilon^1 Q_\tau(\varphi(u,\tau))\, dX_\tau(\tau - \varphi(u,\tau)) \to \eta.$$

Let, in addition, the following condition (B2) *be valid: as $\tau \to \infty$ and $\varepsilon \to 0$,*

$$\text{(B2)} \qquad C_\tau \int_0^{\varphi(\varepsilon,\tau)} \mathbf{E}\{\mu_\tau(t) \mid \mu_\tau(t) \ge 0\}\, dX_\tau(\tau - t) \to 0$$

in probability.

Then $C_\tau Z_\tau(\tau) \to W_2$ in probability as $\tau \to \infty$, and

$$\mathbf{E}e^{-sW_2} = g(1 - h_2(s)),$$

where $h_2(s)$ is the Laplace transform of the distribution function $F_2(x)$ satisfying condition (12) and $g(s)$ is the Laplace transform of the distribution of the random variable η.

The proof of this theorem is similar to the proof of Theorem 1 given above.

REFERENCES

1. B. A. Sevastyanov, *Branching Processes.* Nauka, Moscow, 1971 (in Russian).
2. K. B. Athreya and P. E. Ney, *Branching Processes.* Springer, Berlin, 1972.
3. S. V. Nagaev and M. H. Asadullin, On a scheme of summation of random number of independent random variables with applications to branching processes with immigration. *Soviet Acad. Sci. Dokl.* (1985) **285**, No. 2, 293–296 (in Russian).
4. S. V. Nagaev and R. Mukhamedkhanova, Transient phenomena for discrete-time branching random processes. In: *Limit Theorems and Statistical Theorems.* Fan, Tashkent, 1966, pp. 83–89 (in Russian).
5. S. V. Nagaev, Transient phenomena for discrete-time age-dependent branching processes. I. *Siberian Math. J.* (1974) **15**, 368–394 (in Russian).
6. S. V. Nagaev, Transient phenomena for discrete-time age-dependent branching processes. II. *Siberian Math. J.* (1974) **15**, 570–579 (in Russian).
7. K. B. Athreya, P. R. Parthasarathy, and G. Sankaranarayanan, Supercritical age-dependent branching processes with immigration. *J. Appl. Probab.* (1974) **11**, 695–702.
8. M. H. Asadullin and S. V. Nagaev, Limit theorems for critical branching processes with immigration. *Math. Notes* (1982) **32**, 537–548 (in Russian).
9. I. S. Badalbayev and A. M. Zubkov, A limit theorem for a sequence of branching processes with immigration. *Theory Probab. Appl.* (1983) **28**, 382–388 (in Russian).

Probabilistic Methods in Discrete Mathematics, pp. 111–129
V. F. Kolchin et al. (Eds.)

ASYMPTOTIC BEHAVIOUR OF GENERALIZED NON-ORDINARY COX PROCESSES

V. E. BENING and V. YU. KOROLEV

Moscow

Abstract — We inverstagate asymptotic properties of non-ordinary and generalized non-ordinary Cox processes.

This research was supported by the Russian Foundation for Basic Research, grant 96–01–01919.

1. INTRODUCTION

Let $Y_1, Y_2, \dots$ be independent identically distributed non-negative integer-valued random variables, let $N_0(t)$, $t \geq 0$, be the standard Poisson process (i.e., homogeneous Poisson process with unit intensity) independent of $Y_1, Y_2, \dots$ We set

$$N_1(t) = \sum_{j=1}^{N_0(t)} Y_j, \qquad t \geq 0 \tag{1}$$

(for definiteness we assume that $\sum_{j=1}^{0} = 0$). Let $\Lambda(t)$, $t \geq 0$, be a random process with non-decreasing right-continuous trajectories, $\Lambda(0) = 0$, $\mathsf{P}\{\Lambda(t) < \infty\} = 1$ for all $t > 0$. Assume that the processes $\Lambda(t)$ and $N_1(t)$ are independent, and set

$$N(t) = N_1(\Lambda(t)), \qquad t \geq 0. \tag{2}$$

The process $N(t)$ will be called a non-ordinary doubly stochastic Poisson process (or a non-ordinary Cox process) controlled by the process $\Lambda(t)$ and generated by the sequence $Y_1, Y_2 \dots$

Let $X_1, X_2, \dots$ be independent identically distributed random variables independent of the process $N(t)$. We set

$$S(t) = \sum_{k=1}^{N(t)} X_k, \qquad t \geq 0. \tag{3}$$

The process $S(t)$ will be called a generalized non-ordinary doubly stochastic Poisson process or a generalized non-ordinary Cox process controlled by the process $\Lambda(t)$ and generated by the sequences $Y_1, Y_2, \dots$ and $X_1, X_2, \dots$

The processes $S(t)$ defined by (1)–(3) are quite natural models of the behaviour of some characteristics in financial and actuarial mathematics. Consider some examples.

EXAMPLE 1. Let X_k be the difference between the stock price in the kth contract and that in the $(k-1)$th contract, Y_j be the number of contracts concluded at the jth time instant. Since the actual trade intensity $\Lambda'(t)$ at stock exchanges is variable and even stochastic, the process $N(t)$ defined by (1)–(2) describes the total number of contracts concluded by the time t. Then the process $S(t)$ defined by (1)–(3) describes the difference between the stock price at time t and that at $t = 0$.

EXAMPLE 2. Let Y_j be the number of victims in the jth transport accident. Then the process $N(t)$ defined by (1)–(2) describes the total number of victims by the time t. It is clear that the actual intensity $\Lambda'(t)$ of transport accidents essentially varies depending on the season, weather, the state of roads, etc. If X_k is the insurance payment to the kth victim, then the process $S(t)$ defined by (1)–(3) describes the total loss of an insurance company due to the payments to the victims of transport accidents.

The aim of this paper is to describe the limit laws for $S(t)$ as (in terms of Examples) the number of contracts or victims infinitely increases, to give criteria of convergence of the distributions of $S(t)$ to the limit ones and some estimates of the rate of this convergence, and also to present the asymptotic expansions for the distributions of $S(t)$ as well as the estimates of the concentration functions of these distributions. We continue the researches started in [1–4].

Introduce some additional assumptions. Throughout the paper we will assume that there exist $\mathbf{E}Y_1 = a$ and $\mathbf{D}Y_1 = b^2$. Then it follows from (1) that $\mathbf{E}N_1(t) = at$ and $\mathbf{D}N_1(t) = t(a^2 + b^2)$. For the sake of convenience we assume that $\mathbf{E}\Lambda(t) = t$. This assumption makes it possible for us to investigate the asymptotic behaviour of $S(t)$ as $t \to \infty$ while remaining within the framework of the general formulation of the problem given above.

Assume that there exist $\mathbf{E}X_1 = \alpha$ and $\mathbf{D}X_1 = \sigma^2$. In this paper we consider the case $\alpha = 0$. This is typical for the situation described in Example 1. The case $\alpha \neq 0$ will be considered in another paper.

The symbols $\Rightarrow$, $\xrightarrow{\mathbf{P}}$ and $\stackrel{d}{=}$ denote weak convergence, convergence in probability, and coincidence of distributions respectively. The standard normal distribution function and its density are denoted by $\Phi(x)$ and $\phi(x)$ respectively.

2. SOME ASYMPTOTIC PROPERTIES OF NON-ORDINARY COX PROCESSES

In this section we discuss some properties of the process $N(t)$. The following assertion is rather obvious.

LEMMA 1. *Let $a > 0$. Then $N(t) \xrightarrow{\mathbf{P}} \infty$, as $t \to \infty$, if and only if $\Lambda(t) \xrightarrow{\mathbf{P}} \infty$ as $t \to \infty$.*

PROOF. First assume that $\Lambda(t) \xrightarrow{\mathbf{P}} \infty$ and prove that $N(t) \xrightarrow{\mathbf{P}} \infty$. Let $\varepsilon > 0$ and $n > 0$ be arbitrary. Fix an arbitrary $m > 0$. We have

$$\begin{aligned}\mathbf{P}\{N(t) \le m\} &= \mathbf{P}\{N(t) \le m, \Lambda(t) \le n\} + \mathbf{P}\{N(t) \le m, \Lambda(t) > n\} \\ &\le \mathbf{P}\{\Lambda(t) \le n\} + \int_n^\infty \mathbf{P}\{N_1(\lambda) \le m\} d\mathbf{P}\{\Lambda(t) < \lambda\} \\ &\equiv J_1(t,n) + J_2(t,n,m). \end{aligned} \tag{4}$$

Let $Q_\lambda(x)$ be the concentration function of $N_1(\lambda)$. Since $a > 0$, we have $\mathbf{P}\{Y_1 = 0\} < 1$. Therefore, as follows from [2], there exists a number $C \in (0, \infty)$ depending only on the distribution of Y_1 such that

$$Q_\lambda(x) \le \frac{C(x+1)}{\sqrt{\lambda}}$$

for all $x > 0$ and $\lambda > 0$. Then continuing (4) we obtain

$$J_2(t,n,m) \le C(m+2) \int_n^\infty \frac{1}{\sqrt{\lambda}} d\mathbf{P}(\Lambda < \lambda) \le \frac{C(m+2)}{\sqrt{n}}.$$

Therefore, with m fixed, we can choose $n = n(\varepsilon) = C^2(m+2)^2\varepsilon^{-2}$ so that

$$J_2(t, n(\varepsilon), m) < \varepsilon \tag{5}$$

for all $n > n(\varepsilon)$. Now choose $t = t(\varepsilon)$ so that

$$J_1(t, n(\varepsilon)) < \varepsilon \tag{6}$$

for all $t \ge t(\varepsilon)$ which is possible by virtue of the condition $\Lambda(t) \xrightarrow{\mathbf{P}} \infty$. Relations (4), (5), and (6) imply that $\mathbf{P}\{N(t) \le m\} < 2\varepsilon$ for $t \ge t(\varepsilon)$, that is, $N(t) \xrightarrow{\mathbf{P}} \infty$ as $t \to \infty$.

Now assume that $N(t) \xrightarrow{\mathbf{P}} \infty$ and prove that $\Lambda(t) \xrightarrow{\mathbf{P}} \infty$. For arbitrary positive m and n we have

$$\begin{aligned}\mathbf{P}\{\Lambda(t) \le m\} &= \mathbf{P}\{\Lambda(t) \le m, N(t) \le n\} + \mathbf{P}\{\Lambda(t) \le m, N(t) > n\} \\ &\le \mathbf{P}\{N(t) \le n\} + \int_0^m \mathbf{P}\{N_1(\lambda) > n\} d\mathbf{P}\{\Lambda(t) < \lambda\} \\ &\equiv I_1(t,n) + I_2(t,n,m). \end{aligned} \tag{7}$$

Since $\mathbf{E}N_1(\lambda) = \lambda a$, by the Markov inequality we have

$$I_2(t,n,m) \le \frac{a}{n} \int_0^m \lambda d\mathbf{P}\{\Lambda(t) < \lambda\} \le \frac{am}{n}.$$

Therefore, for an arbitrary $\varepsilon > 0$, with m fixed and $n > n(\varepsilon) = am/\varepsilon$ we have

$$I_2(t, n(\varepsilon), m) < \varepsilon. \tag{8}$$

Now choose $t = t(\varepsilon)$ so that

$$I_1(t, n(\varepsilon)) < \varepsilon \tag{9}$$

for $t \geq t(\varepsilon)$ which is possible due to the condition $N(t) \xrightarrow{\mathsf{P}} \infty$. From (7), (8), and (9) it follows that $\mathsf{P}\{\Lambda(t) \leq m\} < 2\varepsilon$ for $t \geq t(\varepsilon)$ which means that $\Lambda(t) \xrightarrow{\mathsf{P}} \infty$ as $t \to \infty$, since ε and m are arbitrary. The lemma is proved.

In what follows we use the following general assertion concerning the asymptotic behaviour of superpositions of independent random processes proved in [5] (also see [1], Sect. A2.2).

Let $X(t)$ and $M(t)$, $t \geq 0$, be independent random processes such that $X(t)$ is measurable and $\mathsf{P}\{M(t) < \infty\} = 1$ for any $t > 0$. Let $A(t)$, $B(t)$, and $D(t)$ be real functions such that $A(t)$ and $B(t)$ are measurable, $B(t) > 0$, $D(t) > 0$. By L_1 and L_2 we denote metrics which metrize weak convergence in the spaces of one- and two-dimensional random variables respectively. For example, L_1 is the Lévy metric and L_2 is the Lévy–Prokhorov metric.

LEMMA 2. *Assume that $B(t) \to \infty$ and $D(t) \to \infty$ as $t \to \infty$ and the families of random variables*

$$\left\{\frac{X(t) - A(t)}{B(t)}\right\}_{t>0}, \qquad \left\{\frac{B(M(t))}{D(t)}\right\}_{t>0}$$

are tight. Then the one-dimensional distributions of the appropriately centered and normalized superpositions of the processes $X(t)$ and $M(t)$ weakly converge to the distribution of some random variable Z as $t \to \infty$, that is,

$$\frac{X(M(t)) - C(t)}{D(t)} \Rightarrow Z, \qquad t \to \infty,$$

with some real function $C(t)$, if and only if there exists a tight family of triples of random variables $\{W(t), U(t), V(t)\}_{t\geq 0}$ such that

(1) *$Z \stackrel{d}{=} W(t)U(t) + V(t)$ for every $t > 0$, where the random variable $W(t)$ and the pair $U(t)$, $V(t)$ are independent;*

(2) $L_1\left(\dfrac{X(t) - A(t)}{B(t)}, W(t)\right) \to 0$ *as* $t \to \infty$;

(3) $L_2\left(\left(\dfrac{B(M(t))}{D(t)}, \dfrac{A(M(t)) - C(t)}{D(t)}\right), (U(t), V(t))\right) \to 0$ *as* $t \to \infty$.

The following theorem is a sort of the law of large numbers for non-ordinary Cox processes (2).

THEOREM 1. *Let $N(t)$ be a non-ordinary Cox process controlled by the process $\Lambda(t)$. Let $\mathsf{E}Y_1 = a > 0$ and let $D(t) > 0$ be a function such that $D(t) \to \infty$ as $t \to \infty$. Then the following conditions are equivalent.*

(1) *The family of random variables* $\{\Lambda(t)/D(t)\}_{t>0}$ *is tight and the one-dimensional distributions of the normalized non-ordinary Cox process weakly converge to the distribution of some random variable* Z *as* $t \to \infty$, *that is, as* $t \to \infty$,

$$\frac{N(t)}{D(t)} \Rightarrow Z. \tag{10}$$

(2) *The one-dimensional distributions of the controlling process* $\Lambda(t)$, *under the same normalization, weakly converge to the distribution of the random variable* Z/a, *that is, as* $t \to \infty$,

$$\frac{\Lambda(t)}{D(t)} \Rightarrow \frac{Z}{a}. \tag{11}$$

PROOF. To prove this assertion, we use Lemma 2. By virtue of the Chebyshev inequality for an arbitrary $\varepsilon > 0$ we have

$$\mathbf{P}\left\{\left|\frac{N_1(t)}{t} - a\right| > \varepsilon\right\} \le \frac{\mathbf{D}N_1(t)}{\varepsilon^2 t^2} = \frac{a^2 + b^2}{\varepsilon^2 t} \to 0$$

as $t \to \infty$, that is, as $t \to \infty$,

$$\frac{N_1(t)}{t} \Rightarrow a. \tag{12}$$

Therefore in Lemma 2 we put $X(t) \equiv N_1(t)$, $M(t) \equiv \Lambda(t)$, $B(t) \equiv t$, $A(t) \equiv C(t) \equiv 0$. As we have done so, from (12) it follows that any triple of random variables $W(t), U(t), V(t)$ considered in Lemma 2 inevitably has the form $a, U(t), 0$ and hence conditions 2 and 3 of Lemma 2 are reduced to the condition

$$L_1\left(\frac{\Lambda(t)}{D(t)}, U(t)\right) \to 0, \qquad t \to \infty. \tag{13}$$

But due to the form of the triples mentioned above and condition 1 of Lemma 2, for each $t > 0$ the random variables $U(t)$ should satisfy the relation

$$aU(t) \stackrel{d}{=} Z.$$

Therefore in this case (13) is equivalent to (11). The theorem is proved.

Though the following assertion which can be regarded as a central limit theorem for non-ordinary Cox processes will not be used in the rest of this paper, we do present it here. Its proof is based on Lemma 2.

THEOREM 2. *Let* $N(t)$ *be a non-ordinary Cox process controlled by the process* $\Lambda(t)$. *Assume that* $\mathbf{E}Y_1 = a > 0$ *and there exists a function* $D(t) > 0$ *such that* $D(t) \to \infty$

as $t \to \infty$ and the family of random variables $\{\Lambda(t)/D^2(t)\}_{t>0}$ is tight. Then the one-dimensional distributions of the nonrandomly centered and normalized non-ordinary Cox process $N(t)$ weakly converge to the distribution of some random variable Z as $t \to \infty$, that is, as $t \to \infty$,

$$\frac{N(t) - C(t)}{D(t)} \Rightarrow Z \tag{14}$$

with some real function $C(t)$, if and only if

$$\limsup_{t\to\infty} \frac{C(t)}{D^2(t)} = k^2 < \infty \tag{15}$$

and there exists a random variable V such that

$$Z \stackrel{d}{=} k\sqrt{a + b^2/a}W + V,$$

where W is a random variable with the standard normal distribution independent of V and, as $t \to \infty$,

$$L_1\left(\frac{a\Lambda(t) - C(t)}{D(t)}, V(t)\right) \to 0, \tag{16}$$

where for real s

$$\mathsf{E}\exp\{isV(t)\} = \exp\left\{-\frac{s^2}{2}\left(a + \frac{b^2}{a}\right)\left(k^2 - \frac{C(t)}{D^2(t)}\right)\right\}\mathsf{E}\exp\{isV\}.$$

PROOF. We begin with the proof that (14) follows from (15) and the existence of the random variable V pointed out in the theorem. We use Lemma 2. It is well known that the generalized Poisson process $N_1(t)$ is asymptotically normal in the sense that, as $t \to \infty$,

$$\frac{N_1(t) - at}{\sqrt{t(a^2 + b^2)}} \Rightarrow W \tag{17}$$

where W has the standard normal distribution (see, e.g. Theorem 3.4.1 in [1]). Therefore in Lemma 2 we can set $X(t) = N_1(t)$, $M(t) = \Lambda(t)$, $A(t) = at$, $B(t) = \sqrt{t(a^2 + b^2)}$. Then due to the tightness of the family $\{V(t)\}_{t>0}$ and condition 3 of Lemma 2, the family of random variables $\{(a\Lambda(t) - C(t))/D(t)\}_{t\geq 0}$ is tight. Let $l_t(q)$ be the left q-quantile of $\Lambda(t)$. The tightness of the family $\{\Lambda(t)/D^2(t)\}_{t\geq 0}$ implies that

$$\sup_t \frac{l_t(q)}{D^2(t)} = T(q) < \infty \tag{18}$$

for any $q \in (0, 1)$. The tightness of the family $\{(a\Lambda(t) - C(t))/D(t)\}_{t\geq 0}$ implies the boundedness of the function $(al_t(q) - C(t))/D(t)$ in t for each $q \in (0, 1)$. But

$$\frac{al_t(q) - C(t)}{D(t)} = D(t)\left(\frac{al_t(q)}{D^2(t)} - \frac{C(t)}{D^2(t)}\right), \tag{19}$$

so to ensure the boundedness of the left-hand side of (19) for any $q \in (0,1)$, the difference $al_t(q)/D^2(t) - C(t)/D^2(t)$ should tend to zero as $t \to \infty$ for any $q \in (0,1)$, since $D(t) \to \infty$ as $t \to \infty$. But with account of (18) this is possible only if (15) holds.

Let $v_t(q)$ denote the left q-quantile of the random variable

$$B(M(t))/D(t) = \sqrt{(a^2+b^2)\Lambda(t)}/D(t).$$

For each $q \in (0,1)$ we obviously have

$$v_t(q) = \frac{\sqrt{(a^2+b^2)l_t(q)}}{D(t)}$$

whence

$$l_t(q) = \frac{D^2(t)v_t^2(q)}{a^2+b^2}.$$

But as we have seen above,

$$\lim_{t\to\infty}\left|\frac{al_t(q)}{D^2(t)} - \frac{C(t)}{D^2(t)}\right| = 0.$$

Hence

$$\lim_{t\to\infty}\left|\frac{av_t^2(q)}{a^2+b^2} - \frac{C(t)}{D^2(t)}\right| = 0$$

resulting in

$$\lim_{t\to\infty}\left|v_t(q) - \sqrt{\frac{a^2+b^2}{a}\frac{C(t)}{D^2(t)}}\right| = 0$$

irrespective of $q \in (0,1)$. Therefore by virtue of (14) any triple of random variables $W(t), U(t), V(t)$ mentioned in Lemma 2 should necessarily have the form W, $\sqrt{C(t)(a+b^2/a)}/D(t)$, $V(t)$, where W is a standard normal random variable independent of $V(t)$ for each $t \geq 0$. Recall that according to condition 1 of Lemma 2 each of these triples must guarantee the possibility of the representation

$$Z \stackrel{d}{=} \sqrt{a+\frac{b^2}{a}}\frac{\sqrt{C(t)}}{D(t)}W + V \tag{20}$$

for each $t \geq 0$. For convenience we set $k^2(t) = C(t)/D^2(t)$. We consider the random variables satisfying (20) in more detail. The boundedness of $k(t)$ and tightness of $\{V(t)\}_{t\geq 0}$, implied by Lemma 2, allow one to choose a subsequence T_1 from an arbitrary sequence $T = \{t_1, t_2, \ldots\}$ so that $k(t) \to k_0$ and $V(t) \Rightarrow V_0$ as $t \to \infty$, $t \in T_1$, for some number k_0 and some random variable V_0. But then, reasoning as in the proof of the transfer Theorem 3.1.2 in [1], for any real s we obtain

$$\mathbf{E}e^{isZ} = \mathbf{E}e^{isV(t)}\exp\left\{-\frac{s^2k^2(t)}{2}\left(a+\frac{b^2}{a}\right)\right\} \to \mathbf{E}e^{isV_0}\exp\left\{-\frac{s^2k_0^2}{2}\left(a+\frac{b^2}{a}\right)\right\}$$

as $t \to \infty$, $t \in T_1$, hence it follows that the limiting pair (k_0, V_0) also satisfies (20). In other words, the set of pairs $(k(t), V(t))$ satisfying (20) is closed. Let V be the random variable corresponding to the value $k(t) = k$ (see (15)) in representation (20). Then for any $t \geq 0$ we have

$$k\sqrt{a + \frac{b^2}{a}W} + V \stackrel{d}{=} k(t)\sqrt{a + \frac{b^2}{a}W} + V(t), \tag{21}$$

where the summands on both sides are independent. By rewriting (21) in terms of characteristic functions, we obtain

$$\exp\left\{-\frac{s^2k^2}{2}\left(a + \frac{b^2}{a}\right)\right\}\mathbf{E}e^{isV} = \exp\left\{-\frac{s^2k^2(t)}{2}\left(a + \frac{b^2}{a}\right)\right\}\mathbf{E}e^{isV(t)} \tag{22}$$

for all real s and $t \geq 0$. From (22) for the characteristic function of the random variable $V(t)$, we obtain the representation

$$\mathbf{E}\exp\{isV(t)\} = \exp\left\{-\frac{s^2}{2}\left(k^2 - k^2(t)\right)\left(a + \frac{b^2}{a}\right)\right\}\mathbf{E}e^{isV}.$$

Finally, relation (16) with $V(t)$ just described follows from Lemma 2. The proof of the first part of the theorem is completed.

Let us derive (14) from (15) and the existence of the random variable V with the properties given in the theorem. In the same way as we have done in the proof of the first part, we make sure that condition (16) and the tightness of the family $\{\Lambda(t)/D^2(t)\}_{t\geq 0}$ imply that, as $t \to \infty$,

$$L_1\left(\frac{a\Lambda(t)}{D^2(t)}, \frac{C(t)}{D^2(t)}\right) \to 0.$$

Now it remains to refer to Lemma 2 with the account of (16). The theorem is proved.

Theorem 2 allows us to obtain the following criterion of the asymptotic normality of non-ordinary Cox processes.

COROLLARY 1. *Under the conditions of Theorem 2, the non-randomly centered and normalized non-ordinary Cox process $N(t)$ is asymptotically normal with some variance $\delta^2 > 0$, that is, as $t \to \infty$,*

$$\mathbf{P}\left\{\frac{N(t) - C(t)}{D(t)} < x\right\} \Rightarrow \Phi\left(\frac{x}{\delta}\right)$$

if and only if

$$\sup_{t\geq 0}\frac{C(t)}{D^2(t)} \leq \frac{a\delta^2}{a^2 + b^2}$$

and

$$\lim_{t\to\infty} L_1\left(\mathbf{P}\left\{\frac{a\Lambda(t) - C(t)}{D(t)} < x\right\}, \Phi\left(\frac{\sqrt{a}xD(t)}{\sqrt{a\delta^2D^2(t) - (a^2 + b^2)C(t)}}\right)\right) = 0.$$

PROOF. The assertion follows from Theorem 2 and the Cramér–Lévy theorem on the decomposability of a normal law only into normal components, according to which any random variable $V(t)$ satisfying (16) must necessarily be normal with zero mean and variance

$$\delta^2 - \left(a + \frac{b^2}{a}\right)\frac{C(t)}{D^2(t)}.$$

In other words, a non-randomly centered non-ordinary Cox process is asymptotically normal if so is its controlling process.

REMARK 1. Theorems 1 and 2 hold not only for non-ordinary Cox processes, but for all processes of the form (2) with N_1 being an asymptotically degenerate and asymptotically normal counting process.

3. LIMIT THEOREMS FOR NON-CENTERED GENERALIZED NON-ORDINARY COX PROCESSES

To prove the validity of the criterion of convergence of non-centered generalized non-ordinary Cox processes, we need the following criterion of convergence of random sums proved in [6] (see also [1], Theorem 3.2.1).

Let $N_1, N_2, \ldots$ be non-negative integer-valued random variables such that for each $k \geq 1$ the random variables $N_k, X_1, X_2, \ldots$ are independent. For $k \geq 1$ we set

$$S_k = X_1 + \ldots + X_k.$$

LEMMA 3. *Assume that* $N_k \xrightarrow{\mathsf{P}} \infty$ *as* $k \to \infty$ *and the sequences of numbers* $\{a_k\}$ *and* $\{b_k\}$, $b_k > 0$, $b_k \to \infty$ *as* $k \to \infty$, *are such that the family of random variables* $\{(S_k - a_k)/b_k\}_{k\geq 1}$ *is tight. The weak convergence of random sums*

$$\frac{1}{d_k}\left(\sum_{j=1}^{N_k} X_j - c_k\right) \Rightarrow Z,$$

as $k \to \infty$, *to some random variable* Z *takes place with some real* c_k *and* $d_k > 0$, $d_k \to \infty$, *as* $k \to \infty$, *if and only if there exists a tight sequence of triples of random variables* $\{W_k, U_k, V_k\}_{k\geq 1}$ *such that*

(1) $Z \stackrel{d}{=} W_k U_k + V_k$, *where* W_k *and the pair* U_k, V_k *are independent for each* $k \geq 1$;

(2) $L_1\left(\frac{S_k - a_k}{b_k}, W_k\right) \to 0$ *as* $k \to \infty$;

(3) $L_2\left(\left(\frac{b_{N_k}}{d_k}, \frac{a_{N_k} - c_k}{d_k}\right), (U_k, V_k)\right) \to 0$ *as* $k \to \infty$.

This assertion can be regarded as a modification of Lemma 2 in which the condition of the tightness of the family $\{B(M(t))/D(t)\}_{t>0}$ is replaced by the condition $N_k \xrightarrow{\mathbf{P}} \infty$.

Recall that here we assume that $\alpha = \mathbf{E}X_1 = 0$ which results in $\mathbf{E}S(t) = 0$ and $\mathbf{D}S(t) = a\sigma^2 t$.

THEOREM 3. *Assume that $a > 0$ and $\Lambda(t) \xrightarrow{\mathbf{P}} \infty$ as $t \to \infty$. Then one-dimensional distributions of a normalized generalized non-ordinary Cox process weakly converge to the distribution of some random variable Z:*

$$\frac{S(t)}{\sigma\sqrt{at}} \Rightarrow Z,$$

as $t \to \infty$, if and only if there exists a non-negative random variable U such that

(1) $\mathbf{P}\{Z < x\} = \int_0^\infty \Phi(x/\sqrt{u})\, d\mathbf{P}\{U < u\}, \quad x \in \mathbf{R};$

(2) $\frac{\Lambda(t)}{t} \Rightarrow U$ *as* $t \to \infty$.

PROOF. From Lemma 1 it follows that the conditions $\Lambda(t) \xrightarrow{\mathbf{P}} \infty$ and $N(t) \xrightarrow{\mathbf{P}} \infty$ are equivalent. The properties of the random variables $X_1, X_2, \ldots$ guarantee that, as $k \to \infty$,

$$\mathbf{P}\left\{\frac{1}{\sigma\sqrt{k}}\sum_{j=1}^{k} X_j < x\right\} \Rightarrow \Phi(x). \tag{23}$$

Let $t_1, t_2, \ldots$ be an arbitrary infinitely increasing sequence of time instants. By virtue of (23) we can put in Lemma 3 $N_k = N(t_k)$, $b_k = \sigma\sqrt{k}$, $d_k = \sigma\sqrt{at_k}$. Then by Lemma 3 for

$$\frac{S(t_k)}{\sigma\sqrt{at_k}} = \frac{1}{\sigma\sqrt{at_k}}\sum_{j=1}^{N_k} X_j \Rightarrow Z,$$

as $k \to \infty$, it is necessary and sufficient that there exists a tight sequence of random variables $\{U_k\}_{k\geq 1}$ such that for any real x

$$\mathbf{P}\{Z < x\} = \mathbf{E}\Phi\left(\frac{x}{\sqrt{U_k}}\right) \tag{24}$$

and, as $k \to \infty$,

$$L_1\left(\frac{N_k}{at_k}, U_k\right) \to 0. \tag{25}$$

But the family of scale mixtures of normal laws is identifiable, therefore the distributions of all U_k coincide, so that instead of the family of random variables $\{U_k\}$ satisfying (24) there exists a random variable U such that for any real x

$$\mathbf{P}\{Z < x\} = \mathbf{E}\Phi\left(\frac{x}{\sqrt{U}}\right). \tag{26}$$

Hence (25) reduces to

$$\frac{N_k}{at_k} \Rightarrow U. \tag{27}$$

But the family of random variables $\{\Lambda(t_k)/(at_k)\}_{k\geq 1}$ is tight, since by the Markov inequality and the assumption $\mathsf{E}\Lambda(t) = t$ we have

$$\lim_{R\to\infty}\sup_k \mathsf{P}\left\{\frac{\Lambda(t_k)}{at_k} > R\right\} \leq \lim_{R\to\infty}\sup_k \frac{\mathsf{E}\Lambda(t_k)}{at_k R} = \lim_{R\to\infty}\frac{1}{aR} = 0.$$

Therefore by Theorem 1 for (27) it is necessary and sufficient that, as $k \to \infty$,

$$\frac{\Lambda(t_k)}{t_k} \Rightarrow U.$$

Now noting that the sequence $\{t_k\}$ is arbitrary and the distributions of the random variable U in (26) for all sequences $\{t_k\}$ coincide due to the identifiability of the family of scale mixtures of normal laws, we complete the proof.

It should be noted that the class of scale mixtures of normal laws which are the limit distributions in Theorem 3 is very rich and contains, for example, Cauchy, Student, Laplace (double exponential) distributions, symmetric stable distributions. All scale mixtures of normal laws have heavier tails than the normal distribution itself (see Lemma 3.5.2 in [1]).

COROLLARY 2. *Under the conditions of Theorem 3,*

$$\mathsf{P}\left\{\frac{S(t)}{\sigma\sqrt{at}} < x\right\} \Rightarrow \Phi(x),$$

as $t \to \infty$, if and only if

$$\frac{\Lambda(t)}{t} \Rightarrow 1,$$

as $t \to \infty$.

This assertion is an immediate consequence of Theorem 3 and the identifiability of scale mixtures of normal laws.

In other words, a non-centered generalized non-ordinary Cox process with finite variance is asymptotically normal if and only if its controlling process is asymptotically degenerate.

REMARK 2. As is seen from the proof of Theorem 3, we did not use representation (1) of $N_1(t)$. Therefore Theorem 3 and Corollary 2 hold for any process $N(t)$ satisfying (2) with $N_1(t)$ being asymptotically degenerate in the sense (12) with some positive a.

4. ESTIMATES OF THE CONVERGENCE RATE

From Theorem 3 it follows that the distribution of the random variable $S(t)/(\sigma\sqrt{at})$ is close to the limit one if and only if the distributions of $\Lambda(t)/t$ and U are close to each other, or which is in some sense the same, the distribution of $\Lambda(t)$ is close to that of Ut. Let $N^*(t)$ be the non-ordinary Cox process generated by the same sequence $Y_1, Y_2, \ldots$ as $N(t)$ but controlled by the process $\Lambda^*(t) = Ut$ instead of $\Lambda(t)$. Note that if there exists $\mathbf{E}U$, then $\mathbf{E}U = 1$, since $\mathbf{E}\Lambda(t)/t = 1$. We set

$$S^*(t) = \sum_{j=1}^{N^*(t)} X_j. \tag{28}$$

By $F_t(x)$ and $F_t^*(x)$, respectively, we denote the distribution functions of $S(t)/(\sigma\sqrt{at})$ and $S^*(t)/(\sigma\sqrt{at})$ respectively. Then from the identity

$$F_t(x) = (F_t(x) - F_t^*(x)) + F_t^*(x) \tag{29}$$

it follows that if we know an appropriate estimate of the accuracy of the approximation of the distribution of $S(t)$ by that of $S^*(t)$ and a convergence rate estimate for $S^*(t)$, then we can construct an appropriate convergence rate estimate for $S(t)$. Therefore our nearest aim is to construct an estimate for

$$\sup_x |F_t(x) - F_t^*(x)|.$$

By the direct analysis of the characteristic function of $S(t)$ we easily obtain the following simple assertion (see [7]).

LEMMA 4. *For each $t > 0$*

$$\sum_{j=1}^{N_1(t)} X_j \stackrel{d}{=} \sum_{j=1}^{N_0(t)} Z_j,$$

where the random variables $N_0(t), Z_1, Z_2, \ldots$ are independent for each $t > 0$ and

$$Z_j \stackrel{d}{=} \sum_{k=1}^{Y_1} X_k, \qquad j \geq 1.$$

From Lemma 4 it follows that one-dimensional distributions of the generalized non-ordinary Cox process $S(t)$ controlled by the process $\Lambda(t)$ and generated by the sequences $Y_1, Y_2, \ldots$ and $X_1, X_2, \ldots$ coincide with those of the generalized ordinary Cox process controlled by the same process $\Lambda(t)$ and generated by the sequence $Z_1, Z_2, \ldots$ defined above.

LEMMA 5. *Let $S_1(t)$ and $S_2(t)$ be two generalized non-ordinary Cox processes generated by the same sequences $Y_1, Y_2, \ldots$ and $X_1, X_2, \ldots$ but controlled by the processes $\Lambda_1(t)$ and*

$\Lambda_2(t)$ *respectively. Then*

$$\sup_x |\mathsf{P}\{S_1(t) < x\} - \mathsf{P}\{S_2(t) < x\}| \le \int_0^\infty \min\left(2, \frac{1}{\sqrt{\lambda}}\right) |\mathsf{P}\{\Lambda_1(t) < \lambda\} - \mathsf{P}\{\Lambda_2(t) < \lambda\}| d\lambda.$$

PROOF. Set $\Psi_t(\lambda) = \mathsf{P}\{\Lambda_1(t) < \lambda\} - \mathsf{P}\{\Lambda_2(t) < \lambda\}$. Then by Lemma 4 we have

$$\begin{aligned}\mathsf{P}\{S_1(t) < x\} - \mathsf{P}\{S_2(t) < x\} &= \int_0^\infty \mathsf{P}\left\{\sum_{j=1}^{N_1(\lambda)} X_j < x\right\} d\Psi_t(\lambda) \\ &= \int_0^\infty \mathsf{P}\left\{\sum_{j=1}^{N_0(\lambda)} Z_j < x\right\} d\Psi_t(\lambda) \\ &= \int_0^\infty \sum_{k=0}^\infty \mathsf{P}\left\{\sum_{j=1}^{k} Z_j < x\right\} \frac{\lambda^k e^{-\lambda}}{k!} d\Psi_t(\lambda).\end{aligned}$$

The further reasoning word for word repeats the proof of Lemma A2.5.1 in [1] with X_j replaced by Z_j. The lemma is proved.

We set

$$\delta(t) = \int_0^\infty \min\left(2, \frac{1}{\sqrt{\lambda}}\right) |\mathsf{P}\{\Lambda(t) < \lambda\} - \mathsf{P}\{Ut < \lambda\}| d\lambda. \tag{30}$$

It was demonstrated in [1, 3] that if, for example, $\Lambda(t)$ is geometrically distributed with the parameter $1/t$ (in this case U has the standard exponential distribution), then

$$\delta(t) \le \frac{2\sqrt{\pi t}}{t-1}.$$

The following lemma presents a convergence rate estimate for random sums with the number of summands representable as $N_1(t)$ given by (1). In this case according to Lemma 4 the distribution of the random sum is nothing else but generalized Poisson and therefore the limit law of this random sum, as $t \to \infty$, is normal (see, e.g., Theorem 3.4.2 in [1]).

LEMMA 6. *Let* $a > 0$, $\alpha = \mathsf{E}X_1 = 0$, $\mathsf{E}Y_1^3 < \infty$, $\mathsf{E}|X_1|^3 < \infty$. *Then*

$$\sup_x \left| \mathsf{P}\left\{\frac{1}{\sigma\sqrt{at}} \sum_{j=1}^{N_1(t)} X_j < x\right\} - \Phi(x)\right| \le \frac{C_0}{\sqrt{t}} \frac{\mathsf{E}Y_1^3 \mathsf{E}|X_1|^3}{a^{3/2}\sigma^3},$$

where C_0 *is the absolute constant in the Berry-Esseen inequality,* $C_0 \le 0.7655$.

This assertion is a particular case of the estimate proved in [7] with the constant refined in [8].

THEOREM 4. *Let* $a > 0$, $\alpha = 0$, $\mathbf{E}Y_1^3 < \infty$, $\mathbf{E}|X_1|^3 < \infty$, $\varkappa = \mathbf{E}U^{-1/2} < \infty$. *Then*

$$\sup_x \left| \mathbf{P}\left\{ \frac{S(t)}{\sigma\sqrt{at}} < x \right\} - \mathbf{E}\Phi\left(\frac{x}{\sqrt{U}}\right) \right| \leq \frac{C_0}{\sqrt{t}} \frac{\varkappa \mathbf{E}Y_1^3 \mathbf{E}|X_1|^3}{a^{3/2}\sigma^3} + \delta(t) \tag{31}$$

with $\delta(t)$ *defined in* (30).

PROOF. With the account of (28) we have

$$\begin{aligned}
\Delta(t) &= \sup_x \left| \mathbf{P}\left\{ \frac{S^*(t)}{\sigma\sqrt{at}} < x \right\} - \mathbf{E}\Phi\left(\frac{x}{\sqrt{U}}\right) \right| \\
&= \sup_x \left| \mathbf{P}\left\{ \frac{1}{\sigma\sqrt{at}} \sum_{j=1}^{N_1(Ut)} X_j < x \right\} - \mathbf{E}\Phi\left(\frac{x}{\sqrt{U}}\right) \right| \\
&= \sup_x \left| \int_0^\infty \left(\mathbf{P}\left\{ \frac{1}{\sigma\sqrt{at\lambda}} \sum_{j=1}^{N_1(\lambda t)} X_j < \frac{x}{\sqrt{\lambda}} \right\} - \Phi\left(\frac{x}{\sqrt{\lambda}}\right) \right) d\mathbf{P}\{U < \lambda\} \right| \\
&\leq \int_0^\infty \sup_x \left| \mathbf{P}\left\{ \frac{1}{\sigma\sqrt{at\lambda}} \sum_{j=1}^{N_1(\lambda t)} X_j < x \right\} - \Phi(x) \right| d\mathbf{P}\{U < \lambda\}.
\end{aligned}$$

By estimating the integrand in the latter expression with the help of Lemma 6, we obtain

$$\Delta(t) \leq \frac{C_0}{\sqrt{t}} \frac{\mathbf{E}Y_1^3 \mathbf{E}|X_1|^3}{a^{3/2}\sigma^3} \int_0^\infty \frac{1}{\sqrt{\lambda}} d\mathbf{P}\{U < \lambda\} = \frac{C_0}{\sqrt{t}} \frac{\varkappa \mathbf{E}Y_1^3 \mathbf{E}|X_1|^3}{a^{3/2}\sigma^3}.$$

The theorem is proved.

REMARK 3. In some statistical applications, if the observations are equidistant in time (i.e., represent a time series) with zero trend, then it is resonable to assume from the very beginning that $\Lambda(t) = Ut$ which results in $\delta(t) = 0$.

5. ASYMPTOTIC EXPANSIONS FOR THE ONE-DIMENSIONAL DISTRIBUTIONS

Relation (29) makes it possible to obtain an appropriate asymptotic expansion for $F_t(x)$ from the asymptotic expension for $F_t^*(x)$ and the estimate for the difference $F_t(x) - F_t^*(x)$ (for example, from the one given in Lemma 5).

We say that a random variable X satisfies Cramér's (C) condition (see, e.g. [9], Chapter 7, Section 5), if

$$\text{(C)} \qquad \limsup_{|s|\to\infty} |\mathbf{E}\exp\{isX\}| < 1.$$

The following proposition is well known (see, e.g. [10], Theorem 6.3.1). Let, as above,

$$S_n = X_1 + \ldots + X_n.$$

PROPOSITION 1. *Let in addition to the conditions introduced in Section 1, the random variables* $X_1, X_2, \dots$ *satisfy* (C) *condition and* $\mathbf{E}|X_1|^k < \infty$ *for some integer* $k \geq 3$. *Then*

$$\sup_x \left| \mathbf{P}\{S_n < x\sigma\sqrt{n}\} - \Phi(x) - \sum_{j=1}^{k-2} Q_j(x) n^{-j/2} \right| = o(n^{-(k-2)/2}),$$

where

$$Q_j(x) = -\varphi(x) \sum H_{j+2l-1}(x) \prod_{m=1}^{j} \frac{1}{k_m!} \left(\frac{\gamma_{m+2}}{(m+2)!\sigma^{m+2}} \right)^{k_m}, \quad j = 1, 2, \dots, \tag{32}$$

with summation over all integer non-negative solutions $(k_1, \dots, k_j)$ *of the equation* $k_1 + 2k_2 + \dots + jk_j = j$, $l = k_1 + \dots + k_j$, γ_{m+2} *is the semi-invariant of order* $m + 2$ *of the random variable* X_1, *and* $H_m(x)$ *are the Chebyshev–Hermite polynomials of degree* m, *i.e.,*

$$H_m(x)\varphi(x) = (-1)^m \varphi^{(m)}(x).$$

In particular, if $\alpha_l = \mathbf{E}X_1^l$, $l = 1, 2, \dots$, then

$$Q_1(x) = -\varphi(x)(x^2 - 1)\frac{\alpha_3}{6\sigma^3}, \tag{33}$$

$$Q_2(x) = -\varphi(x) \left((x^3 - 3x)\frac{\alpha_4 - 3\sigma^4}{24\sigma^4} + (x^5 - 10x^3 + 15x)\frac{\alpha_3^2}{72\sigma^6} \right). \tag{34}$$

Now we prove an analogue of this statement for the process $S(t)$. Let

$$v_l = \mathbf{E}Y_1(Y_1 - 1)\dots(Y_1 - l + 1), \quad l = 1, 2, \dots,$$

$$Z_1 = \sum_{k=1}^{Y_1} X_k,$$

$$(\log t)_k = \begin{cases} i, & k = 3, \\ \log t, & k > 3. \end{cases}$$

THEOREM 5. *Let in addition to the conditions introduced in Section 1,* $\mathbf{E}|X_1|^k < \infty$ *and* $\mathbf{E}Y_1^k < \infty$ *for some integer* $k \geq 3$. *Assume also that* X_1 *satisfies* (C) *condition,* $a = \mathbf{E}Y_1 > 0$, $\mathbf{E}U = 1$, $\mathbf{E}U^{-(k-2)/2} < \infty$ *and, as* $t \to \infty$,

$$(\log t)_k t^{(k-2)/2} \mathbf{E}q^{Ut} \to 0 \tag{35}$$

for any $q \in (0, 1)$. *Then*

$$\sup_x \left| F_t(x) - \mathbf{E}\Phi(x/\sqrt{U}) - \sum_{j=1}^{k-2} t^{-j/2} q_j(x) \right| \leq \varepsilon(t),$$

where

$$\varepsilon(t) = \delta(t) + o(t^{-(k-2)/2}),$$
$$q_j(x) = \mathbf{E}U^{j/2}Q_j^*(x/\sqrt{U}), \quad j = 1, \dots, k-2,$$

the functions $Q_j^*(\cdot)$ are defined by formulae (32) *with σ and the semi-invariants γ_{m+2} replaced by $\sigma\sqrt{a}$ and the moments $\mathbf{E}Z_1^{m+2}$, respectively, $m = 1, \dots, j, j = 1, \dots, k-2$, and $\delta(t)$ is defined in* (30).

Before proving this theorem we make some remarks.

REMARK 4. Condition (35) means that the Laplace–Stieltjes transform $\psi_U(s)$ of U must decrease faster than $s^{(2-k)/2}$ as $s \to \infty$. This property is inherent in the gamma-distribution with the shape parameter exceeding $(k-2)/2$ including the exponential distribution for which the only k providing (35) is $k = 3$, the Lévy distribution (the stable distribution with the characteristic exponent $\alpha = 1/2$), the uniform distribution on $[0, 2]$ for which $k = 3$ is also the only k providing (35), as well as in all distributions concetrated on finite intervals not containing the origin (see [1], p. 240).

REMARK 5. The first two functions $Q_1^*(x)$ and $Q_2^*(x)$ (see (33) and (34)) have the form

$$Q_1^*(x) = -\varphi(x)(x^2-1)\frac{\alpha_3}{6\sigma^3\sqrt{a}},$$
$$Q_2^*(x) = -\varphi(x)\left((x^3-3x)\frac{a\alpha_4 + 6v_2\sigma^4}{24\sigma^4a^2} + (x^5 - 10x^3 + 15x)\frac{\alpha_3^2}{72\sigma^6 a}\right).$$

REMARK 6. The moments of the random variable Z_1 can be calculated by the formulas

$$\mathbf{E}Z_1^m = m!\sum v_d \prod_{r=1}^{m}\frac{1}{k_r!}\left(\frac{\alpha_r}{r!}\right)^{k_r}, \qquad m = 1, 2, \dots,$$

where the summation is over all non-negative integer solutions $(k_1, \dots, k_m)$ of the equation $k_1 + 2k_2 + \dots + mk_m = m$ and $d = k_1 + \dots + k_m$. Since $\alpha_1 = \alpha = 0$, we can assume here that $k_1 = 0$. This formula follows from the definition of Z_1 and Lemma 6.1.1 in [10].

PROOF OF THEOREM 5. With regard to relations (29), (30), and Lemma 5 it suffices to show that

$$\sup_x \left| F_t^*(x) - \mathbf{E}\Phi(x/\sqrt{U}) - \sum_{j=1}^{k-2} t^{-j/2}q_j(x)\right| = o(t^{-(k-2)/2}).$$

Notice that according to Lemma 4 the characteristic function $f_t^*(s)$ of the random variable $S^*(t)/(\sigma\sqrt{at})$ can be represented as

$$f_t^*(s) = \mathbf{E}\exp\left\{is\frac{S^*(t)}{\sigma\sqrt{at}}\right\} = \mathbf{E}\exp\left\{Ut\left(f_{Z_1}\left(\frac{s}{\sigma\sqrt{at}}\right) - 1\right)\right\}$$
$$= \mathbf{E}(v(s/(\sigma\sqrt{at})))^{Ut}, \tag{36}$$

where

$$f_{Z_1}(s) = \mathbf{E}(f_{X_1}(s))^{Y_1},$$
$$f_{X_1}(s) = \mathbf{E}e^{isX_1},$$
$$v(s) = \exp\{f_{Z_1}(s) - 1\}, \qquad s \in \mathbf{R},$$

since

$$S^*(t) \stackrel{d}{=} \sum_{j=1}^{N_0(Ut)} Z_j$$

with $Z_1, Z_2, \ldots$ being independent identically distributed random variables. The further proof follows from the proof of Theorem A2.6.1 in [1] concerning the asymptotic expensions for generalized ordinary Cox processes with zero means with the random variable X_1 replaced by Z_1, the characteristic function $f_{X_1}(s)$ replaced by $f_{Z_1}(s)$ and from the inequalities

$$\limsup_{|s|\to\infty} |v(s)| \leq \limsup_{|s|\to\infty} \exp\{|f_{Z_1}(s)| - 1\} < 1,$$

since there exists $q_0 \in (0, 1)$ such that

$$\limsup_{|s|\to\infty} |f_{Z_1}(s)| \leq \mathbf{E}q_0^{Y_1}$$

with the left-hand side being less than 1 due to the condition $\mathbf{P}\{Y_1 = 0\} < 1$. The theorem is proved.

REMARK 7. If we compare the asymptotic expension for the one-dimensional distributions of generalized non-ordinary Cox processes with that for generalized ordinary Cox processes, then we notice that the first terms (i.e., the limit distributions) coincide up to a non-random linear scale transformation so that the difference of the problems manifests itself only in the second terms. Moreover, if $a = 1$, then the normalizations as well as the second terms also coincide and the distinction is seen only in the third terms! In other words, if for the random variables $Y_1, Y_2, \ldots$ we substitute unities, then this simplification of the problem will reveal itself only in the second (if $a \neq 1$) or even in the third (if $a = 1$) term of the asymptotic expension.

6. ESTIMATES OF THE CONCENTRATION FUNCTIONS

Recall that the concentration function of a random variable X is the function

$$Q_X(l) = \sup_x \mathbf{P}\{x \leq X \leq x + l\}, \qquad l \geq 0.$$

It is well known (see, e.g. [10], Theorem 1.2.2) that if the random variables $Z_1, Z_2, \ldots$ are independent and have one and the same non-degenerate distribution, then there exist $\varepsilon > 0$ and $\delta > 0$ such that

$$|f_{Z_1}(s)| \leq 1 - \varepsilon s^2, \qquad |s| \leq \delta, \tag{37}$$

where $f_{Z_1}(s)$ is the characteristic function of the random variable Z_1 and the concentration function of their sum

$$T_n = Z_1 + \ldots + Z_n$$

satisfies the inequality

$$Q_{T_n}(l) \leq C(\varepsilon, \delta)\frac{l+1}{\sqrt{n}}, \qquad l \geq 0, \tag{38}$$

with

$$C(\varepsilon, \delta) = \left(\frac{96}{95}\right) \max(1, 1/\delta)\sqrt{\pi/\varepsilon}, \tag{39}$$

where ε and δ are the same as in (37).

Now we present an analogue of (38) for the processes $S(t)$ defined by (3) under the conditions of Section 4. We set

$$Q_t(l) = Q_{S(t)}(l), \qquad l \geq 0.$$

Let, as above,

$$Z_1 = \sum_{k=1}^{Y_1} X_k.$$

THEOREM 6. *Assume that* $\varkappa = \mathbf{E}U^{-1/2} < \infty$.

If the random variable Z_1 *is non-degenerate, then*

$$Q_t(l) \leq \varkappa C(\varepsilon, \delta)\frac{l+1}{\sqrt{t}} + \delta(t),$$

where $\delta(t)$ *and* $C(\varepsilon, \delta)$ *are defined in* (30) *and* (39) *respectively.*

If the random variable Z_1 *is degenerate at a point* $z \neq 0$, *then*

$$Q_t(l) \leq \varkappa C(11/24, 1)\frac{l|z|^{-1}+1}{\sqrt{t}} + \delta(t).$$

This result directly follows from Theorem A2.7.1 in [1] with regard to (36) and the random variable X_1 replaced by Z_1, since as we have already noted (see Lemma 4),

$$S^*(t) \stackrel{d}{=} \sum_{j=1}^{N_0(Ut)} Z_j.$$

REFERENCES

1. B. V. Gnedenko and V. Yu. Korolev, *Random Summation: Limit Theorems and Applications*. CRC Press, Boca Raton, Florida, 1996.
2. V. E. Bening, V. Yu. Korolev, and S. Ya. Shorgin, On approximations to generalized Poisson distributions. *J. Math. Sci.* (1997) **83**, 360–373.
3. V. E. Bening and V. Yu. Korolev, Asymptotic behaviour of generalized Cox processes. *Bull. Moscow Univ., Ser. 15, Comput. Math. and Cybern.* (1996), No. 3, 53–66 (in Russian).
4. V. E. Bening and V. Yu. Korolev, Asymptotic behaviour of non-randomly centered generalized Cox processes. *Fundamental and Applied Mathematics* (1996) **2**, 957–976 (in Russian).
5. V. Yu. Korolev, A general theorem on the limit behavior of superpositions of independent random processes with applications to Cox processes. *J. Math. Sci.* (1996) **81**, 2951–2956.
6. V. Yu. Korolev, Convergence of random sequences with independent random indices. I. *Theory Probab. Appl.* (1994) **39**, 282–297 (in Russian).
7. S. Ya. Shorgin, On the estimate of the accuracy of the normal approximation to the distributions of random sums with infinitely divisible indices. *Theory Probab. Appl.* (1996) **41**, 920–926 (in Russian).
8. V. Yu. Korolev and S. Ya. Shorgin, On the absolute constant in the remainder term estimate in the central limit theorem for Poisson random sums. *These Proceedings* pp. 305–308.
9. H. Cramér, *Random Variables and Probability Distributions*. Cambridge Univ. Press, Cambridge, 1970.
10. V. V. Petrov, *Sums of Independent Random Variables*. Springer, Berlin, 1975.

Probabilistic Methods in Discrete Mathematics, pp. 131–136
V. F. Kolchin et al. (Eds.)

THE STEADY STATE DISTRIBUTION OF THE QUEUE LENGTH FOR A QUEUE WITH BULK ARRIVAL AND PROCESSOR SHARING DISCIPLINE

O. I. BOGOYAVLENSKAYA

Petrozavodsk

Abstract — A service system $M|G|1|m$ is considered. The service time distribution is supposed to be of the form corresponding to the bulk arrival structure. Using Erlang phase method, we construct a Markov process to describe the behaviour of the analyzed queue. On the base of the analytic solution of the equilibrium equations, we find the steady state distribution of the queue length in this system.

1. INTRODUCTION

A service system with a single processor, possible loss of demands, and processor sharing discipline is considered. As usual, in Kendall's notation, this system is denoted by $M|G|1|m$. The service time distribution is supposed to be of the form corresponding to the bulk arrival structure. There are many papers devoted to both bulk arrival models and processor sharing models. In [1] the main results on bulk arrival models can be found both for a fixed and random bulk size. Some problems on processor sharing disciplines are discussed in [1] as well. Service systems with recurrent non-ordinary flow is analyzed in [2]. We point out the paper [1] on the processor sharing systems, where the sojourn time distribution under heavy traffic are studied for $M|M|1$ and $M|M|1|m$ with a processor sharing discipline. The last paper contains an extensive bibliography on the subject.

Many results on $M|G|1|m$ queue under FIFO discipline and its generalizations are discussed in [4] in detail.

In this paper we construct a discontinuous Markov process to describe the behaviour of the service system under consideration using the Erlang phase method. We find an analytical solution of the equilibrium equations and obtain the stable distribution of queue length.

2. DESCRIPTION OF THE SYSTEM

We consider a queueing system with a single processor, a finite buffer with m positions for waiting, and a random bulk flow of demands. The bulks arrived are placed in the queue buffer. Each bulk occupies one buffer position. Bulks are served in accordance with a processor sharing discipline. This means that all bulks standing in the queue are served simultaneously and the service intensity decreases in proportion to the number of bulks served at the same time. Single demands in a bulk are served one by one in order of coming. The bulk is lost if all positions in the buffer are busy at the moment of bulk arriving.

The scheme of service lets us to simplify the model essentially, considering each bulk as one long demand came for service. Bulk interarrivals and service times of single demands are supposed to be independent and distributed exponentially with parameters λ and μ respectively.

3. EQUILIBRIUM SYSTEM OF EQUATIONS

The queue length (the queue length here is the number of busy positions in the buffer) is one of the most important parameters of the considered queueing system. However, the stochastic process equal to the number of busy buffer position $\{n(t)\}_{t\geq 0}$ is not a Markov process in the general case, since the bulk service time distribution which is a sum of the service times of single demands has the after-effect as a convolution of exponential distributions. In order to have a possibility to use the analytical instruments of the theory of Markov processes, we consider the random vector $N(t) = (n_1(t), \dots, n_m(t))$, where m is the buffer capacity and $n_i(t)$, $i = 1, \dots, m$, is the number of demands waiting for service and placed in the ith position of the buffer at an instance of time t. By the assumption the service times of a single demand has the exponential distribution, therefore the random process $\zeta(t) = N(t)$, $t > 0$, is a Markov process with countable number of states from the direct product $Q = \mathbf{N} \times \dots \times \mathbf{N}$ of m sets $\mathbf{N} = \{0, 1, \dots\}$.

We define the function $p(t, k_1, \dots, k_m)$ as the probability that the system is in a state $k_1, \dots, k_m$ at an instant of time t, i.e., we set

$$p(t, k_1, \dots, k_m) = \mathbf{P}\{n_1(t) = k_1, \dots, n_m(t) = k_m\}.$$

All states of the Markov process $\zeta(t)$ are mutually communicating. According to [1] we can build the Erlang system of equations (or equilibrium system of equations) for the steady state probabilities

$$p_{k_1,\dots,k_m} = \lim_{t\to\infty} p(t, k_1, \dots, k_m)$$

of the process $\zeta(t)$.

Let the number of demands in a bulk be a random variable ξ with a distribution

$$\mathbf{P}\{\xi = i\} = \varphi_i, \quad i = 1, 2 \dots, \quad \sum_{i=1}^{\infty} \varphi_i = 1.$$

We assume that $E\xi < \infty$.

We use the following notation. Let $I = \{1, \dots, m\}$ be the set of numbers of the buffer positions, $J_k = \{i \in I \colon k_i \neq 0\}$ be the set of numbers of the busy buffer positions in a state $K = \{k_1, \ldots, k_m\}$, and let $B(k_1, \ldots, k_m)$ be the function equal to the number of busy buffer positions in the state K.

The system can appear in a state $(k_1, \ldots, k_m)$ either from the states $(k_1, \ldots, k_{i-1}, 0, k_{i+1}, \ldots, k_m)$, $i \in J_K$, after arrival of a new bulk of size k_i to the position i or from the states $(k_1, \ldots, k_i + 1, \ldots, k_m)$, $i \in I$, if a single demand from the position $i \in I$ is serviced. The third possibility is to stay in the state $(k_1, \ldots, k_m)$ without transition.

The equilibrium system of equations contains three groups of equations: for the zero state

$$-\lambda p_{0\ldots 0} + \sum_{i=1}^{m} p_{0\ldots 01_i 0\ldots 0} = 0, \tag{1}$$

for the case $0 < B(k_1, \ldots, k_m) < m$

$$\sum_{i\in J_K} \lambda\varphi_{k_i} \frac{p_{k_1,\ldots,k_{i-1},0,k_{i+1},\ldots,k_m}}{m - B(k_1, \ldots, k_{i-1}, 0, k_{i+1}, \ldots, k_m)} - (\lambda + \mu)p_{k_1,\ldots,k_m} + \sum_{i\in I} \mu \frac{p_{k_1,\ldots,k_i+1,\ldots,k_m}}{B(k_1, \ldots, k_i + 1, \ldots, k_m)} = 0, \tag{2}$$

and

$$\sum_{i\in J_K} \lambda\varphi_{k_i} \frac{p_{k_1,\ldots,k_{i-1},0,k_{i+1},\ldots,k_m}}{m - B(k_1, \ldots, k_{i-1}, 0, k_{i+1}, \ldots, k_m)} - \mu p_{k_1,\ldots,k_m} + \sum_{i\in I} \mu \frac{p_{k_1,\ldots,k_i+1,\ldots,k_m}}{B(k_1, \ldots, k_i + 1, \ldots, k_m)} = 0, \tag{3}$$

if all positions in the buffer are busy.

Note that the factor $1/(m - B(k_1, \ldots, k_{i-1}, 0, k_{i+1}, \ldots, k_m))$ corresponds to the fact that a new demand appeared has the equal access to all position non-occupied at the moment of arrival.

The system is complemented by the normalizing condition

$$\sum_{Q} p_{k_1,\ldots,k_m} = 1, \tag{4}$$

where Q is the set of possible states of the system.

4. THE SOLUTION OF THE EQUILIBRIUM SYSTEM OF EQUATIONS

THEOREM 1. *The solution of the system* (1)–(4) *is of the form*

$$p_{k_1,\ldots,k_m} = p_0 \frac{n!\,(m-n)!}{m!} \left(\frac{\lambda}{\mu}\right)^n \prod_{J_K} F_{k_j}, \tag{5}$$

where

$$F_0 = 1, \quad F_k = \sum_{i=k}^{\infty} \varphi_i, \quad p_0 = p_{0\ldots 0}, \quad n = B(k_1, \ldots, k_m).$$

PROOF. We demonstrate that expression (5) satisfies the system of equations (1)–(4). We write the system in a more convenient form and simultaneously substitute expression (5) into the left-hand side of the system.

We begin with the case $0 < n < m$. In this case the left-hand side takes the form

$$\frac{1}{m-n+1}\left(\frac{\lambda}{\mu}\right)^{n-1}\frac{(n-1)!(m-n+1)!}{m!}\sum_{i\in J_K}\lambda\varphi_{k_i}\prod_{j\in J_K\setminus i}F_{k_j}$$
$$-(\lambda+\mu)\left(\frac{\lambda}{\mu}\right)^{n}\frac{(n)!(m-n)!}{m!}\prod_{j\in J_K}F_{k_j}$$
$$+\frac{1}{n+1}\left(\frac{\lambda}{\mu}\right)^{n+1}\frac{(n+1)!(m-n-1)!}{m!}\sum_{i\in I\setminus J_K}\prod_{j\in J\cup i}F_{k_j}$$
$$+\frac{\mu}{n}\left(\frac{\lambda}{\mu}\right)^{n}\frac{(n)!(m-n)!}{m!}\sum\_i\in J_K F_{k_i+1}\prod_{j\in J_K\setminus i}F_{k_j}. \qquad (6)$$

Note that $F_1 = 1$. Therefore if the coordinates of a state vector which are equal to zero are replaced by 1, the product $\prod_{J_K} F_{k_j}$ does not change. Note also that the cardinality of the set $|I \setminus J|$ is equal to $m - n$. These two facts let us group the similar terms, divide the left-hand side into two parts, and obtain

$$\frac{\lambda^n}{\mu^{n-1}}\left(\frac{(n-1)!(m-n)!}{m!}\sum_{i\in J_K}\varphi_{k_i}\prod_{j\in J_K\setminus i}F_{k_j}-\frac{(n)!(m-n)!}{m!}\prod_{j\in J_K}F_{k_j}\right.$$
$$\left.+\frac{(n-1)!(m-n)!}{m!}\sum_{i\in J_K}F_{k_i+1}\prod_{j\in J_K\setminus i}F_{k_j}\right)$$
$$+\frac{\lambda^{n+1}}{\mu^n}\prod_{J_K}F_{k_j}\left(\frac{m-n}{n+1}\frac{(n+1)!(m-n-1)!}{m!}-\frac{n!(m-n)!}{m!}\right). \qquad (7)$$

It is easily seen that this expression is equal to zero, since $F_{k+1} + \varphi_k = F_k$.

In the case of $n = 0$ it is obvious that $p_{0,\ldots,1,\ldots,0} = p_0\lambda/(m\mu)$ satisfies equation (6). In the case $m = n$ the arrival of new groups is impossible, therefore the corresponding probabilities are equal to zero. Hence it follows that the last term in (7) is equal to zero and, consequently, probabilities (5) satisfy equation (3).

COROLLARY 1. *The distribution of the queue length in the queueing system considered is given by the formulae*

$$p_n = \left(\frac{\lambda}{\mu}\mathbf{E}\xi\right)^n p_0, \qquad n = 0, 1, \ldots, m. \qquad (8)$$

PROOF. It is clear that using (5) we can represent the probability p_n in the form

$$p_n = p_0 \sum_{K\in Q:\ B(k_1,\ldots,k_m)=n} p_{k_1,\ldots,k_m}; \qquad (9)$$

consequently,

$$p_n = p_0 \left(\frac{\lambda}{\mu}\right)^n \sum_{|J_K|=n} \frac{n!(m-n)!}{m!} \prod_{J_K} F_{k_j}. \tag{10}$$

Note that the probabilities $p_{k_1,\ldots,k_m}$ do not depend on permutations of the indices. Therefore, we can arrange the indices in the increasing order, i.e., assume that $k_i \leq k_{i+1}$, $i = 1, \ldots, m-1$. Then the first $m-n$ elements of the vector are equal to zero. We denote by s_1 the number of the first identical non-zero elements of the vector K, by s_2 the number of the next identical non-zero elements s_2 and so on. Obviously, the number s_i of elements in the ith group may take values $0, \ldots, n$ depending on the vector K, but the number of groups of non-zero elements can not be greater than n, and $s_1 + \ldots + s_n = n$ (if the number of groups k of elements of the vector K is such that $k < n$, then we put $s_i = 0$ for $i = k, \ldots, n$). We denote by b_i the value of the elements of the ith group, $i = 1, \ldots, n$. Then, summing the identical products, we obtain

$$p_n = p_0 \left(\frac{\lambda}{\mu}\right)^n \sum_{s_1+\ldots+s_n=n} \frac{n!\,(m-n)!}{m!} \frac{m!}{s_1!\ldots s_n!(m-n)!} F_{s_1}^{b_1} \ldots F_{s_n}^{b_n}. \tag{11}$$

Thus,

$$\begin{aligned} p_n &= p_0 \left(\frac{\lambda}{\mu}\right)^n \sum_{s_1+\ldots+s_n=n} \frac{n!}{s_1!\ldots s_n!} F_{s_1}^{b_1} \ldots F_{i_n}^{b_n} \\ &= p_0 \left(\frac{\lambda}{\mu}\right)^n \left(\sum_{j=1}^{\infty} F_j\right)^n = p_0 \left(\frac{\lambda}{\mu}\right)^n \left(\sum_{j=1}^{\infty} j\varphi_j\right)^n = p_0 \left(\frac{\lambda}{\mu}\mathbf{E}\xi\right)^n \end{aligned} \tag{12}$$

5. CONCLUSION

The distribution of the queue length let us compute some characteristics of the queueing system under consideration which are important in applications. The probability that the processor is idle is clearly equal to

$$p_0 = \left(\sum_{n=0}^{m} \left(\frac{\lambda}{\mu}\mathbf{E}\xi\right)^n\right)^{-1}.$$

The probability of loss of a bulk is equal to

$$p_m = \left(\frac{\lambda}{\mu}\mathbf{E}\xi\right)^m \left(\sum_{n=0}^{m} \left(\frac{\lambda}{\mu}\mathbf{E}\xi\right)^n\right)^{-1}$$

Note also that the expression $\mathbf{E}\xi/\mu$ is the mathematical expectation of the bulk service time. Hence, the steady state distribution of the queue length is in a good accordance with the well-known results for the queueing systems $M|G|1|\infty$ (see, for example [4]).

The author expresses his deep gratitude to T. Alanko, whose lectures in the Petrozavodsk University stimulated this investigations, and also to V. V. Kalashnikov, E. V. Morozov, and B. M. Shirokov for useful discussions.

This work was fulfilled in the frame of the joint project described in [5].

REFERENCES

1. L. Kleinrock, *Queueing Systems*. Wiley, New York, 1974.
2. M. L. Chaudry and J. G. C. Templton, *A First Course in Bulk Queues*. Wiley, New York, 1983.
3. Tan Xiaoming and C. Knessl, Sojourn time distribution in some processor shared queues. *European J. Appl. Math.* (1993) **4**, 437–448.
4. B. V. Gnedenko and I. N. Kovalenko, *Introduction to Queueing Theory*. Nauka, Moscow, 1987 (in Russian).
5. T. Alanko and I. A. Bogoiavlenski, Performance analysis and capacity planning for a message server in a regional data-communication network. Preprint, Dept. Computer Sci., Univ. Helsinki, Univ. Petrozavodsk, 1995.

Probabilistic Methods in Discrete Mathematics, pp. 137–144
V. F. Kolchin et al. (Eds.)

STATISTICAL ANALYSIS OF RENEWAL PROCESSES

V. I. CHERNETSKII, A. A. ROGOV, and I. V. SHCHEGOLEVA
Petrozavodsk

Abstract — This paper continues the investigations started in [1] and answers some questions stated there.

1. INTRODUCTION

A number of mathematical models of renewal processes, such as classical, alternating, regenerating, branching, etc., are known in the renewal theory. They form a basis for adequate choice of a mathematical model which corresponds to reality with sufficient accuracy. At the same time, the problem of identification of a mathematical model includes the following three tasks concerning directly the statistical analysis of renewal processes:

(1) Evaluation of different probabilistic characteristics of time intervals between events in the renewal process (distribution function, density function, moments).

(2) Developing goodness-of-fit tests to check the agreement between the mathematical model of the renewal process chosen and the statistical data obtained for the corresponding real-life process.

(3) Evaluation of characteristics of the renewal process (the renewal function and its density, the values of overshooting and undershooting) on the basis of the statistical data obtained.

In this article the emphasis is on the classical renewal process and non-parametric estimates of the renewal function. In addition, we consider the problem of choosing a mathematical model of the renewal process on the basis of statistical data.

2. ANALYSIS OF TESTING PLANS

When observing n independent realizations of a renewal process, the following designs of experiment are considered [2, 7]:

(1) The intervals of observation are fixed a priori and are equal to T. This plan is denoted by $[n, B, T]$.

(2) The observation stops after r failures. This plan is denoted by $[n, B, r]$.

(3) The observation stops either after r failures or at the moment of time T depending on what event occurs earlier. This plan is denoted by $[n, B, (r, T)]$.

(4) The ith realization of the renewal process is observed during the time interval T_j and, maybe, not all of T_j are equal to each other. This plan is denoted by $[n, B, T_1, \ldots, T_n]$.

(5) If in the plan $[n, B, T_1, \ldots, T_n]$ the moments of time $T_1, \ldots, T_n$ are independent identically distributed random variables independent of the number of failures and having a common distribution function $G(t)$, then the plan is denoted by $[n, B, T_G]$.

(6) The observation of the ith realization of the process stops after r_i failures. This plan is denoted by $[n, B, r_1, \ldots, r_n]$.

We denote by x_{ij} the time between $j-1$ and j failures in the ith sample, $i = 1, 2, \ldots, n$, $j = 1, 2, \ldots, r_i$; the 0th failure corresponds to the beginning of the observation, the r_ith failure corresponds to the end of the ith observation of the renewal process.

The plan $[n, B, r_1, \ldots, r_n]$ is the most general, i.e., it contains as particular cases the other plans given above.

We suggest the following procedure of transition from the plan $[n, B, T_1, \ldots, T_n]$ to the plan $[n, B, r_1, \ldots, r_n]$. Suppose that

$$T_i = \sum_{j=1}^{r_i} x_{ij}, \qquad i = 1, 2, \ldots, n.$$

The intervals with the indices i_{r_i} are not complete, since not all tests are stopped at the time of failure. To estimate the length of the last interval till the failure, the following method can be used. Among x_{ij}, $i = 1, 2, \ldots, n, j = 1, 2, \ldots, r_i - 1$, we choose the intervals which are greater than the intervals x_{ir_i} and denote them by $y_1, y_2, \ldots, y_m$, where m is the number of such intervals. Substituting all the incomplete intervals with indices ir_i by their estimates

$$\bar{x}_{ir_i} = \frac{1}{m} \sum_{j=1}^{m} y_j, \qquad i = 1, 2, \ldots, n,$$

we obtain the plan $[n, B, r_1, \ldots, r_n]$. This approach lets us use the statistical data obtained under other plans in statistical estimates of renewal functions constructed for the plan $[n, b, r_1, \ldots, r_n]$.

Consider the plan $[1, B, T]$. Let

$$\sum_{i=1}^{r} x_i = T.$$

If t is essentially less than T, then the transition from the plan $[1, B, T]$ to the plan $[n, B, (r_1, \ldots, r_{n-1}, T)]$ can be performed in the following way. Using the recurrence relation

$$r_j = \max\left\{p\colon \sum_{i=r_{j-1}+1}^{p-1} x_i \le t\right\}, \quad r_0 = 1, \quad j = 1, 2, \ldots, n-1,$$

we find $r_1, \ldots, r_{n-1}$; we set

$$T_n = \sum_{i=r_{n-1}+1}^{n} x_i,$$

and determine n from the condition $T_n \le t$. Then the initial sample can be represented as n realizations of the renewal process, i.e.,

$$x_{11} = x_1, \ldots, x_{1r_1} = x_{r_1},$$
$$x_{21} = x_{r_1+1}, \ldots, x_{2r_2} = x_{r_1+r_2},$$
$$\ldots$$
$$x_{n1} = x_{r_1+\ldots+r_{n-1}+1}, \ldots, x_{nr_n} = x_r.$$

Thus, we obtain the plan $[n, B, r_1, \ldots, r_n]$. This approach let us estimate the characteristics of a renewal process using the formulas given above if we have one sufficiently large realization of this process.

Using the classification given above, we can note that the most studied is the first problem for the plans $[n, B, r]$ and $[n, B, T]$ (see [3,7]). The second problem is considered as a rule for Poisson processes [2,3].

In this paper the authors consider the third problem, namely, the statistical estimates of the renewal function. We point out two papers on this theme [5, 7].

3. RENEWAL FUNCTION ESTIMATES

For the classical renewal model, we distinguish three methods of estimating the renewal function.

(1) On the basis of the statistical data, we estimate the distribution function $F(t)$ of failures and then solve (usually by numerical methods) the renewal equation

$$B(t) = F(t) + \int_0^t B(t-\tau)F'(\tau)\,d\tau.$$

The major disadvantage of this method is its difficulty and the complexity of precision evaluation. In [5], possible approaches to this problem are discussed and examples of statistical estimation of the density of renewal function are given. In [6], a theorem is proved which simplifies the solution of the renewal equation in the case where the reliability function is approximated by a mixture of exponential distributions.

(2) On the basis of theoretical data, we choose a parametric representation of the renewal function and then on the basis of statistical data estimate the unknown parameters and apply some goodness-of-fit test. It should be noted that whereas the characterization theorems for renewal function exist [4, 6], the methods of parameter estimating and hypothesis testing are not sufficiently developed except for the Poisson process.

(3) The third method consists in using non-parametric approach to estimation of the renewal function with regard to its properties.

For the plan $[n, B, T_1, \dots, T_n]$ we suggest a non-parametric estimate of the renewal function $B^*(t)$ whose construction takes into account its monotonicity and the fact that the relative frequencies converge to the unknown probabilities. Without loss of generality we may assume that $T_1 \le T_2 \le \dots \le T_n = T$.

We set

$$B^*(t) = B^*(T_k) + \frac{1}{N(t)} \sum_{i=k}^{n} y_i^k(t), \qquad 0 \le t \le T,$$

where $B^*(0) = 0$, $T_0 = 0$, $N(t) = n - k$,

$$k = \max\{i\colon\ T_i < t,\ i = 0, 1, \dots, n\},$$

$y_i^k(t)$ is the number of failures on the interval $(T_k, t]$ in the ith sample, $i = k+1, k+2, \dots, n$.

THEOREM 1. *The estimate $B^*(t)$ is non-decreasing, consistent, and unbiased, i.e.,*

$$\begin{aligned} B^*(t_1) &\le B^*(t_2), && t_1 \le t_2, \\ \mathbf{E}B^*(t) &= B(t), && t \in [0, T], \\ \lim_{N(t)\to\infty} \mathbf{P}\{|B^*(t) - B(t)| > \varepsilon\} &= 0, && \forall \varepsilon > 0. \end{aligned}$$

If new statistical data are obtained, the estimate $B^*(t)$ may be simply corrected unlike the estimates of the renewal function obtained by the first and second methods. To do this, it is sufficient to know $N(t)$ and the previous estimate. Suppose we observe the $(n+1)$th realization of the renewal process during time T_{n+1}. We set $N(t) = 0$ if $t > T$ and

$$N^*(t) = \begin{cases} N(t), & t > T_{n+1}, \\ N(t) + 1, & t \le T_{n+1}. \end{cases}$$

The corrected estimate $B^{**}(t)$ is calculated using the recurrence relation

$$B^{**}(t) = B^{**}(T_k) + \frac{B * (t)N(t) + y_{n+1}^k(t)}{N^*(t)}, \qquad 0 \le t \le \max\{T, T_{n+1}\},$$

where $B^{**}(0) = 0$, $T_0 = 0$,

$$k = \max\{i\colon\ T_i < t,\ i = 0, 1, \dots, n+1\},$$

$y_{n+1}^k(t)$ is the number of failures on the interval $(T_k, t]$ for the $(n+1)$th sample. If $T_{n+1} < T_k$, then $y_{n+1}^k(t) = 0$.

These formulas may be used for the plan $[n, B, r_1, r_2, \dots, r_n]$, if we set

$$T_i = \sum_{j=1}^{r_i} x_{ij}.$$

In [7] a statistical estimate of the renewal function for the plan $[n, B, T_1, \dots, T_n]$ is given. However, this estimate has a deficiency which consists in the fact that it can decrease on some intervals. It is clear that this property contradicts the property of renewal functions. For example, let us consider two experiments. If $T_1 < T_2$ and the number of failures in the first experiment is greater than the number of failures in the second experiment, then the estimate of the renewal function decreases on the interval $[T_1, T_1 + \varepsilon]$ for some $\varepsilon > 0$. Besides, if $t > \min\{T_1, \dots, T_n\}$, this estimate does not use the information about some failures.

As a result of 100 experiments on a renewal process with Poisson flow of failures performed by the statistical simulation method, we obtain that in 60 cases the variance of the estimate $B^*(t)$ is less than the variance of the estimate given in [7]. Thus, we believe that the estimate $B^*(t)$ is more effective.

For practical solution of some problems, it is required to extrapolate the estimate of the renewal function into the points $t > T$. Taking into account one of the properties of renewal functions, we suggest to use for the extrapolation the linear function $y = \lambda t$, where $\lambda = 1/\mathbf{E}\xi$ and realizations of the random variable ξ form the renewal process observed. As the estimate of $\mathbf{E}\xi$ we take the sampling mean, i.e.,

$$\lambda = \sum_{i=1}^{n} (r_i - 1) \left(\sum_{i=1}^{n} \sum_{j=1}^{r_i - 1} x_{ij} \right)^{-1}.$$

Then the problem of finding the point t, from which we may use the approximation described above, arises. To evaluate this number, the following procedure can be used. We choose a sequence of points $\{t^0, t^1, \dots, t^s\}$, $t^0 = 0$, in the interval $[0, T]$. Using the least squares method, we construct sequentially the regressions of the kind $y = \lambda t$ on the base of the samples from the interval $[t^i, T)$ until the sequence of λ obtained stabilizes.

Let us give a formula for calculating λ. Let $T_1, \dots, T_n$ be the jumps of the function $N(t)$, $T_1 < T_2 < \dots < T_n$, and $k = \min\{i\colon T_k > t, i = 1, \dots, n\}$.

We denote by $\tau_1^i, \tau_2^i, \dots, \tau_{N_i}^i$ the sequence of the renewal instances in the total sample for the interval $[t, T_k]$, $i = k - 1$, and by $v_1^i, v_2^i, \dots, v_{N_i}^i$ the corresponding sequence for the interval $(T_i, T_{i+1}]$, $k \le i < n$. The jth renewal instance is the time from the beginning of the observation $t = 0$ till the appearance of the jth renewal. Then to evaluate λ, we need to minimize

$$f(\lambda) = \sum_{i=1}^{N_{k-1}} \left(B(t) + \frac{v_i^{k-1}}{N(T_k)} - \lambda \tau_i^{k-1} \right)^2 + \sum_{j=k}^{n} \sum_{i=1}^{N_j} \left(B(T_{j-1}) + \frac{v_i^j}{N(T_j)} - \lambda \tau_i^j \right)^2. \quad (1)$$

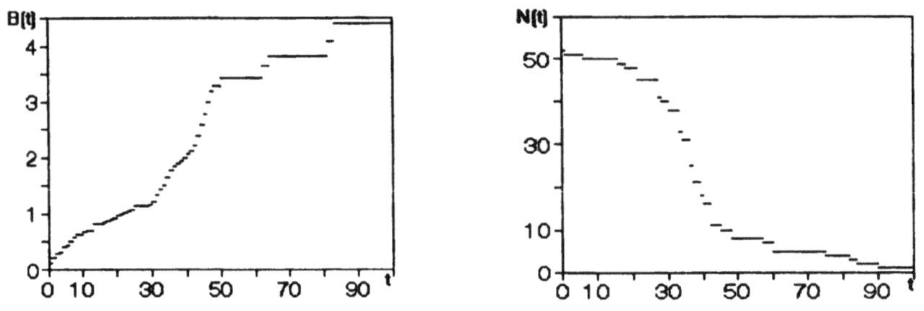

Figure 1. 1990–91, machine 10, $n = 52$, $T = 100$

It is not difficult to see that the root of the equation $f'(\lambda) = 0$ is equal to

$$\lambda = \sum_{i=1}^{N_{k-1}} \left(B(t) + \frac{v_i^{k-1}}{N(T_k)} \right) \tau_i^{k-1} + \sum_{j=k}^{n} \sum_{i=1}^{N_j} \left(B(T_{j-1}) + \frac{v_i^j}{N(T_j)} \right) \tau_i^j \left(\sum_{j=k}^{n} \sum_{i=1}^{N_j} (\tau_i^j)^2 \right)^{-1} .$$

It is easy to see that $f''(\lambda) > 0$, that is, the point found gives the minimum of (1).

For the plan $[n, B, T]$ in the book [6] there is the following result proved by Belyaev. If the renewal function $B(t)$ is continuous, then for all t and $x > 0$, as $n \to \infty$,

$$\mathbf{P}\{\max_t |B^*(t) - B(t)| < x\sqrt{B^*(t)/n}\} \to L(x),$$

where

$$L(x) = \int_0^x \frac{4}{\sqrt{2\pi}} \sum_{k=0}^{\infty} (-1)^k (2k+1) e^{-(2k+1)^2 x^2/2} dx$$

Using this result, we can estimate the accuracy of the estimation obtained for $B^*(t)$ depending on the number of observations n.

4. AN EXAMPLE

In practice a renewal process is met more seldom than an alternating process, that is, a process with a non-zero reconstruction time. In the statistic analysis of such a process, the problems similar to described above arise.

The problem of estimation of the average number of failures or the average number of renewals is reduced to the problems for the classical renewal process and the formulae described above can be used. The statistical analysis of such a type was carried out for the failures of paper-machines on the Segezhabumprom factory. We processed the data obtained in 1990 and 1991. As the instances of the end of observation the instances of the planned preventive maintenance were chosen. The scale of t is one term (8 working hours).

From the analysis of the given graphs it is clear that the values of the renewal function stop being certain when the number of observations $N(t)$ is less than 10.

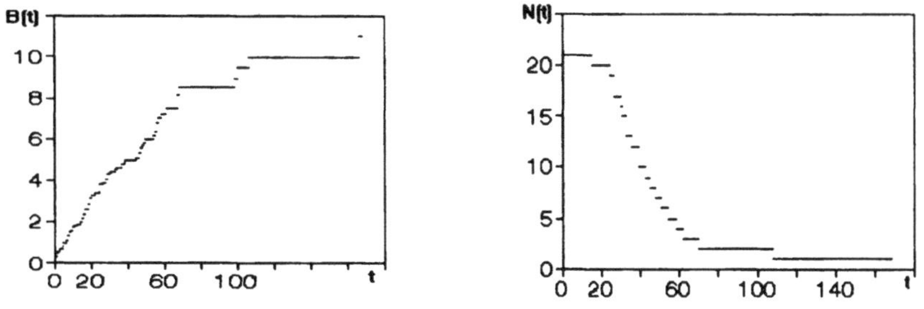

Figure 2. 1991, machine 1, $n = 21$, $T = 168$

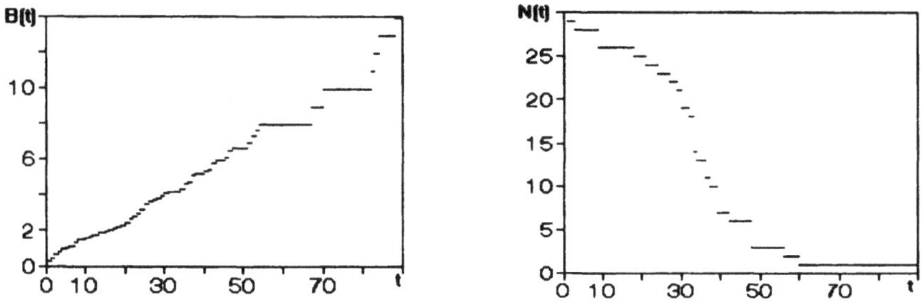

Figure 3. 1991, machine 11, $n = 30$, $T = 90$

REFERENCES

1. A. A. Rogov and V. I. Tchernetskii, On problems of renewal system theory for finite time interval. *Probabilistic Methods in Discrete Mathematics, Proc. III Intern. Petrozavodsk Conf.* VSP/TVP, Utrecht/Moscow, 1993, pp. 386–394

2. B. V. Gnedenko, Yu. K. Belyaev, and A. D. Solovyev, *Mathematical Methods of Reliability Theory.* Nauka, Moscow, 1965 (in Russian).

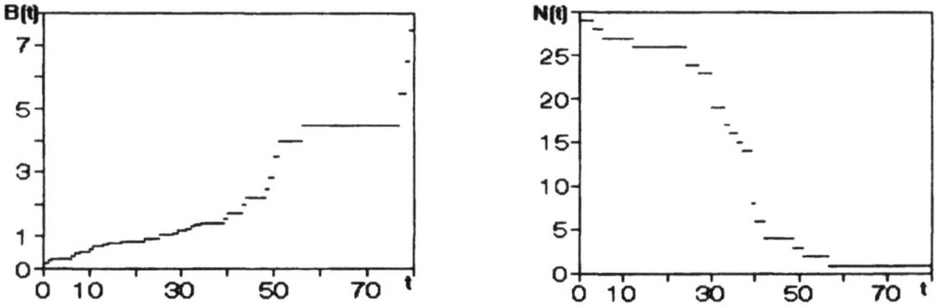

Figure 4. 1991, machine 9, $n = 29$, $T = 80$

3. D. R. Cox and P. A. W. Lewis, *The Statistical Analysis of Series of Events.* Wiley, New York, 1966.
4. A. A. Rogov, On the identification of renewal functions. Dep. VINITI, No. 2979–B92, 1992, p. 15 (in Russian).
5. V. M. Skripnik, A. E. Nazin, Yu. G. Prikhodko, and Yu. N. Blagoveshchenskii, *Analysis of Reliability of Technical Systems on Censored Samples.* Radio i Svyaz, Moscow, 1988 (in Russian).
6. E. Yu. Barzilovich, *Technical Maintenance Models.* Vysshaya Shkola, Moscow, 1982 (in Russian).
7. E. V. Chepurin, On statistical inference for renewal processes. In: *Statist. Methods in Reliability Theory and Quality Control.* Moscow Univ. Press, Moscow, 1973, pp. 9–25 (in Russian).

Probabilistic Methods in Discrete Mathematics, pp. 145–155
V. F. Kolchin et al. (Eds.)

FUNCTIONAL LIMIT THEOREMS FOR OBSERVATIONS OF STOCHASTIC PROCESSES AT A RANDOM TIME POINT

A. N. CHUPRUNOV

Kazan

Abstract — This paper came into existence from author's works connected with applications of probability theory to statistics. The author dealt with samples which were observations of stochastic processes at random time points. In order to perform the statistical analysis of such samples, the author needed some limit theorems. These limit theorems are presented here.

1. LIMIT THEOREMS FOR RANDOM STEP AND BROKEN LINES

We denote by $\xrightarrow{d}$ the convergence in probability and by $\overset{d}{=}$, the equality in distribution.

Let $(\Omega, \mathfrak{A}, \mathbf{P})$ be a probability space and $(T, \mathfrak{B})$ be a measurable space. Let $X_n(t)$, $t \in T, n \in \mathbf{N}$, be stochastic processes defined on $(\Omega, \mathfrak{A}, \mathbf{P})$ and measurable with respect to $\mathfrak{B}$. Let X_{nk}, $1 \le k \le k_n$, $n \in \mathbf{N}$, be independent copies of X_n. Let ξ and ξ_n, $n \in \mathbf{N}$, be independent identically distributed random variables with values in $(T, \mathfrak{B})$ defined on another probability space, $(\Omega_1, \mathfrak{A}_1, \mathbf{P}_1)$. We assume that $k_n, a_n \in \mathbf{N}, n \in \mathbf{N}$, are sequences such that $k_n \to \infty$, $n \to \infty$, and $a_1 \le a_2 \le a_3 \le \dots$ We set

$$S_n(t) = \sum_{k=1}^{k_n} X_{nk}(t), \qquad t \in T,$$

and

$$S_n^*(\omega_1)(x) = \sum_{k=a_n+1}^{a_n+[k_n x]} X_{nk}(\xi_k(\omega_1)), \quad n \in \mathbf{N}, \quad \omega_1 \in \Omega_1, \quad 0 \le x \le 1.$$

Obviously, $S_n(t)$ is a sequence of stochastic processes on T and $S_n^*(\omega_1)(x)$ is a sequence of stochastic processes on $[0, 1]$ with the parameter ω_1.

Let e be a random variable with an infinitely divisible distribution. Then its characteristic function $\hat{e}$ has the form

$$\hat{e}(y) = \hat{e}(b, \sigma^2, L)(y) = \exp\left\{ iby - \frac{\sigma^2 y^2}{2} + \int_{|x|>0} \left(e^{iyz} - 1 - \frac{iyz}{1+z^2} \right) dL(z) \right\},$$

where $y, b \in \mathbf{R}$, $\sigma^2 \geq 0$, and L is a function defined both on $(-\infty, 0)$ and $(0, \infty)$ and increasing on each of them. The set of such functions will be denoted by $\mathfrak{C}(\mathbf{R})$.

Recall that $e(x)$, $x \in [0,1]$, is a homogeneous random process with independent increments if and only if

$$e(x) \stackrel{d}{=} e(xb, x\sigma^2, xL), \qquad x \in [0,1],$$

for some $b \in \mathbf{R}$, $\sigma^2 \geq 0$, and $L \in \mathfrak{C}(\mathbf{R})$.

Let $\tau > 0$. We set

$$A(t) = \sup_{n \in \mathbf{N}} k_n \left| \mathbf{E} X_n(t) \, \mathbf{I}_{|X_n(t)| \leq \tau} \right|,$$

$$B(t) = \sup_{n \in \mathbf{N}} k_n \mathbf{E} \frac{(X_n(t))^2}{1 + (X_n(t))^2}, \qquad t \in T,$$

$$b_k = \min(a_{k+1} - a_k, k), \qquad k \in \mathbf{N}.$$

Let $\mathbf{E}_1$ stand for the expectation with respect to $\mathbf{P}_1$.

We consider the following conditions:

(A) there exists $C > 0$ such that

$$\sum_{k=1}^{n} a_k \leq C n^p, \quad n \in \mathbf{N}, \quad p < 2,$$

and a_k are arbitrary for $p = 2$;

(B) there exists $C > 0$ such that $Cn \leq k_n$ for all $n \in \mathbf{N}$.

The following analogue of the Skorokhod theorem [7] is valid.

THEOREM 1. *Let $1 \leq p \leq 2$, and, as $n \to \infty$,*

$$S_n(t) \stackrel{d}{\to} e(b(t), \sigma^2(t), L(t)).$$

Let (A) and (B) be valid, and $\mathbf{E}_1(A(\xi))^p < \infty$, $\mathbf{E}_1(B(\xi))^p < \infty$.

Then, as $n \to \infty$,

$$S_n^*(\omega_1)(x) \stackrel{d}{\to} e(b_1, \sigma_1^2, L_1)$$

in the Skorokhod space $D[0,1]$ for almost all $\omega_1 \in \Omega_1$, where $b_1 = \mathbf{E}_1 b(\xi)$, $\sigma_1^2 = \mathbf{E}_1 \sigma^2(\xi)$, and $L_1(x) = \mathbf{E}_1 L(\xi)(x)$, $x \in \mathbf{R}$.

The proof of Theorem 1 is based on a criterion for the convergence in distribution of sums of independent random variables (see [8]) and the following lemmas.

Let $f, f_n \colon T \to \mathbf{R}$, $n \in \mathbf{N}$, be measurable functions. We set

$$H(t) = \sup_{n \in \mathbf{N}} k_n |f_n(t)|, \qquad t \in T.$$

LEMMA 1. *Let* $1 \le p \le 2$, $\mathbf{E}_1(H(\xi))^p < \infty$, *and let* (A) *and* (B) *be valid. Then, as* $n \to \infty$,

$$\max_{a_n < i \le a_n + k_n} |f_n(\xi_i(\omega_1))| \to 0,$$

for almost all $\omega_1 \in \Omega_1$.

Lemma 1 ensures the property of the uniform negligibility for the array $X_{nk}(\xi_k(\omega_1))$, $a_n + 1 \le k \le a_n + k_n$, $n \in \mathbf{N}$, for almost all $\omega_1 \in \Omega_1$.

LEMMA 2. *Let* $k_n f_n(t) \to f(t)$, *let* $n \to \infty$ *for all* $t \in T$, *let* (A) *be valid, and let* $\mathbf{E}_1(H(\xi))^p < \infty$. *Then, as* $n \to \infty$,

$$\sum_{i=a_n+1}^{a_n+k_n} f_n(\xi_i(\omega_1)) \to \mathbf{E}_1 f(\xi)$$

for almost all $\omega_1 \in \Omega_1$.

We consider the stochastic processes

$$S_n^{**}(\omega_1)(x) = \sum_{k=a_n+1}^{a_n+[k_n x]} (X_{nk}(\xi_k(\omega_1)) + \{k_n x\} X_{nk}(\xi_{a_n+[k_n x]+1}(\omega_1))),$$

where $n \in \mathbf{N}$, $\omega_1 \in \Omega_1$, $0 \le x \le 1$. We denote by $\gamma(\sigma)$ the centered Gaussian random variable with the variance σ^2 and by W, the Wiener process.

The following analogue of the Donsker theorem is true.

THEOREM 2. *Let the conditions of Theorem* 1 *be valid*, $S_n(t) \xrightarrow{d} \gamma(\sigma(t))$, *as* $n \to \infty$, *for all* $t \in T$ *and* $\sigma_1^2 = 1$.

Then, as $n \to \infty$,

$$S_n^{**}(\omega_1)(x) \xrightarrow{d} W$$

in $C[0, 1]$ *for almost all* $\omega_1 \in \Omega_1$.

Let $Y(t)$ and $Y_i(t)$, $i \in \mathbf{N}$, $t \in T$, be independent identically distributed stochastic processes defined on $(\Omega, \mathfrak{A}, \mathbf{P})$ and measurable with respect to $\mathfrak{B}$.

We consider the stochastic processes

$$S_n^{**}(\omega_1)(x) = n^{-1/2} \sum_{k=a_n+1}^{a_n+[nx]} (Y_{nk}(\xi_k(\omega_1)) + \{k_n x\} Y_k(\xi_{k(a_n+[nx]+1)}(\omega_1))),$$

where $n \in \mathbf{N}$, $\omega_1 \in \Omega_1$, $0 \le x \le 1$.

COROLLARY 1. *Let* $1 \le p \le 2$, $\mathbf{E}Y(t) = 0$ *for all* $t \in T$, $\mathbf{E}_1\mathbf{E}(Y(\xi))^2 = 1$, *and* $\mathbf{E}_1(\mathbf{E}(Y(\xi)^2)^p < \infty$. *Let* (A) *and* (B) *be valid.*

Then, as $n \to \infty$,

$$S_n^{**}(\omega_1)(x) \xrightarrow{d} W$$

in $C[0,1]$ for almost all $\omega_1 \in \Omega_1$.

For $x = 1$ Corollary 1 implies the following assertion.

COROLLARY 2. *Suppose that $\sigma_1^2 = \mathbf{E}_1(\mathbf{E}(Y(\xi)^2) < \infty$.*

(1) *Let the conditions of Corollary 1 be valid. Let*

$$S_n^{\#}(\omega_1)(x) = n^{-1/2} \sum_{k=a_n+1}^{a_n+n} Y_k(\xi_k(\omega_1)).$$

Then, as $n \to \infty$,

$$S_n^{\#}(\omega_1)(x) \xrightarrow{d} \gamma(\sigma_1)$$

for almost all $\omega_1 \in \Omega_1$.

(2) *Let*

$$S_n^{\#\#}(\omega_1)(x) = n^{-1/2} \sum_{k=1}^{n} Y_k(\xi_k(\omega_1)).$$

Then, as $n \to \infty$,

$$S_n^{\#\#}(\omega_1)(x) \xrightarrow{d} \gamma(\sigma_1)$$

for almost all $\omega_1 \in \Omega_1$.

(3) *Let*

$$\xi_{ij} : \Omega_1 \to T, \qquad i, j \in \mathbf{N},$$

be random variables such that

$$\xi_{ij} \stackrel{d}{=} \xi$$

and for all $i \in \mathbf{N}$, ξ_{ij} are independent. Let $\mathbf{E}_1(\mathbf{E}(Y(\xi)^2)^2 < \infty$ and let

$$S_n^{a}(\omega_1)(x) = n^{-1/2} \sum_{j=1}^{n} Y_j(\xi_{nj}(\omega_1)).$$

Then, as $n \to \infty$,

$$S_n^{a}(\omega_1)(x) \xrightarrow{d} \gamma(\sigma_1)$$

for almost all $\omega_1 \in \Omega_1$.

COROLLARY 3. *Let the conditions of Corollary 2 (3) be valid. Let $k_n : \Omega \to \mathbf{N}, n \in \mathbf{N}$, be integer-valued random variables such that the families $\{k_n\}$ and $\{Y, Y_n\}$ are independent and, as $n \to \infty$, $k_n/n \to C$ in probability, where $0 \le C < \infty$.*

Let

$$S_n(\omega_1) = n^{-1/2} \sum_{i=a_n+1}^{a_n+k_n} Y_i(\xi_{ni}(\omega_1)),$$

where $\omega_1 \in \Omega_1, n \in \mathbf{N}$.

Then, as $n \to \infty$,

$$S_n(\omega_1) \xrightarrow{d} C^{1/2}\gamma(\sigma_1)$$

for almost all $\omega_1 \in \Omega_1$.

Corollary 3 is easily derived from Corollary 2 and Theorem 2.4.2 in [10]. Also, by using the methods of the proof of Theorem 2.4.2 in [10], we obtain the following assertion.

COROLLARY 4. *Let the conditions of Corollary 2 (3) be valid. Let $r_{ni} : \Omega \to \{0, 1\}$, $n, i \in \mathbf{N}$, be random variables such that the families $\{r_{ni}\}$ and $\{Y, Y_n\}$ are independent,*

$$\sum_{i \in \mathbf{N}} r_{ni} < \infty$$

for almost all $\omega \in \Omega$, and, as $n \to \infty$,

$$\frac{1}{n} \sum_{i \in \mathbf{N}} r_{ni} \to C$$

in probability, where $0 \le C < \infty$.

Let

$$S_n(\omega_1) = n^{-1/2} \sum_{i \in \mathbf{N}} r_{ni} Y_i(\xi_{ni}(\omega_1)), \quad \omega_1 \in \Omega_1, \quad n \in \mathbf{N}.$$

Then, as $n \to \infty$,

$$S_n(\omega_1) \xrightarrow{d} C^{1/2}\gamma(\sigma)$$

for almost all $\omega_1 \in \Omega_1$.

We use Corollary 4 in Section 2.

Let $G(t)$ be the distribution function of $Y(t)$, $t \in T$. We consider the following analogue of empirical processes:

$$F_n(\omega_1) = n^{-1/2} \sum_{k=a_n}^{a_n+n} (\mathbf{I}_{\{X_k(\xi_k(\omega_1)) \le x\}} - G(\xi_k(\omega_1))(x))$$

where $n \in \mathbf{N}, x \in \mathbf{R}, \omega_1 \in \Omega_1$.

THEOREM 3. *Let $G(t)(x) - G(t)(y) \le M(x) - M(y)$ for all $t \in T, x \ge y$, for some bounded increasing function $M: \mathbf{R} \to \mathbf{R}$.*

Then, as $n \to \infty$,

$$F_n(\omega_1) \xrightarrow{d} W_1$$

in the Skorokhod space $D(\mathbf{R})$ for almost all $\omega_1 \in \Omega_1$, where W_1 is a centered Gaussian process defined on $\mathbf{R}$ with the covariance function

$$\mathbf{E}W_1(x)W_1(y) = \mathbf{E}_1(1 - F(\xi)(x))F(\xi)(y), \quad x, y \in \mathbf{R}, \quad y \le x.$$

The proofs of the results of this section coincide with those of the corresponding results in [2, 3] and [4]. The only difference is that instead of the Kolmogorov law of large numbers we use this law in the form given in [6].

REMARK 1. Consider some statistical applications. Corollary 2 to Theorem 2 can be applied to some statistical problems. For example, we examine certain parameters of health for people working at the same factory. With each man, we connect a pair (t, x), where t is the time that the man has worked at the factory and x is the value of the parameter of health of the man.

Consider the stochastic process $X(t)$, $t \in T$, $T \subset \mathbf{R}$, such that for all $t \in T$, $X(t)$ is the random variable characterizing this parameter of health of the man who has worked at the factory during t years. Thus, we have a sample for the stochastic process $X(t)$, $t \in T$, at random time points.

For such samples we can find the confidence interval for an estimate of the expectation and compare such samples using the Student test.

Corollary 2 to Theorem 2 is applicable in the design of experiments (in this case, we choose a distribution of ξ) and in the regression analysis as well.

We may utilize Corollary 2 to Theorem 2 in the empirical Bayes approach to estimating a parameter. In this case, we consider the parameter as a random variable ξ.

Theorem 3 can be applied to obtain analogues to the von Mises test and the Kolmogorov–Smirnov test for such samples.

2. LIMIT THEOREMS FOR Q-SCHEMES

Limit theorems for Q-schemes for independent identically distributed random variables were obtained in [11]. Here, we generalize Rusakov's results to the random variables $Y_i(\xi_{ij}(\omega_1))$. This generalization is based on the application of Corollary 4 to Theorem 2.

DEFINITION 1. Let ξ_i, ξ_{ij}, $i, j \in \mathbf{N}$, be independent copies of the random variable ξ and let Y_i, Y_{ij}, $i, j \in \mathbf{N}$, be independent copies of the stochastic process Y. We assume that $\mathbf{E}Y(t) = 0$ for all $t \in T$ and $\mathbf{E}_1(\mathbf{E}(Y(\xi)^2))^2 < \infty$, and $m \in \mathbf{N}$ is fixed. Let $K \in \mathbf{N}$, $K > m$.

(1) Let $Y_i^{(0)}(\omega_1) = Y_i(\xi_i(\omega_1))$, $1 \le i \le K$.

(2) Let $Y_i^{(n)}(\omega_1)$, $1 \le i \le K$, be the nth row of a Q-scheme. We choose randomly numbers $k_i(n)$, $1 \le i \le m$, and replace $Y_{k_i(n)}^{(n)}(\omega_1)$, $1 \le i \le m$, with $Y_{ni}^{(n)}(\xi_{ni}(\omega_1))$. Thus

we obtain the $(n+1)$th row of the Q-scheme. We denote this random variables as $Y_i^{(n+1)}(\omega_1)$, $1 \le i \le K$.

In this construction, the numbers $k_i(n)$, $1 \le i \le m$, do not depend on n or on $Y_i(t)$, $Y_{ij}(t)$, $t \in T$, $i, j \in \mathbf{N}$.

We consider the stochastic processes

$$Z_K(\omega_1)(x) = Z_K(Y)(\omega_1)(x) = K^{-1/2} \sum_{i=1}^{K} Y_i^{([xK])}(\omega_1),$$

where $0 \le x \le 1$, $K \in \mathbf{N}$, $\omega_1 \in \Omega_1$.

THEOREM 4. *Let* $\mathbf{E}_1(\mathbf{E}(Y(\xi))^2)^2 < \infty$ *and* $\mathbf{E}_1\mathbf{E}(Y(\xi))^2 = \sigma_1^2$. *Then, as* $K \to \infty$,

$$Z_K(\omega_1) \xrightarrow{d} Z_{m,\sigma_1^2}$$

in $D[0,1]$ *for almost all* $\omega_1 \in \Omega_1$, *where* Z_{m,σ_1^2} *is the Ornstein–Uhlenbeck process with the parameters* m, σ_1^2.

PROOF. Let us prove that the finite-dimensional distributions of $Z_k(\omega_1)$ converge weakly to the finite-dimensional distributions of Z_{m,σ_1^2} for almost all $\omega_1 \in \Omega_1$.

Let $0 \le x \le 1$, $x \in \mathbf{Q}$. Then

$$Z_K(\omega_1)(x) = K^{-1/2} \sum_{i=1}^{K+m[xK]} r_{Ki} Y_i'(\xi_{Ki}'(\omega_1)).$$

Here the stochastic processes Y_i' are independent and have the same distributions as Y. The random variables

$$\xi_{Ki}' : \Omega_1 \to T, \quad K, i \in \mathbf{N},$$

are independent and have the same distributions as ξ. The random variables $r_{Ki} : \Omega \to \{0, 1\}$ are such that

$$\sum_{i=1}^{K+m[xK]} r_{Ki} = K$$

for almost all $\omega \in \Omega$ and $\{r_{Ki}\}$, $\{Y_i'\}$ are independent families. Therefore, by Corollary 4 to Theorem 2, as $K \to \infty$,

$$Z_K(\omega_1)(x) \xrightarrow{d} \gamma(\sigma_1)$$

for almost all $\omega_1 \in \Omega_1$.

Let $0 \le x < y \le 1$, $x, y \in \mathbf{Q}$. Let, as above,

$$Z_K(\omega_1)(x) = K^{-1/2} \sum_{i=1}^{K+m[xK]} r_{Ki} Y_i'(\xi_{Ki}'(\omega_1)).$$

As above, we have

$$Z_K(\omega_1)(y) = K^{-1/2}\left(\sum_{i=1}^{K+m[xK]} r'_{Ki} r_{Ki} Y'_i(\xi'_{Ki}(\omega_1)) + \sum_{i=K+m[xK]+1}^{K+m[yK]} r'_{Ki} Y'_i(\xi'_{Ki}(\omega_1))\right),$$

where $r'_{Ki}\colon \Omega \to \{0, 1\}$ are such that

$$\sum_{i=1}^{K+m[yK]} r_{Ki} = K$$

for almost all $\omega \in \Omega$ and $\{r'_{Ki}\}$, $\{r_{Ki}\}$, and $\{Y'_i\}$ are independent families.

Then

$$Z_K(\omega_1)(x) \stackrel{d}{=} S_{1K} + S_{2K}$$

and

$$Z_n(\omega_1)(y) = S_{1K} + S_{3K}, \quad K \in \mathbf{N}, \quad \omega_1 \in \Omega_1,$$

where

$$S_{1K} = K^{-1/2} \sum_{1\le i\le K+m[xK], r_{Ki}=1, r'_{Ki}=1} Y'_i(\xi'_i(\omega_1)),$$

$$S_{2K} = K^{-1/2} \sum_{1\le i\le K+m[sK], r_{Ki}=1, r'_{Ki}=0} Y'_i(\xi'_i(\omega_1)),$$

$$S_{3K} = K^{-1/2} \sum_{i=K+m[xK]+1}^{K+m[yK]} r'_{Ki} Y'_i(\xi'_{Ki}(\omega_1)), \qquad K \in \mathbf{N}.$$

By [9]

$$\frac{1}{K}\sum_{i=K+m[xK]+1}^{K+m[yK]} r'_{Ki} \to 1 - e^{-m(y-x)},$$

in probability as $K \to \infty$. Consequently,

$$\frac{1}{K}\sum_{1\le i\le K+m[sK], r_{Ki}=1, r'_{Ki}=1} r_{Ki} \to e^{-m(y-x)},$$

in probability as $K \to \infty$ and

$$\frac{1}{K}\sum_{1\le i\le K+m[sK],\ r_{Ki}=1,\ r'_{Ki}=0} r_{Ki} \to 1 - e^{-m(y-x)},$$

in probability as $K \to \infty$.

Then by Corollary 4 to Theorem 2, as $K \to \infty$,

$$S_{1K} \stackrel{d}{\to} \gamma(\sigma_1 e^{-m(y-x)/2}),$$

$$S_{2K} \stackrel{d}{\to} \gamma(\sigma_1(1 - e^{-m(y-x)})^{1/2}),$$

$$S_{3K} \stackrel{d}{\to} \sigma_1\gamma(\sigma_1 e^{-m(y-x)/2})$$

for almost all $\omega_1 \in \Omega_1$, and for all $K \in \mathbf{N}$ and for almost all $\omega_1 \in \Omega_1$, S_{1K}, S_{2K}, and S_{3K} are independent. Thus, we see that as $K \to \infty$

$$(S_{1K}, S_{2K}, S_{3K}) \xrightarrow{d} (\gamma(\sigma_1 e^{-m(y-x)/2}), \gamma(\sigma_1(1 - e^{-m(y-x)})^{1/2}), \gamma(\sigma_1(1 - e^{-m(y-x)})^{1/2})),$$

where $\gamma(\sigma_1 e^{-m(y-x)/2})$, $\gamma(\sigma_1(1 - e^{-m(y-x)})^{1/2})$, $\gamma(\sigma_1(1 - e^{-m(y-x)})^{1/2})$ are independent. Therefore, as $n \to \infty$,

$$(Z_n(\omega_1)(x), Z_n(\omega_1)(y)) \xrightarrow{d} (Z_{m,\sigma^2}(x), Z_{m,\sigma^2}(y)),$$

for almost all $\omega_1 \in \Omega_1$.

In the same way as in the proof of Theorem 3.1 in [11] we obtain that, as $K \to \infty$,

$$(Z_K(\omega_1)(x_1), \ldots, Z_K(\omega_1)(x_k)) \xrightarrow{d} (Z_{m,\sigma^2}(x_1), \ldots, Z_{m,\sigma^2}(x_k))$$

for all $0 \le x_i \le 1$, $x_i \in \mathbf{Q}$, $1 \le i \le k$, $k \in \mathbf{N}$, $\omega_1 \in \Omega_1'$, where $\Omega_1' \subset \Omega_1$ and $\mathbf{P}(\Omega_1') = 1$. Since Z_{m,σ_1^2} is continuous, we conclude that for almost all $\omega_1 \in \Omega_1'$, the finite-dimensional distributions of Z_n converges weakly to the finite-dimensional distributions of Z_{m,σ^2}.

Let us prove the tightness of the family of the distributions of $Z_n(\omega_1)$ for almost all $\omega_1 \in \Omega_1$. Let first $Y(t) < M$, $t \in T$, for some $M > 0$. In the same way as in the proof of (3.4) in [11], we obtain

$$\begin{aligned} &\mathbf{E}|Z_K(\omega_1)(x) - Z_K(\omega_1)(y)|^4 \\ &\quad \le K^{-2}(m[K(x-y)]M^4 + 12M^4 m[K(x-y)](m[K(x-y)] - 1) \\ &\quad \le C(x-y)^2 \end{aligned}$$

for all $\omega_1 \in \Omega_1$, $K \in \mathbf{N}$, and $x, y \in [0,1]$ such that $x - y > 1/K$, where $C = C(m, L)$.

Moreover, by the Tchebyshev inequality,

$$\mathbf{P}\{|Z_K(\omega_1)(x + 1/K) - Z_K(\omega_1)(x)| > \varepsilon\} \le \frac{2M^2}{\varepsilon^2 K}$$

for all $\omega_1 \in \Omega_1$, $\varepsilon > 0$, $K \in \mathbf{N}$, and $0 \le x \le 1 - 1/K$.

Thus, by Davydov's theorem (see [5]), as $K \to \infty$,

$$Z_K(\omega_1) \xrightarrow{d} Z_{m,\sigma^2}$$

in $D[0,1]$ for almost all $\omega_1 \in \Omega_1$.

Let Y be the stochastic process with the properties given in Theorem 5 and $M > 0$. Let

$$\begin{aligned} Y^M &= Y^M(t) = Y(t)I_{\{|Y(t)| \le M\}} - \mathbf{E}Y(t)I_{\{|Y(t)| \le M\}}, \quad t \in T, \\ Y^{\{M\}} &= Y - Y^M. \end{aligned}$$

Let $\varepsilon, \delta > 0$. Repeating the proof of (2.4) in [11], we obtain

$$\mathbf{P}\{\sup_{0\le x\le 1} |Z_K(Y)(\omega_1)(x) - Z_K(Y^M)(\omega_1)(x)| > \varepsilon\}$$
$$= \mathbf{P}\{\sup_{0\le x\le 1} |Z_K(Y^{\{M\}})(\omega_1)(x)| > \varepsilon\}$$
$$\le \frac{4}{\varepsilon^2 K}\left(2\sum_{i=1}^{m}\sum_{j=1}^{K}\mathbf{E}(Y_{ij}^{\{M\}}(\xi_{ij}(\omega_1)))^2 + \sum_{j=1}^{K}\mathbf{E}(Y_i^{\{M\}}(\xi_j(\omega_1)))^2\right).$$

By the strong law of large numbers, as $K \to \infty$,

$$\frac{4}{\varepsilon^2 K}\left(2\sum_{i=1}^{m}\sum_{j=1}^{K}\mathbf{E}(Y_{ij}^{\{M\}}(\xi_{ij}(\omega_1)))^2 + \sum_{j=1}^{K}\mathbf{E}(Y_i^{\{M\}}(\xi_j(\omega_1)))^2\right)$$
$$\to \frac{4}{\varepsilon^2}(2m+1)\mathbf{E}\mathbf{E}_1(Y^{\{M\}}(\xi))^2$$

for almost all $\omega_1 \in \Omega_1$. Thus, for almost all $\omega_1 \in \Omega_1$ we can find $K_0 \in \mathbf{N}$ and $M > 0$ such that for $K > K_0$

$$\mathbf{P}\{\sup_{0\le x\le 1} |Z_K(Y)(\omega_1)(x) - Z_K(Y^M)(\omega_1)| > \varepsilon\} < \delta.$$

Repeating the proof of Theorem 3.1 in [11], we prove that the latter implies the tightness of the distribution family of $Z_K(Y)(\omega_1)$, $K \in \mathbf{N}$, for almost all $\omega_1 \in \Omega_1$. By the Prokhorov theorem, the tightness of the distribution family of $Z_K(Y)(\omega_1)$, $K \in \mathbf{N}$, and the weak convergence of the finite-dimensional distributions of $Z_K(Y)(\omega_1)$, $K \in \mathbf{N}$, to the finite-dimensional distributions of Z_{m,σ_1^2} imply that, as $K \to \infty$,

$$Z_K(\omega_1) \xrightarrow{d} Z_{m,\sigma_1^2}$$

in $D[0,1]$. The theorem is thus proved.

We consider the stochastic processes

$$Z'_K(\omega_1)(x) = K^{-3/2}\sum_{i=1}^{[Kx]}\sum_{j=1}^{K} Y_i^{(i)}(\omega_1), \quad 0 \le x \le 1, \quad K \in \mathbf{N}, \quad \omega_1 \in \Omega_1.$$

Such as in [11], we obtain the following assertion.

THEOREM 5. *Let the conditions of Theorem 4 be valid. Then, as* $K \to \infty$,

$$Z'_K(\omega_1)(x) \xrightarrow{d} \int_0^x Z_{m,\sigma^2}(y)\,dy$$

in $D[0,1]$ *for almost all* $\omega_1 \in \Omega_1$.

REFERENCES

1. P. Billingsley, *Convergence of Probability Measures.* Wiley, New York, 1968.
2. A. N. Chuprunov, On the convergence in distribution of sums and maxima of independent random variable with a random parameter. *Liet. Matem. Rink.* (1995) **35**, 52–64.
3. A. N. Chuprunov, On the convergence in distribution of empirical processes defined by independent random processes. *Liet. Matem. Rink.* (1995) **35**, 171–180.
4. A. N. Chuprunov, The convergence of the random lines defined by observations of a stochastic process. In: *Proc. XVII Seminar on Stability Problems of Stochastic Models.* Kazan, 1995.
5. Yu. A. Davydov, Weak convergence of discontinuous processes to continuous ones. In: *Proc. Kolmogorov Semester of Probability Theory and Math. Statistics.* Petersburg, 1994, pp. 15–19.
6. V. F. Gaposhkin, On the law of large numbers for moving expectations of independent random variables. *Math. Notes* (1987) **42**, 124–131 (in Russian).
7. I. I. Gihman and A. V. Skorokhod, *Introduction to the Theory of Stochastic Processes.* Nauka, Moscow, 1977 (In Russian).
8. B. V. Gnedenko and A. N. Kolmogorov, *Limit Theorems for Sums of Independent Random Variables.* Addison–Wesley, Reading, MA, 1954.
9. V. F. Kolchin, B. A. Sevastyanov, and V. P. Chistyakov, *Random Allocations.* Wiley, New York, 1978.
10. V. M. Kruglov and V. Yu. Korolev, *Limit Theorems for Random Sums.* Moscow Univ. Press, Moscow, 1990 (In Russian).
11. O. V. Rusakov, A functional limit theorem for random variables with strong rest dependence. *Theory Probab. Appl.* (1995) **40**, 813–832 (in Russian).

Probabilistic Methods in Discrete Mathematics, pp. 157–173
V. F. Kolchin et al. (Eds.)

FUNCTIONAL LIMIT THEOREMS FOR SUMS OF INDEPENDENT RANDOM VARIABLES WITH REPLACEMENTS

A. N. CHUPRUNOV

Kazan

Abstract — We consider an urn model consisting of two boxes with red and white balls. Each of the boxes contains k balls which are red or white. The colour of each ball is red or white with equal probabilities and does not depend on the colours of the other balls. We fulfil k steps consisting in random replacements of the balls from one box to another. We are interested in some characteristics of the model after k steps. For this purpose, we prove a general functional limit theorem with regard to the replacements.

1. INTRODUCTION

We consider the following urn model consisting of two boxes with red and white balls. Each of the boxes contains k balls which are red or white. The colour of each ball is red or white with equal probabilities and does not depend on the colours of the other balls. We fulfil k steps consisting in random replacements of the balls from one box to another.

We describe the process in detail. At the beginning we have two boxes, box 1 and box 2, containing k balls each. The colour of each ball is red or white with equal probabilities and does not depend on the colours of the other balls. In what follows, the balls which initially are in box 1 (box 2) are said to be of type 1 (respectively, type 2). Thus, we connect a Bernoulli random variable with every ball.

At the first step, for all $1 \leq i \leq k$ we make the following operation. With probability $p_1(k)$, we replace the ith ball from box 1 by the ith ball from box 2. At the steps with numbers j, $1 < j \leq k$, for all $1 \leq i \leq k$ we make the following operation. If the ith ball from box 1 is of type 1, then with probability $p_j(k)$ we replace this ball by the ith ball from box 2. If the ball is of type 2, then we leave it in box 1.

In the same way we make k steps and we need to estimate some characteristics of this model. It is clear that this model is determined by the parameter k and the probabilities $p_j(k)$, $1 \leq j \leq k$.

In this paper we prove several functional limit theorems which can be applied to the model described above, since the random variables connected with the balls of type 1 and 2 are independent Bernoulli random variables. According to Theorem 1, if

$$p_j(k) = 1 - \frac{m}{k}, \qquad 1 \leq j \leq k,$$

where $0 < m < k$, then the probabilities related to this model can be estimated for large k by the corresponding probabilities connected with the random process

$$V_1(x) = e^{-mx}W'(e^{mx}) + W''(1) - e^{-mx}W''(e^{mx}), \qquad 0 \le x \le 1.$$

According to Theorem 2, if

$$p_j(k) = 1 - \frac{m}{k + (j-1)m}, \qquad 1 \le j \le k,$$

then the probabilities related to this model can be estimated for large k by the corresponding probabilities connected with the random process

$$V_2(x) = \frac{1}{1+xm}W'(1+xm) + W''(1) - \frac{1}{1+xm}W''(1+xm), \qquad 0 \le x \le 1.$$

Here W' and W'' are independent Wiener processes.

In fact, we deal with a more general case. Instead of Bernoulli random variables we consider independent identically distributed centered random variables with a finite variance. We prove limit theorems on convergence of random step lines processes defined by sums of such random variables multiplied by independent indicators (this is just the method of replacement) to a sum of two independent weighted Wiener processes.

In [6] by the same method of replacement, some functional limit theorems with the Ornstein–Uhlenbeck type limit process is proved. In [5], another version of the method of replacement is utilized and a functional limit theorem with the Ornstein-Uhlenbeck limit process is proved. Our results are similar to the results in [5] and [6]. However, the methods of the proofs of our results coincide with the methods used by Fazekas and Chuprunov in [6] and differ from the methods of [5]. Our proofs as well as the proofs in [6] are based on Rebolledo's theorem (see [4]) on the convergence in distribution of martingales to a Wiener-type process.

2. THE RESULTS

We denote by $\xrightarrow{d}$ the convergence in distribution, by $\xrightarrow{\mathbf{P}}$ the convergence in probability, and by $\stackrel{d}{=}$, the equality in distribution. We denote by $\gamma(\sigma)$ a Gaussian centered random variable with the variance σ^2. The conditional expectation of X with respect to a σ-algebra $\mathfrak{A}$ is denoted by $\mathbf{E}^{\mathfrak{A}}X$. Let $\mathbf{I}_A$ denote the indicator of an event A.

We denote by $\mathbf{N}$ the set of positive integers, $\mathbf{R}^+ = [0,\infty)$, and $\mathbf{R} = (-\infty,\infty)$, $[c]$ denotes the integral part of a real number c.

Throughout the paper, $m > 0$ is fixed, $A_{ij} = A_{ij}(k)$, $i,j,k \in \mathbf{N}$, $k > m$, are events, and $I_{ij} = I_{ij}(k) = \mathbf{I}_{A_{ij}(k)}$.

To define and study our processes, we need the following conditions and notation.

(Y) Y, Y_i, $i \in \mathbf{N}$, and Y', Y'_i, $i \in \mathbf{N}$, are two sequences of independent identically distributed random variables such that

$$\mathbf{E}Y' = \mathbf{E}Y = 0, \quad \mathbf{E}Y^2 = \sigma^2 < \infty, \quad \mathbf{E}Y'^2 = \sigma_1^2 < \infty,$$

and Y_i, Y'_i, $i \in \mathbf{N}$, are independent random variables.

(I) For any fixed $k \in \mathbf{N}$, Y_i', Y_i, $I_{ij}(k)$, $i,j \in \mathbf{N}$, are independent random variables.

We introduce the random processes. We begin with k summands Y_i, and in each step, randomly, we change the summands. More precisely, let

(X) $$X_0^{(k)} = k^{-1/2}\sum_{i=1}^{k} Y_i,$$

$$X_1^{(k)} = k^{-1/2}\left(\sum_{i=1}^{k} I_{1i}(k)Y_i + \sum_{i=1}^{k}(1 - I_{1i}(k))Y_i'\right),$$

$$\dots$$

$$X_l^{(k)} = k^{-1/2}\left(\sum_{i=1}^{k} I_{li}(k)I_{(l-1)i}(k)\dots I_{1i}(k)Y_i + \sum_{i=1}^{k}(1 - I_{li}(k)I_{(l-1)i}(k)\dots I_{1i}(k))Y_i'\right),$$

$$\dots$$

$$X_k^{(k)} = k^{-1/2}\left(\sum_{i=1}^{k} I_{ki}(k)I_{(k-1)i}(k)\dots I_{1i}(k)Y_i + \sum_{i=1}^{k}(1 - I_{ki}(k)I_{(k-1)i}(k)\dots I_{1i}(k))Y_i'\right).$$

We will consider the random processes

(Z) $$Z_k(x) = X_{[kx]}^{(k)}, \quad x \in [0,1], \quad k \in \mathbf{N}.$$

THEOREM 1. *Suppose that* (Y) *and* (I) *are satisfied and*

$$\mathbf{P}(A_{ij}(k)) = 1 - \frac{m}{k}, \quad i,j \in \mathbf{N}, \quad k > m.$$

Then, as $k \to \infty$,

$$Z_k(x) \xrightarrow{d} e^{-mx}\sigma W'(e^{mx}) + \sigma_1 W''(1) - \sigma_1 e^{-mx} W''(e^{mx})$$

in $D[0,1]$, *where* $Z_k(x)$ *is the process defined by* (X) *and* (Z).

THEOREM 2. *Suppose that* (Y) *and* (I) *are satisfied and*

$$\mathbf{P}(A_{ij}(k)) = 1 - \frac{m}{k + (i-1)m}, \quad i,j \in \mathbf{N}, \quad k > m.$$

Then, as $k \to \infty$,

$$Z_k(x) \xrightarrow{d} \frac{\sigma}{1+xm} W'(1+xm) + \sigma_1 W''(1) - \frac{\sigma_1}{1+xm} W''(1+xm)$$

in $D[0,1]$, *where* $Z_k(x)$ *is the process defined by* (X) *and* (Z).

Observe that Theorem 1 and Theorem 2 are true under a weaker assumption on the common distribution of the coefficients of Y_i and Y_i'. More precisely, consider the following generalizations of Theorem 1 and Theorem 2.

(II) Let $A'_{ij}(k)$, $i,j \in \mathbf{N}$ be events such that Y'_i, Y_i, $I'_{ij}(k)$, $i,j \in \mathbf{N}$, are independent random variables, where $I'_{ij} = I'_{ij}(k) = \mathbf{I}_{A'_{ij}(k)}$.

Let

$$(\mathrm{X}')\quad X_0^{(1k)} = k^{-1/2}\sum_{i=1}^{k} Y_i,$$

$$X_1^{(1k)} = k^{-1/2}\left(\sum_{i=1}^{k} I_{1i}(k)Y_i + \sum_{i=1}^{k}(1 - I'_{1i}(k))Y'_i\right),$$

$$\dots$$

$$X_l^{(1k)} = k^{-1/2}\left(\sum_{i=1}^{k} I_{li}(k)I_{(l-1)i}(k)\dots I_{1i}(k)Y_i + \sum_{i=1}^{k}(1 - I'_{li}(k)I'_{(l-1)i}(k)\dots I'_{1i}(k))Y'_i\right),$$

$$\dots$$

$$X_k^{(1k)} = k^{-1/2}\left(\sum_{i=1}^{k} I_{ki}(k)I_{(k-1)i}(k)\dots I_{1i}(k)Y_i + \sum_{i=1}^{k}(1 - I'_{ki}(k)I'_{(k-1)i}(k)\dots I'_{1i}(k))Y'_i\right).$$

We will consider the random processes

$$(\mathrm{Z}')\qquad Z'_k(x) = X_{[kx]}^{(1k)}, \quad x \in [0,1], \quad k \in \mathbf{N}.$$

THEOREM 3. *Suppose that* (Y), (I), *and* (II) *are satisfied and*

$$\mathsf{P}(A'_{ij}(k)) = \mathsf{P}(A_{ij}(k)) = 1 - \frac{m}{k}, \quad i,j \in \mathbf{N}, \quad k > m.$$

Then, as $k \to \infty$,

$$Z'_k(x) \xrightarrow{d} e^{-mx}\sigma W'(e^{mx}) + \sigma_1 W''(1) - \sigma_1 e^{-mx} W''(e^{mx})$$

in $D[0,1]$, *where* $Z'_k(x)$ *is the process defined by* (X′) *and* (Z′).

THEOREM 4. *Suppose that* (Y), (I), *and* (II) *are satisfied and*

$$\mathsf{P}(A'_{ij}(k)) = \mathsf{P}(A_{ij}(k)) = 1 - \frac{m}{k + (i-1)m}, \quad i,j \in \mathbf{N}, \quad k > m.$$

Then, as $k \to \infty$,

$$Z'_k(x) \xrightarrow{d} \frac{\sigma}{1+xm} W'(1+xm) + \sigma_1 W''(1) - \frac{\sigma_1}{1+xm} W''(1+xm)$$

in $D[0,1]$, *where* $Z'_k(x)$ *is the process defined by* (X′) *and* (Z′).

REMARK 1. Observe that Theorem 1 and Theorem 2 may be considered as functional limit theorems for random step lines defined by random sums of independent random variables.

We suppose that $\sum_{i\in\varnothing} \xi_i = 0$, where ξ_i, $i \in \mathbf{N}$, are random variables. Let

$$N_{1k}(l) = \sum_{i=1}^{k} I_{li}(k) I_{(l-1)i}(k) \dots I_{1i}(k),$$

and

$$N_{2k}(l) = k - \sum_{i=1}^{k} I_{li}(k) I_{(l-1)i}(k) \dots I_{1i}(k), \qquad 1 \le l \le k,$$

be random indices of summation. We set

$$\tilde{X}_l^{(k)} = k^{-1/2} \left(\sum_{i=1}^{N_{1k}(l)} Y_i + \sum_{i=1}^{N_{2k}(l)} Y_i' \right), \qquad 0 \le l \le k.$$

Then the finite-dimensional distributions of the stochastic processes

$$\tilde{Z}_k(x) = \tilde{X}_{[kx]}^{(k)}, \qquad 0 \le x \le 1,$$

coincide with the corresponding finite-dimensional distributions of $Z_k(x)$. The sums in the definition of $\tilde{X}_l^{(k)}$ are with random indices of summation. Thus, Theorem 1 and Theorem 2 may be considered as a functional limit theorem for random step lines defined by random sums. It may be shown in the same way that Theorem 3 and 4 may also be considered as functional limit theorems for random step lines defined by random sums. Note that limit theorems on random sums can be found in [7].

3. PROOFS

To prove Theorem 1, we need the following auxiliary lemma. Let

(X″)
$$X_0^{(2k)} = k^{-1/2} \sum_{i=1}^{k} Y_i,$$

$$X_1^{(2k)} = k^{-1/2} \sum_{i=1}^{k} I_{1i}(k) Y_i,$$

$$\dots$$

$$X_l^{(2k)} = k^{-1/2} \sum_{i=1}^{k} I_{1i}(k) I_{2i}(k) \dots I_{li}(k) Y_i,$$

$$\dots$$

$$X_k^{(2k)} = k^{-1/2} \sum_{i=1}^{k} I_{1i}(k) I_{2k}(k) \dots I_{ki}(k) Y_i.$$

We will consider the random processes

(Z″)
$$Z_k^{(1)}(Y, x) = Z_k^{(1)}(x) = X_{[kx]}^{(2k)}, \quad x \in [0, 1], \quad k \in \mathbf{N}.$$

LEMMA 1. *Suppose that (Y) and (I) are satisfied and*

$$\mathbf{P}(A_{ij}(k)) = 1 - \frac{m}{k}, \quad i,j \in \mathbf{N}, \quad k > m.$$

Then, as $k \to \infty$,

$$Z_k^{(1)}(x) \xrightarrow{d} \sigma e^{-mx} W'(e^{mx})$$

in $D[0,1]$ and in the L^∞-norm, where $Z_k^{(1)}$ is the process defined by (X″) and (Z″).

For all $0 \le x_1 \le x_2 \le \ldots \le x_n \le 1$, $b_1, b_2 \ldots b_n \in \mathbf{R}$, $n \in \mathbf{N}$,

$$b_1 Z_k'(x_1) + b_2 Z_k'(x_2) + \ldots + b_n Z_k'(x_n)$$

is the sum of the independent random variables

$$V_i = k^{-1/2}(b_1 I_{1i}(k) \ldots I_{[kx_1]i}(k) Y_i + \ldots + b_n I_{1i}(k) \ldots I_{[kx_n]i}(k) Y_i), \qquad 1 \le i \le k,$$

which satisfy the conditions of the Lindeberg–Feller theorem.

The convergence in $D[0,1]$ in Lemma 1 is proved in [6] (see Theorem 1). According to [1], this convergence is true in the L^∞-norm as well. The second assertion of Lemma 1 follows from the proof of Theorem 1 in [6].

PROOF OF THEOREM 1. If in the definition of $Z_k^{(1)}$ the random variables Y_i' are used instead of Y_i, then the random processes $Z_k^{(1)}$ are denoted by $Z_k^{(2)}$. By the first assertion of Lemma 1, as $k \to \infty$,

$$Z_k^{(2)}(x) \xrightarrow{d} \sigma_1 e^{-mx} W'(e^{mx})$$

in $D[0,1]$ and in the L^∞-norm.

Consider the random processes $Z_k^g = Z_k^{(2)}(0)(x)$, $x \in [0,1]$, $k \in \mathbf{N}$. Then, as $k \to \infty$, $Z_k^g \xrightarrow{d} W''(1)\,\mathbf{I}_{[0,1]}$, in $D[0,1]$ and in the L^∞-norn. Therefore, the distribution family of $\{Z_k^g - Z_k^{(2)}, k \in \mathbf{N}\}$ is tight in L^∞. Since the finite-dimensional distributions of $Z_k^{(2)}$ converge to the finite-dimensional distributions of $\sigma_1 e^{-mx} W''(e^{mx})$, it follows that the finite dimensional distributions of $Z_k^g - Z_k^{(2)}$ converge to the finite-dimensional distributions of $\sigma_1 W''(1) - \sigma_1 e^{-mx} W''(e^{mx})$. Therefore, by the Prokhorov theorem, as $k \to \infty$,

$$Z_k^g(x) - Z_k^{(2)}(x) \xrightarrow{d} \sigma_1 W''(1) - \sigma_1 e^{-mx} W''(e^{mx}) \tag{1}$$

in $D[0,1]$ and in the L^∞-norm.

Hence, the distribution family of $Z_k = Z_k^{(1)} + Z_k^g - Z_k^{(2)}$, $k \in \mathbf{N}$, is tight in the L^∞-norm.

Let $0 \le x_1 \le x_2 \le \ldots \le x_n \le 1$, $b_1, b_2, \ldots, b_n \in \mathbf{R}$, $n \in \mathbf{N}$. Then

$$b_1 Z_k(x_1) + b_2 Z_k(x_2) + \ldots + b_n Z_k(x_n)$$

is the sum of the independent random variables

$$\begin{aligned} U_i = k^{-1/2}(&b_1(I_{1i}(k) I_{2i}(k) \ldots I_{[kx_1]i}(k) Y_i + (1 - I_{1i}(k) I_{2i}(k) \ldots I_{[kx_1]i})(k) Y_i') \\ &+ b_2(I_{1i}(k) I_{2i}(k) \ldots I_{[kx_2]i}(k) Y_i + (1 - I_{1i}(k) I_{2i}(k) \ldots I_{[kx_2]i})(k) Y_i') + \ldots \\ &+ b_n(I_{1i}(k) I_{2i}(k) \ldots I_{[kx_n]i}(k) Y_i + (1 - I_{1i}(k) I_{2i}(k) \ldots I_{[kx_n]i})(k) Y_i')), \qquad 1 \le i \le k. \end{aligned}$$

For all $i, j \in \mathbf{N}$, $1 \le i, j \le k$, $0 \le x_1 \le x_2 \le 1$, the random variables $b_1(I_{1i}(k)I_{2i}(k)\ldots I_{[kx_1]i}(k)Y_i$ and $(1-I_{1j}(k)I_{2j}(k)\ldots I_{[kx_2]j})(k))Y_j'$ are orthogonal. So the random variables U_i, $1 \le i \le k$, $k \in \mathbf{N}$, satisfy the conditions of the Lindeberg–Feller theorem. Hence, by the theorem, Lemma 1, and (1), it follows that

$$b_1 Z_k(x_1) + \ldots + b_n Z_k(x_n) \xrightarrow{d} b_1(e^{-mx_1}\sigma W'(e^{mx_1}) + \sigma_1 W''(1) - \sigma_1 e^{-mx_1} W''(e^{mx_1})) + \ldots$$
$$+ b_n(e^{-mx_n}\sigma W'(e^{mx_n}) + \sigma_1 W''(1) - \sigma_1 e^{-mx_n} W''(e^{mx_n})).$$

Therefore, the finite-dimensional distributions of Z_k converge to the finite-dimensional distributions of $W'(e^{mx}) + \sigma_1 W''(1) - \sigma_1 e^{-mx} W''(e^{mx})$. By the Prokhorov theorem, as $k \to \infty$,

$$Z_k(x) \xrightarrow{d} e^{-mx}\sigma W'(e^{mx}) + \sigma_1 W''(1) - \sigma_1 e^{-mx} W''(e^{mx})$$

in $D[0, 1]$.

LEMMA 2. *Suppose that (Y) and (I) are satisfied and*

$$\mathbf{P}(A_{ij}(k)) = 1 - \frac{m}{k + (i-1)m}, \quad i, j \in \mathbf{N}, \quad k > m.$$

Then, as $k \to \infty$,

$$Z_k^{(1)}(x) \xrightarrow{d} \frac{\sigma}{1 + mx} W'(1 + mx)$$

in $D[0, 1]$ and in L^∞-norm, where $Z_k^{(1)}(x)$ is the process defined by (X″) and (Z″).

For all $0 \le x_1 \le x_2 \le \ldots \le x_n \le 1$, $b_1, b_2, \ldots, b_n \in \mathbf{R}$, $n \in \mathbf{N}$,

$$b_1 Z_k'(x_1) + b_2 Z_k'(x_2) + \ldots + b_n Z_k'(x_n)$$

is the sum of the independent random variables

$$k^{-1/2}(b_1 I_{1i}(k)\ldots I_{[kx_1]i}(k)Y_i + \ldots + b_n I_{1i}(k)\ldots I_{[kx_n]i}(k)Y_i), \qquad 1 \le i \le k,$$

which satisfy the conditions of the Lindeberg–Feller theorem.

PROOF. First we assume that $\mathbf{E}Y^4 = c$, where $c < \infty$.

We use the following notation:

$$b_i = b_i(k) = (\mathbf{P}(A_{ij}(k)))^{-1} = \left(1 - \frac{m}{k + (i-1)m}\right)^{-1} = \frac{k + (i-1)m}{k + (i-2)m},$$
$$b_0(k) = b_0 = 1,$$
$$r_{ij} = r_{ij}(k) = b_i(k) I_{ij}(k),$$
$$V_l^{(k)} = b_l b_{(l-1)} \ldots b_1 b_0 X_l^{(2k)}, \qquad 0 \le l \le k.$$

Then

$$\mathbf{E}r_{ij} = 1, \quad \mathbf{E}r_{ij}^2 = b_i, \quad \mathbf{E}r_{ij}^3 = b_i^2, \quad \mathbf{E}r_{ij}^4 = b_i^3,$$
$$\mathbf{E}(r_{ij} - 1)^2 = b_i - 1 = \frac{m}{k + (i-2)m},$$
$$\mathbf{E}(r_{ij})^2(r_{ij} - 1)^2 = b_i(b_i - 1)^2 = \frac{m + (i-1)m}{k + (i-2)m}\left(\frac{m}{k + (i-2)m}\right)^2.$$

Let

$$\mathfrak{A}_{kl} = \mathfrak{A}\{Y_i,\ r_{ji}(k),\ j \le l,\ 1 \le i \le k\}$$

be the σ-algebra generated by the random variables Y_i, $r_{ji}(k), j \le l$, $1 \le i \le k$.

We consider the random processes

$$Z_k^b(x) = V_{[kx]}^{(k)} - V_0^{(k)}, \quad x \in [0,1], \quad k \in \mathbf{N}.$$

We prove that, as $k \to \infty$,

$$Z_k^b(x) \xrightarrow{d} \sigma W'(xm)$$

in $D[0,1]$.

Observe that

$$\mathbf{E}_{kl}^{\mathfrak{A}} V_{l+1}^{(k)} = V_l^{(k)}.$$

Therefore, $\{V_{l+1}^{(k)} - V_l^{(k)},\ 0 \le l < k\}$ is a martingale-difference and

$$Z_k^b(x) = \sum_{j=1}^{[kx]}(V_j^{(k)} - V_{j-1}^{(k)})$$

is a martingale. As in [4], we denote the quadratic variation by $[\cdot]$:

$$[Z_k^b(x)] = \sum_{j=1}^{[kx]}(V_j^{(k)} - (V_{j-1}^{(k)})^2), \qquad x \in [0,1].$$

We show that, as $k \to \infty$,

$$[Z_k^b(x)] \xrightarrow{\mathbf{P}} \sigma^2 mx \tag{2}$$

for all $x \in [0,1]$. Observe that

$$V_l^{(k)} - V_{l-1}^{(k)} = k^{-1/2}\sum_{i=1}^{k}(r_{li} - 1)r_{(l-1)i}\ldots r_{1i}Y_i.$$

Since the expectation of the product of every two different summands in the expression above is zero, we have

$$\begin{aligned}\mathbf{E}(V_l^{(k)} - V_{l-1}^{(k)})^2 &= \frac{1}{k}\sum_{i=1}^{k}\mathbf{E}((r_{li}-1)r_{(l-1)i}\dots r_{1i}Y_i)^2\\ &= \frac{1}{k}\sum_{i=1}^{k}\frac{m}{k+(l-2)m}\,\frac{k+(l-2)m}{k+(l-3)m}\cdots\frac{k}{k-m}\sigma^2\\ &= \frac{\sigma^2}{k}\frac{km}{k-m}.\end{aligned}$$

Hence,

$$\lim_{k\to\infty}\mathbf{E}[Z_k^b(x)] = \lim_{k\to\infty}\frac{1}{k}\sum_{l=1}^{[kx]}\sigma^2\frac{km}{k-m} = \sigma^2 mx \tag{3}$$

for all $x \in [0,1]$. We consider

$$[Z_k^b(x)] - \mathbf{E}[Z_k^b(x)] = B + C,$$

where

$$B = \frac{1}{k}\sum_{j=1}^{[kx]}\sum_{i=1}^{k}\left(((r_{ji}-1)r_{(j-1)i}\dots r_{1i}Y_i)^2 - \frac{m\sigma^2}{k-m}\right),$$

$$C = \frac{1}{k}\sum_{j=1}^{[kx]}\sum_{\substack{i_1,i_2=1\\ i_1\neq i_2}}^{k}((r_{ji_1}-1)r_{(j-1)i_1}\dots r_{1i_1}Y_{i_1}(r_{ji_2}-1)r_{(j-1)i_2}\dots r_{1i_2}Y_{i_2}).$$

Since the expectation of the product of every two summands in B with different indices i is zero, we have

$$\begin{aligned}0 \le \mathbf{E}B^2 &= \frac{1}{k^2}\sum_{j=1}^{[kx]}\sum_{i=1}^{k}\mathbf{E}\left(((r_{ji}-1)r_{(j-1)i}\dots r_{1i}Y_i)^2 - \frac{m\sigma^2}{k-m}\right)^2\\ &\quad + \frac{2}{k^2}\sum_{\substack{j_1,j_2=1\\ j_1>j_2}}^{[kx]}\sum_{i=1}^{k}\mathbf{E}\left(((r_{j_1i}-1)r_{(j_1-1)i}\dots r_{1i}(r_{j_2i}-1)r_{(j_2-1)i}\dots r_{1i}Y_i^2)^2 - \left(\frac{m\sigma^2}{k-m}\right)^2\right)\\ &\le \frac{1}{k^2}\sum_{i=1}^{[kx]}k\mathbf{E}((r_{ji}-1)r_{(j-1)i}\dots r_{1i}Y_i)^4\\ &\quad + \frac{2}{k^2}\sum_{\substack{j_1,j_2=1\\ j_1>j_2}}^{[kx]}\sum_{i=1}^{k}\mathbf{E}((r_{j_1i}-1)r_{(j_1-1)i}\dots r_{1i}(r_{j_2i}-1)r_{(j_2-1)i}\dots r_{1i}Y_i^2)^2\\ &\le \Sigma_1 + \Sigma_2,\end{aligned}$$

where

$$\Sigma_1 = \frac{1}{k}\sum_{j=1}^{[kx]} \mathbf{E}((r_{ji}-1)r_{(j-1)i}\ldots r_{1i}Y_i)^4,$$

$$\Sigma_2 = \frac{2}{k^2}\sum_{\substack{j_1 j_2=1\\ j_1>j_2}}^{[kx]} \sum_{i=1}^{k} \mathbf{E}((r_{j_1 i}-1)r_{(j_1-1)i}\ldots r_{1i}(r_{j_2 i}-1)r_{(j_2-1)i}\ldots r_{1i}Y_i^2)^2.$$

The moments of r_{ij} have been calculated. Therefore, we obtain

$$\begin{aligned}
\Sigma_1 &= \frac{1}{k}\sum_{j=1}^{[kx]}(b_j^3 - 4b_j^2 + 6b_j - 4 + 1)b_{j-1}^3 b_{j-2}^3\ldots b_2^3 b_1^3 c \\
&= \frac{c}{k}\sum_{j=1}^{[kx]}(b_j - 1)(b_j^2 - 3b_j + 3)b_{j-1}^3 b_{j-2}^3\ldots b_2^3 b_1^3 \\
&= \frac{c}{k}\sum_{j=1}^{[kx]}\frac{m}{k+(j-2)m}\left(\left(1+\frac{m}{k+(j-2)m}\right)^2 - 3\left(1+\frac{m}{k+(j-2)m}\right)+3\right) \\
&\quad\times\left(\frac{k+(j-2)m}{k+(j-3)m}\right)^3\left(\frac{k+(j-3)m}{k+(j-4)m}\right)^3\cdots\left(\frac{k+m}{k}\right)^3\left(\frac{k}{k-m}\right)^3 \\
&\le \frac{c}{k}\sum_{j=1}^{[kx]}\frac{m}{k+(j-2)m}\left(\left(1+\frac{m}{k-m}\right)^2 - 3\left(1+\frac{m}{k-m}\right)+3\right)\frac{(k+(j-2)m)^2}{(k-m)^3} \\
&\le \frac{cm[kx]}{k}\left(\left(1+\frac{m}{k-m}\right)^2+3\right)\frac{(k+km)^2}{(k-m)^3}
\end{aligned}$$

and

$$\begin{aligned}
\Sigma_2 &= \frac{2c}{k}\sum_{\substack{j_1 j_2=1\\ j_1>j_2}}^{[kx]}(b_{j_1}-1)b_{j_1-1}b_{j_1-2}\ldots b_{j_2+1}(b_{j_2}-1)^2 b_{j_2} b_{j_2-1}^3\ldots b_2^3 b_1^3 \\
&= \frac{2c}{k}\sum_{\substack{j_1 j_2=1\\ j_1>j_2}}^{[kx]}\frac{m}{k+(j_1-2)m}\left(\frac{k+(j_1-2)m}{k+(j_1-3)m}\right) \\
&\quad\times\left(\frac{k+(j_1-3)m}{k+(j_1-4)m}\right)\cdots\left(\frac{k+j_2 m}{k+(j_2-1)m}\right)\left(\frac{m}{(k+(j_2-2)m}\right)^2 \\
&\quad\times\left(\frac{k+(j_2-1)m}{k+(j_2-2)m}\right)\left(\frac{k+(j_2-2)m}{k+(j_2-3)m}\right)^3\cdots\left(\frac{k+m}{k}\right)^3\left(\frac{k}{k-m}\right)^3 \\
&= \frac{2c}{k}\sum_{\substack{j_1 j_2=1\\ j_1>j_2}}^{[kx]}\frac{m}{k+(j_2-1)m}\left(\frac{m}{k+(j_2-2)m}\right)^2\left(\frac{k+(j_2-1)m}{k+(j_2-2)m}\right)\left(\frac{k+(j_2-2)m}{k-m}\right)^3
\end{aligned}$$

$$\leq \frac{2c}{k}k^2\left(\frac{m}{k-m}\right)^3\left(\frac{k+km}{k-m}\right)^4.$$

It is not difficult to see that the right-hand sides of the inequalities above tend to zero as $k \to \infty$. Thus, we obtain that for all $x \in [0, 1]$, as $k \to \infty$,

$$\Sigma_1, \Sigma_2 \to 0. \tag{4}$$

Hence, for all $x \in [0, 1]$,

$$\mathbf{E}B^2 \to 0.$$

Using the fact that the expectation of two summands in C having different indices either i or j is zero, we have that for all $x \in [0, 1]$, as $k \to \infty$,

$$\begin{aligned}
\mathbf{E}C^2 &= \frac{1}{k}\sum_{j=1}^{[kx]}\sum_{\substack{i_1,i_2=1\\ i_1\neq i_2}}^{k}\mathbf{E}((r_{ji_1}-1)r_{(j-1)i_1}\ldots r_{1i_1}Y_{i_1}(r_{ji_2}-1)r_{(j-1)i_2}\ldots r_{1i_2}Y_{i_2})^2\\
&= \frac{1}{k^2}\sum_{j=1}^{[kx]}\sum_{\substack{i_1,i_2=1\\ i_1\neq i_2}}^{k}\left(\frac{m\sigma^2}{k-m}\right)^2\\
&= \frac{k-1}{k}[kx]\left(\frac{m\sigma^2}{k-m}\right)^2.
\end{aligned}$$

By virtue of Tchebyshev's inequality, the calculations above imply that for all $x \in [0, 1]$, as $k \to \infty$,

$$[Z_k^b(x)] - \mathbf{E}[Z_k^b(x)] \xrightarrow{\mathbf{P}} 0. \tag{5}$$

From (3) and (5), we arrive at (2).

Let $\varepsilon > 0$. Then we obtain

$$\begin{aligned}
&\mathbf{E}\sum_{j=1}^{[kx]}\mathbf{E}^{\mathfrak{A}_{k(j-1)}}|V_j^{(k)} - V_{j-1}^{(k)}|\,\mathbf{I}_{\{|V_j^{(k)}-V_{j-1}^{(k)}|>\varepsilon\}}\\
&\qquad\qquad \leq \frac{1}{\varepsilon^3}\sum_{j=1}^{[kx]}\mathbf{E}|V_j^{(k)} - V_{j-1}^{(k)}|^4\\
&\qquad\qquad \leq \frac{1}{\varepsilon^3 k^2}\sum_{l=1}^{[kx]}\mathbf{E}\left(\sum_{i=1}^{k}(r_{li}-1)r_{(l-1)i}\ldots r_{1i}Y_i\right)^4.
\end{aligned} \tag{6}$$

By virtue of (4) the right hand side of (6) converges to zero as $k \to \infty$. Therefore, for all $x \in [0, 1]$, as $k \to \infty$,

$$\sum_{j=1}^{[kx]}\mathbf{E}^{\mathfrak{A}_{k(j-1)}}|V_j^{(k)} - V_{j-1}^{(k)}|\,\mathbf{I}_{\{|V_j^{(k)}-V_{(j-1)}^{(k)}|>\varepsilon\}} \xrightarrow{\mathbf{P}} 0, \tag{7}$$

In accordance with [4], it follows from (2) and (7) (see Remark 3 below) that $Z_k^b \xrightarrow{d} W(\sigma^2 mx)$ in $D[0,1]$ as $k \to \infty$.

Denote by M_{1k}, $k \in \mathbf{N}$, the random elements in $D[0,1]$ determined by Z_k^b and set $M_{2k} = V_0^{(k)} \mathbf{I}_{[0,1]}$, $k \in \mathbf{N}$. The sequence of random elements M_{2k} converges in distribution to $\gamma(\sigma) \mathbf{I}_{[0,1]}$ in $D[0,1]$. Let $M_k = M_{1k} + M_{2k}$. Since M_{1k} and M_{2k} converge, the set of distributions of M_k is tight. Since M_k has a finite set of jump points, by [1] (see [1], Chapter 3, Section 14) the set of distributions of M_k is tight in L^∞-norm.

Now we consider the case where $\mathbf{E}Y^4$ may be infinite. Let $D > 0$. We set

$$Y^D = Y \mathbf{I}_{\{|Y| \le D\}} - \mathbf{E} Y \mathbf{I}_{\{|Y| \le D\}}, \qquad Y^{\{D\}} = Y - Y^D.$$

We denote by $M_k(Y^a)$ the random element obtained by replacing Y by Y^a in the definition of M_k. Let $\varepsilon > 0$. By Tchebyshev's inequality and the martingale inequality [2],

$$\begin{aligned}
\mathbf{P}\left\{\sup_{0 \le x \le 1} |M_k(x) - M_k(Y^D)(x)| > \varepsilon\right\} &= \mathbf{P}\left\{\sup_{0 \le x \le 1} |M_k(Y^{\{D\}})(Y^{\{D\}})(x)| > \varepsilon\right\} \\
&\le \frac{1}{\varepsilon^2} \mathbf{E} \sup_{0 \le x \le 1} \left|M_k(Y^{\{D\}})(x)\right|^2 \\
&\le \frac{4}{\varepsilon^2} \mathbf{E}(M_k(Y^{\{D\}})(1))^2 \\
&\le \frac{4}{\varepsilon^2} b_k(k) b_{k-1}(k) \ldots b_2(k) b_1(k) \mathbf{E}(Y^{\{D\}})^2 \\
&\le \frac{4}{\varepsilon^2} \frac{k + mk}{k - m} \mathbf{E}(Y^{\{D\}})^2.
\end{aligned}$$

Therefore, as $D \to \infty$,

$$\sup_{k \in \mathbf{N}} \mathbf{P}\left\{\sup_{0 \le x \le 1} |M_k(x) - M_k(Y^D)(x)| > \varepsilon\right\} \to 0. \tag{8}$$

According to [3] (see Remark 2 below) the tightness of the distribution set of $M_k(Y^D)$, $k \in \mathbf{N}$, in L^∞-norm and (8) imply the tightness of the distribution set of M_k, $k \in \mathbf{N}$, in the L^∞-norm. Hence, the set of the distributions of M_k, $k \in \mathbf{N}$, is tight in $D[0,1]$. We set

$$Z_k^a(x) = Z_k^b(x) + V_0^{(k)}, \quad x \in [0,1], \quad k \in \mathbf{N}.$$

Observe that Z_k^a has the same distribution as M_k.

Let us prove that the finite dimensional distributions of Z_k^a converge to the finite dimensional distributions of $W_1(x) = \sigma W'(1 + xm)$. Let $0 < x_1 < x_2 \le 1$. We show that, as $k \to \infty$,

$$\begin{aligned}
(Z_k^a(0), Z_k^a(x_1) - Z_k^a(0), Z_k^a(x_2) - Z_k^a(x_1)) &\xrightarrow{d} \\
&(W_1(0), W_1(x_1) - W_1(0), W_1(x_2) - W_1(x_1)).
\end{aligned}$$

Let $y_1, y_2 \in \mathbf{R}$. Then $Z_k^a(0) + y_1(Z_k^a(x_1) - Z_k^a(0)) + y_2(Z_k^a(x_2) - Z_k^a(x_1))$ is the sum of k centered independent random variables. Using the same argument as in the proof of (3), for these random variables we obtain the first condition of the Lindeberg–Feller theorem, namely, as $k \to \infty$,

$$\mathbf{E}(Z_k^a(0) + y_1(Z_k^a(x_1) - Z_k^a(0)) + y_2(Z_k^a(x_2) - Z_k^a(x_1)))^2 \to \\ \mathbf{E}(W_1(0))^2 + y_1^2\mathbf{E}(W_1(x_1) - W_1(0))^2 + y_2^2\mathbf{E}(W_1(x_2) - W_1(x_1))^2.$$

Moreover, as in the proof of (7), for these random variables we obtain the second condition of the Lindeberg–Feller theorem. Therefore, by the Lindeberg–Feller theorem, as $k \to \infty$,

$$Z_k^a(0) + y_1(Z_k^a(x_1) - Z_k^a(0)) + y_2(Z_k^a(x_2) - Z_k^a(x_1)) \xrightarrow{d} \\ W_1(0) + y_1(W_1(x_1) - W_1(0)) + y_2(W_1(x_2) - W_1(x_1)).$$

Thus, as $k \to \infty$,

$$(Z_k^a(0), Z_k^a(x_1) - Z_k^a(0), Z_k^a(x_2) - Z_k^a(x_1)) \xrightarrow{d} \\ (W_1(0), W_1(x_1) - W_1(0), W_1(x_2) - W_1(x_1)).$$

Using three or more points in [0, 1] in the same way we prove that the finite dimensional distributions of Z_k^a converge to the finite dimensional distributions of W_1.

By the Prokhorov theorem (see [1], Chapter 1, Section 6), as $k \to \infty$,

$$Z_k^a \xrightarrow{d} W_1$$

in $D[0, 1]$. Observe that

$$Z_k(x) = (b_{[kx]}b_{[kx]-1}\dots b_2 b_1)^{-1} Z_k^a(x), \quad k \in \mathbf{N},$$

and, as $k \to \infty$,

$$(b_{[kx]}b_{[kx]-1}\dots b_2 b_1)^{-1} = \left(\frac{k + ([kx] - 1)m}{k - m}\right)^{-1} \to \frac{1}{1 + xm}$$

uniformly in $x \in [0, 1]$. Consequently, as $k \to \infty$,

$$Z_k(x) \xrightarrow{d} \frac{\sigma}{1 + mx} W(1 + mx)$$

in $D[0, 1]$, and the second assertion of Lemma 2 is true as well. The proof is complete.

PROOF OF THEOREM 2. We obtain the proof of Theorem 2 if we substitute Lemma 2 for Lemma 1 in the proof of Theorem 1.

PROOF OF THEOREM 3. If in the definition of $Z_k^g - Z_k^{(2)}$ we use $I'_{ij}(k)$ instead of $I_{ij}(k)$, then we denote the process $Z_k^g - Z_k^{(2)}$ obtained by $Z_k^{(3)}$. From the proof of Theorem 1, it follows that, as $k \to \infty$,

$$Z_k^{(3)}(x) \xrightarrow{d} \sigma_1 W''(0) - \sigma_1 e^{-mx} W'(e^{mx}),$$

in $D[0,1]$ and in the L^∞-norm. Therefore, by the first assertion of Lemma 1, the distributions family of $\{Z'_k = Z_k^{(1)} + Z_k^{(3)}, k \in \mathbf{N}\}$ is tight in the L^∞-norm and in $D[0,1]$.

We show that the finite-dimensional distributions of Z'_k converge to the finite-dimensional distributions of

$$W'(e^{mx}) + \sigma_1 W''(1) - \sigma_1 e^{-mx} W''(e^{mx}).$$

We denote by φ and φ' the characteristic functions of Y and Y' respectively. Then

$$\varphi(t) = -\exp\{-\sigma^2 t^2/2 + o(t)\}, \qquad t \in \mathbf{R}, \tag{9}$$

and

$$\varphi'(t) = \exp\{-\sigma_1^2 t^2/2 + o(t^2)), \qquad t \in \mathbf{R}. \tag{10}$$

Let $0 \le x_1 \le x_2 \le \ldots \le x_n \le 1$. We consider the characteristic functional

$$\psi_k(b_1, \ldots, b_n) = \mathbf{E}\exp\{i(b_1 Z'_k(x_1) + \ldots + b_n Z'_k(x_n))\},$$

where $b_1, \ldots, b_n \in \mathbf{R}$. Let

$$b = \sum_{i=1}^{n} |b_i| \neq 0.$$

It is not difficult to see that it follows from (9) and (10) that

$$\begin{aligned}\psi_k(b_1, \ldots, b_n) &= \mathbf{E}\exp\{i(b_1 Z'_k(x_1) + \ldots + b_n Z'_k(x_n))\} \\ &= \mathbf{E}\exp\left\{-\frac{\sigma^2}{2k}\sum_{i=1}^{k}(b_1 I_{1i}(k)\ldots I_{[kx_1]i} + \ldots + b_n I_{1i}(k)\ldots I_{[kx_n]i})^2\right. \\ &\quad \left. - \frac{\sigma_1^2}{2k}\sum_{i=1}^{k}(b_1(1 - I'_{1i}(k)\ldots I'_{[kx_1]i}) + \ldots + b_n(1 - I'_{1i}(k)\ldots I'_{[kx_n]i}))^2 + o(1)\right\}\end{aligned}$$

Observe that

$$\begin{aligned}&\frac{1}{k}\mathbf{E}\left(\sum_{i=1}^{k}(b_1 I_{1i}(k)\ldots I_{[kx_1]i} + \ldots + b_n I_{1i}(k)\ldots I_{[kx_n]i})^2\right. \\ &\qquad\qquad \left. - \mathbf{E}(b_1 I_{1i}(k)\ldots I_{[kx_1]i} + \ldots + b_n I_{1i}(k)\ldots I_{[kx_n]i})^2\right)^2 \\ &\qquad = \frac{1}{k}\mathbf{E}\left((b_1 I_{1i}(k)\ldots I_{[kx_1]i} + \ldots + b_n I_{1i}(k)\ldots I_{[kx_n]i})^2\right. \\ &\qquad\quad \left. - \mathbf{E}(b_1 I_{1i}(k)\ldots I_{[kx_1]i} + \ldots + b_n I_{1i}(k)\ldots I_{[kx_n]i})^2\right)^2 \\ &\qquad\qquad \le \frac{b^4}{k} \to 0, \qquad k \to \infty,\end{aligned} \tag{11}$$

and

$$\frac{1}{k}\mathbf{E}\left(\sum_{i=1}^{k}(b_1(1-I'_{1i}(k)\ldots I'_{[kx_1]i}(k))+\ldots+b_n(1-I'_{1i}(k)\ldots I'_{[kx_n]i}(k)))^2\right.$$

$$\left.-\mathbf{E}(b_1(1-I'_{1i}(k)\ldots I'_{[kx_1]i}(k))+\ldots+b_n(1-I'_{1i}(k)\ldots I'_{[kx_n]i}(k)))^2\right)^2$$

$$=\frac{1}{k}\mathbf{E}\left((b_1(1-I'_{1i}(k)\ldots I'_{[kx_1]i}+\ldots+b_n(1-I'_{1i}(k)\ldots I'_{[kx_n]i}))^2\right.$$

$$\left.-\mathbf{E}(b_1(1-I'_{1i}(k)\ldots I'_{[kx_1]i})+\ldots+b_n(1-I'_{1i}(k)\ldots I'_{[kx_n]i}))^2\right)^2$$

$$\leq\frac{b^4}{k}\to 0,\qquad k\to\infty. \tag{12}$$

In the case $Y'=0$ we denote the characteristic functional $\psi_k(b_1,\ldots,b_n)$ by $\psi_k^{(1)}(b_1,\ldots,b_n)$, and in the case $Y=0$ we denote the characteristic functional $\psi_k(b_1,\ldots,b_n)$ by $\psi_k^{(2)}(b_1,\ldots,b_n)$. Then by (9)–(14), the Lebesgue theorem, and Theorem 1, we have

$$\lim_{k\to\infty}\psi_k(b_1,\ldots,b_n)$$

$$=\lim_{k\to\infty}\exp\left\{-\frac{\sigma^2}{2}\mathbf{E}(b_1I_{1i}(k)\ldots I_{[kx_1]i}+\ldots+b_nI_{1i}(k)\ldots I_{[kx_n]i})^2\right.$$

$$\left.-\frac{\sigma_1^2}{2}\mathbf{E}(b_1(1-I'_{1i}(k)\ldots I'_{[kx_1]i})+\ldots+b_n(1-I'_{1i}(k)\ldots I_{[kx_n]i}))^2\right\}$$

$$=\lim_{k\to\infty}\exp\left\{-\frac{\sigma^2}{2}\mathbf{E}(b_1I_{1i}(k)\ldots I_{[kx_1]i}+\ldots+b_nI_{1i}(k)\ldots I_{[kx_n]i})^2\right\}$$

$$\times\lim_{k\to\infty}\exp\left\{-\frac{\sigma_1^2}{2}\mathbf{E}(b_1(1-I'_{1i}(k)\ldots I'_{[kx_1]i})+\ldots+b_n(1-I'_{1i}(k)\ldots I'_{[kx_n]i}))^2\right\}$$

$$=\lim_{k\to\infty}\psi_k^{(1)}(b_1,\ldots,b_n)\lim_{k\to\infty}\psi_k^{(2)}(b_1,\ldots,b_n)$$

$$=\mathbf{E}\exp\left\{i\left(b_1(e^{-mx_1}\sigma\, W'(e^{mx_1})+\sigma_1W''(1)-\sigma_1e^{-mx_1}W''(e^{mx_1}))\right.\right.$$

$$+b_2(e^{-mx_2}\sigma W'(e^{mx_2})+\sigma_1W''(1)-\sigma_1e^{-mx_2}W''(e^{mx_2}))+\ldots$$

$$\left.\left.+b_n(e^{-mx_n}\sigma W'(e^{mx_n})+\sigma_1W''(1)-\sigma_1e^{-mx_n}W''(e^{mx_n}))\right)\right\}.$$

Therefore, the finite-dimensional distributions of Z'_k converge to the finite-dimensional distributions of

$$W'(e^{mx})+\sigma_1W''(1)-\sigma_1e^{-mx}W''(e^{mx}).$$

By the Prokhorov theorem, as $k\to\infty$,

$$Z'_k(x)\xrightarrow{d}e^{-mx}\sigma W'(e^{mx})+\sigma_1W''(1)-\sigma_1e^{-mx}W''(e^{mx})$$

in $D[0,1]$.

PROOF OF THEOREM 4. We obtain the proof of Theorem 4 if in the proof of Theorem 3 we use Theorem 2 instead of Theorem 1.

For the sake of convenience, we recall some theorems used in the previous proofs.

REMARK 2. Let B be a Banach space and X_n, $n \in \mathbf{N}$, be B-valued random elements. If for any $\varepsilon > 0$ there exist B-valued random elements $Y_{\varepsilon n}$, $n \in \mathbf{N}$, such that the family $\{\mathcal{L}(Y_{\varepsilon n}), n \in \mathbf{N}\}$ is tight and for every n

$$\mathbf{P}\{\|X_n - Y_{\varepsilon n}\| > \varepsilon\} < \varepsilon,$$

then the family $\{\mathcal{L}(X_n), n \in \mathbf{N}\}$ is tight.

Here $\mathcal{L}(X)$ stands for the distribution of the random element X. The proof is straightforward; it can be found in [3].

REMARK 3. Let

$$G_n = \{\mathcal{G}_{n,m},\ m \in \mathbf{N}\}, \qquad n \in \mathbf{N},$$

be a sequence of discrete parameter filtering families of σ-algebras. Let $\{\xi_{n,m},\ n, m \in \mathbf{N}\}$ be a double sequence of random variables such that for all $n \in \mathbf{N}$ the process $S_n(m)$ is a uniformly integrable G_n-martingale, where

$$S_n(m) = \sum_{k=0}^{m} \xi_{n,k}, \quad \xi_{n,0} = 0, \quad n, m \in \mathbf{N}.$$

Let

$$M_n(t) = S_n([nt]), \quad n \in \mathbf{N}, \quad t \in \mathbf{R}^+.$$

If, moreover,

$$\sum_{k=1}^{[nt]} \mathbf{E}^{\mathcal{G}_{n,k-1}}(|\xi_{n,k}| I_{\{|\xi_{n,k}|>\varepsilon\}}) \xrightarrow{\mathbf{P}} 0,$$

as $n \to \infty$, and

$$\sum_{k=1}^{[nt]} \xi_{n,k}^2 \xrightarrow{\mathbf{P}} A(t),$$

as $n \to \infty$, for all $\varepsilon > 0$ and $t \in \mathbf{R}^+$, then, as $n \to \infty$,

$$M_n \xrightarrow{d} W(A(t))$$

in the Skorokhod space $D(\mathbf{R}^+)$. The proof can be found in [4].

REFERENCES

1. P. Billingsley, *Convergence of Probability Measures*. Wiley, New York, 1968.
2. A. N. Shiryaev, *Probability*. Nauka, Moscow, 1980 (in Russian).
3. D. H. Mushtari, *Probabilities and Topologies on Linear Spaces*. Kazan Mathematics Foundation, Kazan, 1996.
4. R. Rebolledo, Central limit theorem for local martingales. *Z. Wahrsch. verw. Gebiete* (1980) **51**, 269–286.
5. O. V. Rusakov, A functional limit theorem for random variables with strong remaining dependence. *Theory Probab. Appl.* (1995) **40**, 813–832 (in Russian).
6. I. Fazekas and A. Chuprunov, Convergence of random step lines to Ornstein–Uhlenbeck type processes. Techn. Report of the Debrecen Univ., No. 24/1996, p. 15.
7. V. M. Kruglov and V. Yu. Korolev, *Limit Theorems for Random Sums*. Moscow Univ. Press, Moscow, 1990 (in Russian).

Probabilistic Methods in Discrete Mathematics, pp. 175–178
V. F. Kolchin et al. (Eds.)

ON THE LIMIT DISTRIBUTION OF THE ASYMMETRY OF RANDOM GRAPHS

O. V. DENISOV

Moscow

Abstract — The asymmetry $A(G)$ of a graph G is the minimal number such that removal or adding such a number of edges transform the graph G to a graph with non-trivial automorphism group. We find the limit (as $n \to \infty$) distribution of $A(G)$ assuming that the uniform distribution is given on the set of all graphs with n vertices.

Let $\mathcal{G}_n$ be the set of all undirected simple graphs with n vertices and S_n be the group of all substitutions of this vertices. The automorphism group of a graph G is the set of all $g \in S_n$ such that $g(G) = G$. Then the asymmetry $A(G)$ of G can be defined as follows:

$$A(G) = \min_{g \in S_n \setminus \{e\}} a(G, g),$$

where $a(G, g)$ is the minimal number of modifications of the graph G, each of which consists of removing or adding an edge, transforming the graph G to a graph H with the property $g(H) = H$.

It is proved in [2, 3] that $A(G) \le (n-1)/2$ for all $G \in \mathcal{G}_n$, but for any constant $c > 1$ for almost all $G \in \mathcal{G}_n$

$$A(G) \ge \frac{n}{2} - c\sqrt{n \ln n}.$$

We assume that any two vertices are connected by an edge with probability 1/2 independently of others pairs of vertices so that the uniform measure is induced on $\mathcal{G}_n$:

$$\mathbf{P}(G) = 2^{-n(n-1)/2}.$$

In [2] it is in fact proved that for almost all G

$$A(G) = \min_{g \in T_n} a(G, g),$$

where T_n is the set of all transpositions from S_n.

For $g = (b, c) \in T_n$

$$a(G, g) = \sum_{d \ne b, c} \eta(g, d), \tag{1}$$

where

$$\eta(g,d) = I(\{\{b,d\} \in G, \{c,d\} \in G\} \cup \{\{b,d\} \notin G, \{c,d\} \notin G\})$$

is the indicator of the event that the edges $\{b,d\}, \{c,d\}$ are simultaneously present or absent in G. It is easy to see that $\mathbf{E}\eta(g,d) = 1/2$, $\mathbf{D}\eta(g,d) = 1/4$, and the random variables $\eta(g_1,d_1)$ and $\eta(g_2,d_2)$ are independent if $g_1 \neq g_2$.

We use the method of moments to prove the following theorem about the limit distribution of the random variable

$$\xi(n) = \sum_{g\in T_n} I(a(G,g) < n/2 - u(n)\sqrt{n}/2),$$

which is equal to the number of transpositions $g \in T_n$ such that the random variable $a(G,g)$ is less than $n/2 - u(n)\sqrt{n}/2$, where $u(n)$ is a specially chosen sequence tending to infinity.

THEOREM 1. *Let*

$$\lim_{n\to\infty} \binom{n}{2} \Phi(-u_\lambda(n)) = \lambda > 0,$$

where

$$\Phi(x) = \frac{1}{\sqrt{2\pi}} \int_{-\infty}^{x} e^{-y^2/2}\, dy.$$

Then for any fixed $k = 0, 1, \ldots$

$$\lim_{n\to\infty} \mathbf{P}\{\xi(n) = k\} = \frac{\lambda^k}{k!} e^{-\lambda}.$$

PROOF. According to Sevastyanov's theorem [5], it is sufficient to prove that for any fixed $r \geq 1$ the asymptotic equality

$$\mathbf{P}\left\{\bigcap_{i=1}^{r} \{a(G,g_i) < n/2 - u_\lambda(n)\sqrt{n}/2\}\right\}$$
$$= \prod_{i=1}^{r} \mathbf{P}\{a(G,g_i) < n/2 - u_\lambda(n)\sqrt{n}/2\}(1 + o(1)) \quad (2)$$

holds uniformly in $\{g_1, \ldots, g_r\} \subset T_n$ as $n \to \infty$.

Let $g_i = g_i^{(n)} = (b_i, c_i)$ for $i = 1, \ldots, r$ and let $m \leq 2r$ be the cardinality of the set

$$M = M^{(n)}(g_1, \ldots, g_r) = \bigcup_{i=1}^{r} \{b_i, c_i\}.$$

Taking (1) into account, the left-hand side of (2) can be represented as

$$\mathbf{P}\left\{\bigcap_{i=1}^{r}\left\{\frac{1}{\sqrt{(n-m)/4}}\sum_{d\notin M}(\eta(g_i,d)-1/2)\right.\right.$$
$$\left.\left. < -u_\lambda(n)\sqrt{n/(n-m)}-\frac{1}{\sqrt{n-m}}\left(\sum_{d\in M\setminus\{b_i,c_i\}}(\eta(g_i,d)-1/2)-1\right)\right\}\right\}$$
$$=\mathbf{P}\left\{\bigcap_{i=1}^{r}\left\{\frac{1}{\sqrt{n-m}}\sum_{d\notin M}(-1)^{\eta(g_i,d)+1} < -u_\lambda(n)(1+O(r/\sqrt{n})\right\}\right\} \qquad (3)$$

uniformly in $\{g_1,\dots,g_r\}\subset T_n$.

In the normalized sum $\sum_{d\notin M}X^{(n)}(d)$ the random vectors $X^{(n)}(d) = ((-1)^{\eta(g_1,d)+1},\dots,(-1)^{\eta(g_r,d)+1})$, $d\notin M$ are independent because the value of the dth vector is defined by presence or absence of edges of the graph G of the form $\{b_i,d\},\{c_i,d\}$, $i=1,\dots,r$, and the sets $\bigcup_{i=1}^{r}\{\{b_i,d\},\{c_i,d\}\}$ are mutually disjoint for distinct $d\notin M$ by the construction of M. The distribution of the random variable $X_i^{(n)}(d)=(-1)^{\eta(g_i,d)+1}$ does not depend on n, d, i and satisfies the Cramer condition. The mathematical expectation of the vector $X^{(n)}(d)$ is equal to zero, and its covariance matrix is the identity matrix. Since, in addition,

$$u_\lambda(n)(1+O(r/\sqrt{n})=O(\ln n)=o((n-m)^{1/6}),$$

all the conditions of the central limit theorem for large deviations [4] are satisfied. Therefore probability (3) can be estimated as

$$\Phi^r(-u_\lambda(n)(1+O(n^{-1/2}))(1+o(1))=\Phi^r(-u_\lambda(n))(1+o(1))$$

uniformly in $\{g_1,\dots,g_r\}\subset T_n$. Hence (2) follows. Theorem is proved.

This theorem and estimates in [3] allow us to derive the limit distribution of $A(G)$.

COROLLARY 1. *For any fixed* x

$$\lim_{n\to\infty}\mathbf{P}\left\{\frac{A(G)-n/2+\sqrt{n\ln n}(1-(\ln(32\pi\ln n))/(8\ln n))}{\sqrt{n/(16\ln n)}}\geq x\right\}=e^{-e^{x}}.$$

PROOF. We set

$$u_x(n)=2\sqrt{\ln n}\left(1-\frac{\ln(32\pi\ln n)+2x}{8\ln n}\right)$$

and use the estimate [3]

$$\mathbf{P}\left\{\min_{g\notin T_n\cup\{e\}}a(G,g)\leq n/2\right\}=o(1).$$

Then we have

$$\mathbf{P}\left\{A(G) < n/2 + \sqrt{n \ln n}\left(1 - \frac{\ln(32\pi \ln n)}{8 \ln n} + \frac{x}{4 \ln n}\right)\right\}$$
$$+ \mathbf{P}\left\{\min_{g \in T_n} a(G, g) < n/2 - \sqrt{n} u_x(n)/2\right\} + o(1)$$
$$= 1 - \mathbf{P}\{\xi(n) = 0\} + o(1)$$

Applying the theorem just proved, we see that, as $n \to \infty$,

$$1 - \mathbf{P}\{\xi(n) = 0\} \to 1 - \exp\left\{-\lim_{n\to\infty} \binom{n}{2} \frac{\exp(-u_x(n)^2/2)}{\sqrt{2\pi} u_x(n)}\right\} = 1 - e^{-e^x}.$$

The corollary is proved.

Note that each hypergraph with n vertices can be represented as a binary function of n variables. Results on the limit distribution of the asymmetry of binary functions are presented in [1].

REFERENCES

1. O. V. Denisov, On the asymmetry of binary functions. In: *Proc. Second All-Russian Conference on Stochastic Methods*. TVP, Moscow, 1995, pp. 45–46 (in Russian).
2. P Erdős and A. Renyi, Asymmetric graphs. *Acta Math. Acad. Sci. Hung.* (1963) **14**, 295–315.
3. P Erdős and J. Spencer, *Probabilistic Methods in Combinatorics*. Academic Press, New York, 1974.
4. B. V. Ryasanov and S. I. Chechota, On the approximation of a random Boolean function by the set of all quadratic forms. *Discrete Math. Appl.* (1995) **5**, 473–490.
5. B. A. Sevastyanov, Limit Poisson law in a scheme of summing dependent random variables. *Theory Probab. Appl.* (1972) **17**, 733–738 (in Russian).

Probabilistic Methods in Discrete Mathematics, pp. 179–188
V. F. Kolchin et al. (Eds.)

THE DISTRIBUTION OF VERTICES IN STRATA OF PLANE PLANTED FOREST

I. A. EGOROVA

Petrozavodsk

Abstract — The distribution of the numbers of vertices in strata of a random plane planted forest which consists of N trees and n non-root vertices is considered for various domains of variation of N, n, t, where t is the strata number.

In [3] a number of theorems describing the limit behaviour of the number of vertices in strata of random forest which consists of N rooted trees and n non-root labelled vertices were obtained as $n \to \infty$. In this article the same problem for the set $M(N, n)$ of all forests consisting of N plane planted trees with n non-root vertices is considered. In [4] Pavlov proved that the number of different forests in the set $M(N,n)$ is equal to

$$\frac{N}{n+N}\binom{2n+N-1}{n}.$$

Let the uniform probability distribution on the set $M(N, n)$ be given and let $\mu(t, N, n)$ be the number of vertices in the tth strata of a random forest from $M(N, n)$.

Let $\xi_1, \xi_2, \dots$ be independent identically distributed random variables such that

$$\mathbf{P}\{\xi_1 = k\} = \left(\frac{1}{2}\right)^{k+1}, \qquad k = 0, 1, 2, \dots \tag{1}$$

In this paper we prove the following assertions.

THEOREM 1. *For any fixed positive integers $N, t, k_1, \dots, k_t$, as $n \to \infty$,*

$$\mathbf{P}\{\mu(1, N, n) = k_1, \dots, \mu(t, N, n) = k_t\}$$
$$\to \mathbf{P}\{1 + \xi_1 + \dots + \xi_{N+1} = k_1\}\mathbf{P}\{1 + \xi_1 + \dots + \xi_{k_1+1} = k_2\}\dots\mathbf{P}\{1 + \xi_1 + \dots + \xi_{k_{t-1}+1} = k_t\}.$$

THEOREM 2. *Let $n, t \to \infty$, $N/\sqrt{n} \to 0$, $N/t \to 0$, $t/\sqrt{n} \to 0$. Then for any fixed $x > 0$*

$$\mathbf{P}\{\mu(t, N, n)/t \le x\} \to 1 - e^{-x} - xe^{-x}.$$

THEOREM 3. *Let $n, t, N \to \infty$, $N/\sqrt{n} \to 0$, $N/t \to \alpha$, where α is a positive constant. Then for any fixed $x > 0$*

$$\mathbf{P}\{\mu(t,N,n)/t \le x\} \to e^{-\alpha} \sum_{k=0}^{\infty} \frac{\alpha^k}{k!} K_{2k+4}(2x),$$

where $K_{2k+4}(2x)$ is the distribution function of the χ^2-distribution with $2k+4$ degrees of freedom at the point $2x$.

THEOREM 4. *Let $n, N \to \infty$, $N/\sqrt{n} \le \gamma < \infty$, $N/t \to \infty$. Then for any fixed x*

$$\mathbf{P}\left\{\frac{\mu(t,N,n) - N}{\sqrt{Nt}} \le x\right\} \to \frac{1}{\sqrt{2\pi}} \int_{-\infty}^{x} e^{-u^2/2} du.$$

THEOREM 5. *Let $n, t \to \infty$, $N/\sqrt{n} \to 0$, $t\sqrt{2/n} \to \beta$, where β is a positive constant. Then for any fixed $x_2 > x_1 > 0$*

$$\mathbf{P}\{x_1 \le \mu(t,N,n)/t \le x_2\} \to F(x_1, x_2),$$

where

$$F(x_1, x_2) = \int_{x_1}^{x_2} \int_0^1 \frac{x}{(1-y)^{3/2}} \exp\left\{-\frac{x^2\beta^2}{8(1-y)}\right\} dG_\beta(x,y) \tag{2}$$

and the distribution function $G_\beta(x,y)$ has the characteristic function

$$\psi(\theta_1, \theta_2) = \left(\frac{\sinh u}{u} - i\theta_1 \left(\frac{2\sinh(u/2)}{u}\right)^2\right)^{-1},$$
$$u = \beta\sqrt{-2i\theta_2}. \tag{3}$$

THEOREM 6. *Let $n, N, t \to \infty$, $N/\sqrt{n} \to \gamma$, $t\sqrt{2/n} \to \beta$, where β and γ are positive constants. Then for any fixed $x_2 > x_1 > 0$*

$$\mathbf{P}\{x_1 \le \mu(t,N,n)/t \le x_2\} \to F_\gamma(x_1, x_2),$$

where

$$F_\gamma(x_1, x_2) = \int_{x_1}^{x_2} \int_0^1 \frac{x\beta(1 - e^{-\gamma\sqrt{2}/\beta})}{\sqrt{2}\gamma(1-y)^{3/2}} \exp\left\{-\frac{x^2\beta^2}{8(1-y)}\right\} dG_{\beta,\gamma}(x,y) \tag{4}$$

and the distribution function $G_{\beta,\gamma}(x,y)$ has the characteristic function

$$\psi_\gamma(\theta_1, \theta_2) = \left(1 - e^{-\gamma\sqrt{2}/\beta}\right)^{-1} e^{-\gamma\sqrt{-i\theta_2}} \left(e^{-\gamma f(\theta_1,\theta_2)\sqrt{2}/\beta} - e^{-\gamma f(\theta_2)\sqrt{2}/\beta}\right), \tag{5}$$
$$f(\theta_2) = ue^{-u}\left(1 - e^{-u}\right)^{-1}, \tag{6}$$
$$f(\theta_1, \theta_2) = e^{-u}\left((-i\theta_1 - u/2)^{-1} + (1 - e^{-u})/u\right)^{-1}. \tag{7}$$

To prove Theorems 1–6, we consider the Galton–Watson homogeneous branching process G beginning with N particles such that the number of offspring of one particle has distribution (1). Let ν_N denote the total progeny of the process G, $\mu_N(t)$ be the number of particles at instant t, $\mu_N^{(r)}(t)$ be the number of particles at instant t having exactly r direct descendants, and let $\nu_N(t) = \mu_N(0) + \ldots + \mu_N(t-1)$.

Let $\mu_r(t, N, n)$ be the number of vertices of height t and of degree r in the random forest from $M(N, n)$, $r = 0, 1, \ldots, n$. Let $M = \|m_r^{(t)}\|$, $r, t = 0, 1, \ldots, n$, denote a matrix composed of non-negative integers. In [4] Pavlov showed that the distributions of the random variables $\mu_r(t, N, n)$ and $\mu_N^{(r)}(t)$ are related in the following way.

LEMMA 1. *The equality*

$$\mathbf{P}\{\|\mu_r(t, N, n)\| = M\} = \mathbf{P}\{\|\mu_N^{(r)}(t)\| = M \mid \nu_N = n + N\}$$

holds.

Since the random variables $\mu(t, N, n)$ and $\mu_N(t)$ are equally expressed by $\mu_r(t, N, n)$ and $\mu_N^{(r)}(t)$, respectively, the following assertion is a consequence of Lemma 1.

LEMMA 2. *For any x and t*

$$\mathbf{P}\{\mu(t, N, n) \leq x\} = \mathbf{P}\{\mu_N(t) \leq x \mid \nu_N = n + N\}.$$

The branching process G introduced can be considered as the union of N independent processes, each of which begins with one particle. Then

$$\mu_N(t) = \mu^{(1)}(t) + \ldots + \mu^{(N)}(t),$$
$$\nu_N = \nu^{(1)} + \ldots + \nu^{(N)},$$
$$\nu_N(t) = \nu^{(1)}(t) + \ldots + \nu^{(N)}(t),$$

where $\mu^{(i)}(t)$ is the number of particles of the tth generation, and $\nu^{(i)}$ is the total progeny of the process which begins with the ith initial particle, $i = 1, \ldots, N$,

$$\nu^{(i)}(t) = \mu^{(i)}(0) + \ldots + \mu^{(i)}(t-1).$$

We will consider the limit behaviour of $\mu_N(t)$ under the condition $\mu_N(t) > 0$. By analogy with Lemmas 4.5.2–4.5.5 from [4], we can arrive at the following results.

LEMMA 3. *Let $t \to \infty$, $N/t \to 0$. Then for any fixed $x > 0$*

$$\mathbf{P}\{\mu_N(t)/t \leq x \mid \mu_N(t) > 0\} \to 1 - e^{-x}.$$

LEMMA 4. *Let $N, t \to \infty$, $N/t \to \alpha$, where α is a positive constant. Then for any fixed $x > 0$ the conditional distribution of $\mu_N(t)/t$ under the condition $\mu_N(t) > 0$ converges weakly to the distribution with the density*

$$\alpha \frac{e^{-(\alpha+x)}}{1 - e^{-\alpha}} \sum_{k=1}^{\infty} \frac{(\alpha x)^{k-1}}{k!\,(k-1)!}.$$

LEMMA 5. *Let $n, t \to \infty$, $N/\sqrt{n} \to 0$, $t\sqrt{2/n} \to \beta$, where β is a positive constant. Then the joint distribution of $(\mu_N(t)/t, \nu_N(t)/n)$ under the condition $\mu_N(t) > 0$ converges weakly to the distribution with the characteristic function* (3).

LEMMA 6. *If $n, N, t \to \infty$, $t\sqrt{2/n} \to \beta$, $N/\sqrt{n} \to \gamma$, where β and γ are positive constants, then the joint distribution of $(\mu_N(t)/t, \nu_N(t)/n)$ under the condition $\mu_N(t) > 0$ converges weakly to the distribution with the characteristic function* (5).

We will consider the limit behaviour of $\mu_N(t)$ under the condition $\nu_N = n + N$. By analogy with Lemma 4.5.6 from [4] we arrive at the following lemma.

LEMMA 7. *For any fixed positive integers $N, t, k_1, \dots, k_t$, as $n \to \infty$,*

$$N\mathbf{P}\{\mu_N(1) = k_1, \dots, \mu_N(t) = k_t \mid \nu_N = n + N\} \to k_t\mathbf{P}\{\mu_N(1) = k_1, \dots, \mu_N(t) = k_t\}.$$

LEMMA 8. *Let $n, t \to \infty$, $N/\sqrt{n} \to 0$, $N/t \to 0$, $t/\sqrt{n} \to 0$. Then for any positive fixed x*

$$\mathbf{P}\{\mu_N(t)/t \le x \mid \nu_N = n + N\} \to 1 - e^{-x} - xe^{-x}.$$

Lemma 8 is proved by analogy with Theorem 2.3.3 from [2], using Lemma 3 and equality

$$\mathbf{P}\{\mu_N(t) > 0\} = (N/t)(1 + o(1)). \tag{8}$$

LEMMA 9. *Let $n, t, N \to \infty$, $N/\sqrt{n} \to 0$, $N/t \to \alpha$, where α is a positive constant. Then for any fixed $x > 0$*

$$\mathbf{P}\{\mu_N(t)/t \le x \mid \nu_N = n + N\} \to e^{-\alpha} \sum_{k=0}^{\infty} \frac{\alpha^k}{k!} K_{2k+4}(2x),$$

where $K_{2k+4}(2x)$ is the distribution function of the χ^2-distribution with $2k + 4$ degrees of freedom at the point $2x$.

PROOF. Lemma 9 is proved by analogy with Theorem 2.3.3 from [2]. Using Lemma 4 and equality

$$\mathbf{P}\{\mu_N(t) > 0\} = 1 - (1 - \mathbf{P}\{\mu^{(1)}(t) > 0\})^N, \tag{9}$$

where for $\mathbf{P}\{\mu^{(1)}(t) > 0\}$ the equality from [2]

$$\mathbf{P}\{\mu^{(1)}(t) > 0\} = (1/t)(1 + o(1)) \tag{10}$$

holds, we obtain

$$\mathbf{P}\{\mu_N(t)/t \le x \mid \nu_N = n + N\} \to e^{-\alpha} \sum_{k=1}^{\infty} \frac{\alpha^{k-1}}{k!\,(k-1)!} \int_0^x e^{-u} u^k du.$$

Since

$$e^{-\alpha}\sum_{k=1}^{\infty}\frac{\alpha^{k-1}}{k!\,(k-1)!}\int_0^x e^{-u}u^k du = e^{-\alpha}\sum_{k=0}^{\infty}\frac{\alpha^k}{k!}K_{2k+4}(2x),$$

we obtain the assertion of Lemma 9.

LEMMA 10. *Let* $n, N \to \infty$, $N/\sqrt{n} \to 0$, $N/t \to \infty$. *Then for any fixed* x

$$\mathbf{P}\left\{\frac{\mu_N(t)-N}{\sqrt{Nt}} \le x \mid \nu_N = n+N\right\} \to \frac{1}{\sqrt{2\pi}}\int_{-\infty}^{x} e^{-u^2/2}du.$$

PROOF. This lemma can be proved by the method described in the proof of Theorem 5.4 from [1]. For arbitrary fixed $x_1 < x_2$ we set

$$k_1 = N + x_1\sqrt{Nt}, \qquad k_2 = N + x_2\sqrt{Nt}.$$

Then

$$\mathbf{P}\left\{x_1 \le \frac{\mu_N(t)-N}{\sqrt{Nt}} \le x_2 \mid \nu_N = n+N\right\}$$
$$= \sum_{k_1\le k\le k_2}\sum_{s=N}^{n+N-k}\frac{\mathbf{P}\{\mu_N(t)=k,\, \nu_N(t)=s\}\,\mathbf{P}\{\nu_k = n+N-s\}}{\mathbf{P}\{\nu_N = n+N\}}. \quad (11)$$

We divide the domain of summation into two parts, $M_1 = \{s\colon N \le s \le n\sqrt{t/N}\}$ and $M_2 = \{s\colon \sqrt{t/N} < s \le n+N-k\}$, and denote the sums over these parts by S_1 and S_2 respectively. Then

$$\mathbf{P}\left\{x_1 \le \frac{\mu_N(t)-N}{\sqrt{Nt}} \le x_2 \mid \nu_N = n+N\right\} = S_1 + S_2.$$

We first consider S_1. It is known [2] that for $n \ge k \ge 0$, $n \ge 1$,

$$\mathbf{P}\{\nu_k = n\} = (k/n)\mathbf{P}\{\xi_1 + \ldots + \xi_n = n-k\}, \quad (12)$$

where $\xi_1, \ldots, \xi_n$ are independent identically distributed random variables with distribution (1). According to the local limit theorem on convergence to the normal law

$$\sigma\sqrt{N}\mathbf{P}\{\xi_1 + \ldots + \xi_N = n\} - \frac{1}{\sqrt{2\pi}}\exp\left\{-\frac{(n-aN)^2}{2\sigma^2 N}\right\} \to 0, \quad (13)$$

where σ^2 and a are the variance and the expectation of ξ_1, respectively, we obtain

$$\mathbf{P}\{\nu_k = n+N-s\} = \frac{N(1+o(1))}{2n\sqrt{\pi n}} \quad (14)$$

and

$$\mathbf{P}\{\nu_N = n + N\} = \frac{N(1 + o(1))}{2n\sqrt{\pi n}}. \tag{15}$$

Hence

$$\begin{aligned} S_1 &= \sum_{k_1 \le k \le k_2} \sum_{s \in M_1} \mathbf{P}\{\mu_N(t) = k, \nu_N(t) = s\}(1 + o(1)) \\ &= (1 + o(1)) \left(\mathbf{P}\{k_1 \le \mu_N(t) \le k_2\} - \mathbf{P}\{k_1 \le \mu_N(t) \le k_2, \nu_N(t) > n\sqrt{t/N}\} \right). \end{aligned} \tag{16}$$

It is easy to see that

$$\mathbf{E}\mu^{(1)}(t) = 1, \quad \mathbf{D}\mu^{(1)}(t) = 2t, \quad \mathbf{E}(\mu^{(1)}(t))^3 \le c_1 t^2;$$

here and below $c, c_1, c_2, \ldots$ are some positive constants. Using these relations, we can check that for the sequence of random variables $(\mu^{(i)}(t) - 1)/\sqrt{2Nt}$, $i = 1, 2, \ldots, N$, Lyapunov's condition is satisfied, i.e.,

$$N\mathbf{E}\left|\frac{\mu^{(1)}(t) - 1}{\sqrt{2Nt}}\right|^3 \le c_2\sqrt{t/N} \to 0.$$

Thus

$$\mathbf{P}\{k_1 \le \mu_N(t) \le k_2\} = \frac{1}{\sqrt{2\pi}} \int_{x_1}^{x_2} e^{-u^2/2} du + o(1). \tag{17}$$

It can be shown that

$$\mathbf{D}\nu^{(1)}(t) = t(t - 1)(2t - 1)/3.$$

Taking into account the Chebyshev inequality, we obtain

$$\begin{aligned} \mathbf{P}\{k_1 \le \mu_N(t) \le k_2, \nu_N(t) > n\sqrt{t/N}\} &\le \mathbf{P}\{\nu_N(t) > n\sqrt{t/N}\} \\ &\le \mathbf{P}\{|\nu_N(t) - Nt| > n\sqrt{t/N} - Nt\} \\ &\le \mathbf{P}\{|\nu_N(t) - Nt| > n\sqrt{t/N}/2\} \\ &\le \frac{4\mathbf{D}\nu_N(t)}{n^2(t/N)} \le c_3\frac{N^2t^2}{n^2} = o(1). \end{aligned} \tag{18}$$

Using (16)–(18), we obtain

$$S_1 = (1 + o(1))\frac{1}{\sqrt{2\pi}} \int_{x_1}^{x_2} e^{-u^2/2} du. \tag{19}$$

Now let us estimate S_2. Let $s = n + N - x$. Consider $\mathbf{P}\{\nu_k = n + N - s\}$. If $x \ge cN^2$, then with the use of (12) and (13) it is easy to show that

$$\mathbf{P}\{\nu_k = n + N - s\} \le c_4/N^2. \tag{20}$$

If $x < cN^2, N/x \to 0$, then, using (12), we obtain

$$\mathbf{P}\{\nu_k = n + N - s\} \le c_5 \frac{k}{x^{3/2}} e^{-k^2/(8x)},$$

since the sum of the geometrically distributed independent random variables has the negative binomial distribution.

Let $x = N^2 y$. Then

$$\mathbf{P}\{\nu_k = n + N - s\} \le c_5 \frac{1}{N^2 y^{3/2}} e^{-1/(8y)}.$$

Let $z = 1/y$. If $y \to 0$, then $z^{3/2} e^{-1/(8y)} \to 0$. Therefore (20) is also true in this case.

Let $x = ck$, where $c > 1$. Using (12) and equality

$$\mathbf{P}\{\xi_1 + \ldots + \xi_m = r\} = \binom{m + r - 1}{r} \frac{1}{2^{m+r}}, \tag{21}$$

we obtain

$$\mathbf{P}\{\nu_k = n + N - s\} \le c_6 (g(c))^k,$$

where the function $g(c)$ is given by

$$g(c) = \left(1 + \frac{1}{2c - 2}\right)^{c-1} \left(1 - \frac{1}{2c}\right)^c, \qquad c > 1.$$

It is easy to show that $g(c) \le 1$ for any c, hence (20) is true.

Let $x = k + m$, where $m \ge 0$. Using (12) and (21), we obtain

$$\mathbf{P}\{\nu_k = n + N - s\} = \mathbf{P}\{\nu_k = m + k\} \le c_6 \frac{(2m + k)^m}{2^{k+2m}} \le \frac{c_7}{N^2}.$$

Therefore (20) is true for $s \in M_2$. Combining (11), (15), (18), and (20), we obtain

$$S_2 \le c_8 \frac{t^2}{N\sqrt{n}}. \tag{22}$$

Using (11), (19), and (22), we obtain the assertion of Lemma 10.

LEMMA 11. *Let $n, t \to \infty$, $N/\sqrt{n} \to 0$, $t\sqrt{2/n} \to \beta$, where β is a positive constant. Then for any fixed $x_2 > x_1 > 0$*

$$\mathbf{P}\{x_1 \le \mu_N(t)/t \le x_2 \mid \nu_N = n + N\} \to F(x_1, x_2),$$

where the distribution function $F(x_1, x_2)$ is given by (2).

PROOF. The proof of this lemma is analogous to the proof of Lemma 4.5.10 from [4]. Let arbitrary positive x_1 and x_2, $x_1 < x_2$, be fixed and

$$k_1 = x_1 t, \qquad k_2 = x_2 t.$$

Then

$$\mathbf{P}\{x_1 \le \frac{\mu_N(t)}{t} \le x_2 \mid \nu_N = n+N\}$$
$$= \sum_{k_1 \le k \le k_2} \sum_{s=N}^{n+N-k} \frac{\mathbf{P}\{\mu_N(t) = k, \nu_N(t) = s\}\mathbf{P}\{\nu_k = n+N-s\}}{\mathbf{P}\{\nu_N = n+N\}}. \tag{23}$$

For any δ (such that $0 < \delta < 1$), we divide the sum over s into the sum from 1 to $(1-\delta)n$ and the sum from $(1-\delta)n$ to $n+N-k$. We denote these sums by S_1 and S_2 respectively. Then

$$\mathbf{P}\{x_1 \le \mu_N(t)/t \le x_2 \mid \nu_N = n+N\} = S_1 + S_2. \tag{24}$$

Consider S_2. Using (12) and (13), we obtain

$$\mathbf{P}\{\nu_k = n+N-s\} = \frac{k}{n-s}\frac{1}{\sqrt{4\pi(n-s)}}\exp\left\{-\frac{k^2}{4(n-s)}\right\}(1+o(1)),$$
$$\mathbf{P}\{\nu_N = n+N\} = \frac{N}{n\sqrt{4\pi n}}(1+o(1)).$$

Let $y = s/n$ and $k = xt$. For x and y such that $x_1 \le x \le x_2$ and $y \le 1-\delta$

$$\frac{\mathbf{P}\{\nu_k = n+N-s\}}{\mathbf{P}\{\nu_N = n+N\}} = \frac{k\exp\left\{-\frac{x^2\beta^2}{8(1-y)}\right\}}{N(1-y)^{3/2}}(1+o(1)), \tag{25}$$

uniformly with respect to the integers k and s. By virtue of (13) and (25),

$$S_1 = \sum_{k_1 \le k \le k_2} \sum_{N \le s \le (1-\delta)n} \mathbf{P}\{\mu_N(t) = k, \nu_N(t) = s \mid \mu_N(t) > 0\}\frac{x}{(1-y)^{3/2}}\exp\left\{-\frac{x^2\beta^2}{8(1-y)}\right\}$$
$$\times(1+o(1)).$$

Let $G_\beta(x, y)$ denote the distribution function with the characteristic function $\psi(\theta_1, \theta_2)$. Using Lemma 5, we obtain

$$S_1 \to \int_{x_1}^{x_2}\int_0^{1-\delta} \frac{x}{(1-y)^{3/2}}\exp\left\{-\frac{x^2\beta^2}{8(1-y)}\right\} dG_\beta(x, y). \tag{26}$$

Now we consider S_2. Combining (12) and (21), we obtain

$$\mathbf{P}\{\nu_k = n+N-s-1\} < \mathbf{P}\{\nu_k = n+N-s\}.$$

For $s > (1-\delta)n$

$$\mathbf{P}\{\nu_k = n + N - s\} \le \frac{c_9}{n\delta^{3/2}} e^{-c_{10}/\delta}.$$

Therefore from (23) and (24) it follows that

$$S_2 \le \frac{c_9}{N\delta^{3/2}} e^{-c_{10}/\delta} \mathbf{P}\{\nu_N(t) > (1-\delta)n\}.$$

By the Chebyshev inequality,

$$\mathbf{P}\{\nu_N(t) > (1-\delta)n\} \le c_{11} N/\sqrt{n}.$$

Then

$$S_2 \le c_{12} e^{-c_{10}/\delta}/\delta^{3/2}. \tag{27}$$

Using (24), (26), and (27) we obtain the assertion of Lemma 11.

LEMMA 12. *Let $n, N, t \to \infty$, $N/\sqrt{n} \to \gamma$, $t\sqrt{2/n} \to \beta$, where β and γ are positive constants. Then for any fixed $x_2 > x_1 > 0$*

$$\mathbf{P}\{x_1 \le \mu_N(t)/t \le x_2 \mid \nu_N = n + N\} \to F_\gamma(x_1, x_2),$$

where the distribution function $F_\gamma(x_1, x_2)$ is given by (4).

By virtue of Lemma 6, the proof of this lemma is analogous to the proof of Lemma 11.
By analogy with Lemma 10, we can prove the following lemma.

LEMMA 13. *Let $n, N \to \infty$, $N/\sqrt{n} \to z$, $N/t \to \infty$, where z is a positive constant. Then for any fixed x*

$$\mathbf{P}\left\{\frac{\mu_N(t) - N}{\sqrt{Nt}} \le x \mid \nu_N = n + N\right\} \to \frac{1}{\sqrt{2\pi}} \int_{-\infty}^{x} e^{-u^2/2} du.$$

Now we can prove Theorems 1–6. Using (1), we obtain

$$\begin{aligned} k_t \mathbf{P}\{\mu_N(1) = k_1, \dots, \mu_N(t) = k_t\} \\ &= k_t \mathbf{P}\{\xi_1 + \dots + \xi_N = k_1\} \mathbf{P}\{\xi_1 + \dots + \xi_{k_1} = k_2\} \times \dots \\ &\quad \times \mathbf{P}\{\xi_1 + \dots + \xi_{k_{t-1}} = k_t\} \\ &= N \mathbf{P}\{1 + \xi_1 + \dots + \xi_{N+1} = k_1\} \mathbf{P}\{1 + \xi_1 + \dots + \xi_{k_1+1} = k_2\} \times \dots \\ &\quad \times \mathbf{P}\{1 + \xi_1 + \dots + \xi_{k_{t-1}+1} = k_t\}. \end{aligned} \tag{28}$$

From Lemmas 2, 7 and (28) we obtain the assertion of Theorem 1.

Theorems 2, 3, 5, and 6 are proved with the use of Lemmas 8, 9, 11, and 12, respectively, and of Lemma 2.

From Lemmas 2, 10, and 13 we obtain the assertion of Theorem 4.

REFERENCES

1. I. B. Kalugin, A class of random mappings. *Proc. Steklov Inst. Math.* (1986) **177**, 79-110.
2. V. F. Kolchin, *Random Mappings*. Springer, New York, 1986.
3. Yu. L. Pavlov, On the distribution of the number of vertices in strata of a random forest. In: *Proc. 1st World Congress Bernulli Society, Vol.* 1. VNU Science Press, Utrecht, 1987, pp. 239–241.
4. Yu. L. Pavlov, *Random Forests*. Karelian Sci. Centre of the Russian Acad. Sci., Petrozavodsk, 1996 (in Russian).

Probabilistic Methods in Discrete Mathematics, pp. 189–198
V. F. Kolchin et al. (Eds.)

THE LIMIT DISTRIBUTION OF THE NUMBERS OF EMPTY CELLS IN THE SCHEME OF ALLOCATING GROUPED PARTICLES TO GROUPED CELLS

N. YU. ENATSKAYA and E. R. KHAKIMULLIN

Moscow

Abstract — We consider a scheme of allocating grouped particles to groups of cells. We prove a theorem on asymptotic normality of the numbers of empty cells in the chosen groups of cells.

1. FORMULATION OF THE PROBLEM

Let N cells be divided into $k+1$ groups and let N_i be the number of cells in the ith group, $i = 1, \dots, k+1$. We choose the first k groups and set

$$M = \sum_{i=1}^{k} N_i.$$

We allocate sequentially n groups of particles. We suppose that the jth group contains m_j particles, these m_j particles are allocated in different cells in such a way that any of $\binom{n}{m_j}$ possible allocations is occupied with equal probabilities, $j = 1, \dots, n$; we assume also that the allocations of these n groups are independent.

We study the asymptotic behaviour, as $N \to \infty$, of the random vector of the numbers of empty cells in the chosen k groups of cells

$$\bar{\mu}_0(n, N_1, \dots, N_k) = \{\mu_0(n, N_i),\ i = 1, \dots, k\},$$

where $\mu_0(n, N_i)$ is the number of empty cells in the ith group of cells after allocating n groups of particles. In particular, $\mu_0(n, M)$ is the total number of empty cells among the M chosen after allocating n groups of particles.

The case where $N = M$, $N \to \infty$, and $k = 1$ was considered in detail in [1], the case where $k = 1$ and $M \le N$ was considered in [3]. The limit theorem for the vector $\bar{\mu}_0(n, N_1, \dots, N_k)$ was proved in [2] in the case where $N \to \infty$, $M = N$, and

$$\frac{N_i}{N} = \beta_i, \quad 0 < c_1 \le \beta_i \le c_2 < \infty, \quad i = 1, \dots, k,$$

where c_1 and c_2 are constants.

In [4, 5] we investigated the case where $N \to \infty$, $M \le N$, $k \ge 1$,

$$\frac{N_i}{N} \to \beta_i, \quad 0 < \beta_i < 1, \quad i = 1, \dots, k, \quad \sum_{i=1}^{k} \beta_i = 1, \quad m_j = m, \quad j = 1, 2, \dots$$

In this paper we consider the case where the sizes of the chosen groups can be of different orders with respect to the total number of cells N, as $N \to \infty$.

2. NOTATION

We introduce the following notation. Let $\mathbf{E}\xi$ and $\mathbf{D}\xi$ be the mathematical expectation and the variance of the random variable ξ, c_i, $i = 1, 2, \dots$, be positive constants,

$$p = \prod_{i=1}^{n} (1 - m_i/N),$$

$$q = 1 - p,$$

$$\varphi(x) = e^{-x^2/2}/\sqrt{2\pi},$$

$$Q = \sqrt{\frac{M}{2\pi}} \left((2\pi pq)^{k-1} \mathbf{D}\mu_0(n, M) \prod_{i=1}^{k} N_i \right)^{-1/2},$$

$$M_l = \sum_{i=1}^{l} N_i, \quad l = 1, \dots, k-1, \quad M_k = M,$$

$$\sigma_1 = \sqrt{M_l pq},$$

$$\sigma_2 = \sqrt{(M - M_l) pq},$$

$$\alpha = \sum_{i=1}^{k} \alpha_i,$$

$$Q_1 = \sqrt{\frac{M}{2\pi}} \left((2\pi pq)^{k-l-1} \mathbf{D}\mu_0(n, M) \prod_{i=l+1}^{k} N_i \prod_{i=1}^{l} \mathbf{D}\mu_0(n, N_i) \right)^{-1/2},$$

$$N(r, s) = \min_{r \le i \le s} N_i, \quad r, s = 1, \dots, k, \quad r \le s.$$

3. RESULTS

THEOREM 1. *Let the integers*

$$\alpha_i = \mathbf{E}\mu_0(n, N_i) + u_i \sqrt{\mathbf{D}\mu_0(n, N_i)}, \quad i = 1, \dots, k,$$

be such that $|u_i| \le c$, *where* c *is a constant, and let*

$$\min_{1 \le i \le k} N_i \to \infty, \qquad \min_{1 \le i \le k} \mathbf{D}\mu_0(n, N_i) \to \infty.$$

Then

$$\mathbf{P}\{\mu_0(n,N_i)=\alpha_i, i=1,\dots,k\} = Q\exp\left\{-\frac{1}{2}\left(\sum_{i=1}^{k}\frac{u_i^2\mathbf{D}\mu_0(n,N_i)}{N_ipq} - \left(\sum_{i=1}^{k}\frac{u_i\sqrt{\mathbf{D}\mu_0(n,N_i)}}{\sqrt{\mathbf{D}\mu_0(n,M)}}\right)^2\left(\frac{\mathbf{D}\mu_0(n,M)}{Mpq}-1\right)\right)\right\} \times\left(1+O\left(\frac{1}{\sqrt{N(1,k)}}+\frac{1}{\sqrt{\mathbf{D}\mu_0(n,M)}}\right)\right). \quad (1)$$

COROLLARY 1. *Let the conditions of Theorem 1 be satisfied, $M_l = o(M)$ for some l, $1 \le l \le k-1$, and $\mathbf{D}\mu_0(n,M) = O(Mpq)$.*

Then

$$\mathbf{P}\{\mu_0(n,N_i)=\alpha_i,\ i=1,\dots,k\} = Q\exp\left\{-\frac{1}{2}\left(\sum_{i=1}^{k}\frac{u_i^2\mathbf{D}\mu_0(n,N_i)}{N_ipq} - \left(\sum_{i=l+1}^{k}\frac{u_i\sqrt{\mathbf{D}\mu_0(n,N_i)}}{\sqrt{\mathbf{D}\mu_0(n,M)}}\right)^2\left(\frac{\mathbf{D}\mu_0(n,M)}{Mpq}-1\right)\right)\right\} \times\left(1+O\left(\sum_{i=1}^{l}\sqrt{\mathbf{D}\mu_0(n,N_i)/\mathbf{D}\mu_0(n,M)}\right)\right). \quad (2)$$

COROLLARY 2. *Let the conditions of Theorem 1 be satisfied, let $N_1 < N_2 < \dots < N_k$, let $N_i = o(N_{i+1})$, $i = 1,\dots,k-1$, and $\mathbf{D}\mu_0(n,M) = O(Mpq)$.*

Then

$$\mathbf{P}\{\mu_0(n,N_i)=\alpha_i,\ i=1,\dots,k\} = Q\exp\left\{-\frac{1}{2}\left(\sum_{i=1}^{k}\frac{u_i^2\mathbf{D}\mu_0(n,N_i)}{N_ipq} + u_k^2\frac{\mathbf{D}\mu_0(n,N_i)}{\mathbf{D}\mu_0(n,M)}\left(\frac{\mathbf{D}\mu_0(n,M)}{Mpq}-1\right)\right)\right\} \times\left(1+O\left(\sum_{i=1}^{k-1}\sqrt{\mathbf{D}\mu_0(n,N_i)/\mathbf{D}\mu_0(n,M)}\right)\right). \quad (3)$$

COROLLARY 3. *Let the conditions of Corollary 1 be satisfied and*

$$\begin{aligned}
\alpha_i &= \mathbf{E}\mu_0(n,N_i) + u_i\sqrt{M_lpq}, && i = 1,\dots,l,\\
\alpha_i &= \mathbf{E}\mu_0(n,N_i) + u_i\sqrt{Mpq}, && i = l+1,\dots,k,\\
\frac{N_i}{M_l} &\to \beta_i^{(1)}, && i = 1,\dots,l,\\
\frac{N_i}{M} &\to \beta_i^{(2)}, && i = l+1,\dots,k,
\end{aligned}$$

where $0 \le \beta_i^{(1)} \le 1$, $i = 1, \dots, l$, *and* $0 \le \beta_i^{(2)} \le 1$, $i = l+1, \dots, k$.

Then

$$\mathbf{P}\{\mu_0(n, N_i) = \alpha_i,\ i = 1, \dots, k\} = Q_1 \exp\left\{ -\frac{1}{2} \left(\sum_{i=1}^{l} \frac{u_i^2}{M_l pq} + \sum_{i=l+1}^{k} \frac{u_i^2 M}{N_i} + \left(\sum_{l+1}^{k} u_i^2 \right)^2 \left(\frac{Mpq}{\mathbf{D}\mu_0(n, M)} - 1 \right) \right) \right\} \times \left(1 + O\left(\frac{1}{\sqrt{N(l+1, k)pq}} \right) \right). \quad (4)$$

4. METHOD OF THE PROOFS

We use the almost obvious representation

$$\mathbf{P}\{\mu_0(n, N_i) = \alpha_i, i = 1, \dots, k\} = \mathbf{P}\{\mu_0(n, M) = \alpha\} \binom{N}{\alpha}^{-1} \prod_{i=1}^{k} \binom{N_i}{\alpha_i}. \quad (5)$$

This representation was used in [4] in the case $M \le N$ and in [2] for $M = N$.

The asymptotic behaviour of the second factor in (5) is analyzed in the same way as it was done in [2]. The first factor is treated by the method based on the idea used in [1], which leads to the representation of the components of the vector $\bar{\mu}_0(n, N_1, \dots, N_k)$ as sums of independent random variable taking the values 0 and 1.

5. AUXILIARY RESULTS

For the sake of convenience we give here in addition to representation (5) two known results which are used in the proofs below.

Let $(x)_j = x(x-1)\dots(x-j+1)$ for positive integers $j = 1, 2, \dots$

LEMMA 1. *For* $j = 1, 2, \dots,$

$$\mathbf{E}(\mu_0(n, M))_j = (M)_j \prod_{i=1}^{n} \binom{N-j}{m_j} \Big/ \binom{N}{m_i}. \quad (6)$$

PROOF. The proof of the lemma repeats word by word the proof of Lemma 1 in [3] with only substitution of

$$\prod_{i=1}^{n} \binom{N-j}{m_j} \Big/ \binom{N}{m_i}$$

for

$$\left(\binom{N-j}{m} \Big/ \binom{N}{m} \right)^n.$$

THEOREM 2. *There exists $C > 0$ such that*

$$\max_{0\le k\le N-m}\left|\sqrt{\mathbf{D}\mu_0(n,M)}\mathbf{P}\{\mu_0(n,M)=k\}-\varphi\left(\frac{k-\mathbf{E}\mu_0(n,M)}{\sqrt{\mathbf{D}\mu_0(n,M)}}\right)\right|\le\frac{C}{\sqrt{\mathbf{D}\mu_0(n,M)}}.$$

The proof of Theorem 2 is based on the following lemmas given below.

LEMMA 2. *The generating function of the random variable $\mu_0(n, M)$ is equal to*

$$\mathbf{E}z^{\mu_0(n,M)} = A^{-1}R(z-1),$$

where

$$A=\prod_{i=1}^{n}(N)_{m_i},$$

$$R(z)=\sum_{j=0}^{s}\binom{M}{j}z^j\prod_{i=1}^{n}(N-j)_{m_i},\qquad s=\min_i(M,N-m_i).$$

PROOF. The assertion is a consequence of (6) and the following chain of equalities:

$$\begin{aligned}\mathbf{E}z^{\mu_0(n,M)} &= \mathbf{E}\sum_{j=0}^{M}\binom{\mu_0(n,M)}{j}(z-1)^j\\ &=\sum_{j=0}^{M}(z-1)^j\binom{M}{j}\prod_{i=1}^{n}\binom{N-j}{m_i}\Big/\binom{N}{m_i}\\ &=\sum_{j=0}^{M}(z-1)^j\binom{M}{j}\prod_{i=1}^{n}\frac{(N-j)_{m_i}}{(N)_{m_i}}\\ &=\sum_{j=0}^{M}(z-1)^j\binom{M}{j}\prod_{i=1}^{n}(N-j)_{m_i}\Big/\prod_{i=1}^{n}(N)_{m_i}\\ &=A^{-1}R(z-1),\end{aligned}$$

since

$$\mathbf{E}\binom{\mu_0(n,M)}{j}=\mathbf{E}(\mu_0(n,M))_j/j!.$$

LEMMA 3. *The polynomial $R(z)$ has only real roots.*

PROOF. The proof is similar to the proof of Lemma 2 in [1]. It is clear that if the roots of a polynomial $f(z)$ of degree n are real, then all the roots of the polynomials

$$z^nf(z^{-1}),\qquad z^jf(z),\quad j=1,2,\ldots,\qquad f'(z)$$

and consequently, of the polynomial

$$D_j f(z) = z^j \frac{d^j}{dz^j} f(z), \qquad j = 1, 2, \ldots,$$

are real as well. To prove the lemma, it suffices to see that

$$D_j z^n = (n)_j z^n,$$
$$D_{m_1} \ldots D_{m_n}(z^{N-M}(1+z)^M) = z^N R(z^{-1}).$$

The first relation is obvious, and the second one follows from the equalities

$$z^{N-M}(1+z)^M = \sum_{j=0}^{M} \binom{M}{j} z^{N-M+j},$$

$$\begin{aligned} D_{m_n}(z^{N-M}(1+z)^M) &= z^{m_n} \frac{d^{m_n}}{dz^{m_n}} \sum_{j=1}^{M} \binom{M}{j} z^{N-M+j} \\ &= z^{m_n} \sum_{j=0}^{M} \binom{M}{j} z^{N-M+j-m_n}(N-M+j)_{m_n} \\ &= \sum_{j=0}^{M} \binom{M}{j} z^{N-M+j}(N-M+j)_{m_n}, \end{aligned}$$

$$\begin{aligned} D_{m_{n-1}} D_{m_n}(z^{N-M}(1+z)^M) &= z^{m_{n-1}} \frac{d^{m_{n-1}}}{dz^{m_{n-1}}} \sum_{j=0}^{M} \binom{M}{j} z^{N-M+j}(N-M+j)_{m_n} \\ &= \sum_{j=0}^{M} \binom{M}{j} z^{N-M+j}(N-M+j)_{m_{n-1}}(N-M+j)_{m_n}, \end{aligned}$$

therefore

$$\begin{aligned} &D_{m_1} D_{m_2} \ldots D_{m_n}(z^{N-M}(1+z)^M) \\ &\quad = \sum_{j=0}^{M} \binom{M}{j} z^{N-M+j} \prod_{i=1}^{n} (N-M+j)_{m_i} \sum_{l=0}^{M} \binom{M}{l} z^{N-l} \prod_{i=1}^{n} (N-l)_{m_i}. \qquad (7) \end{aligned}$$

The lemma is proved.

The specific character of the case $M \leq N$ in comparison with the case $M = N$ considered in [1] has no effect on the proof of Theorem 2. This proof is based on the fact that the distribution of the random variable $\mu_0(n, M)$ coincides with the distribution of a random variable representable as a sum of independent random variables taking the values 0 and 1, which was used in [1], and in our case is presented in Lemmas 2 and 3. The proof is very similar to the corresponding reasoning in [1], and we omit it.

The following theorem is a direct corollary to Theorem 2.

THEOREM 3. *If* $\mathbf{D}m_0(n, M) \to \infty$, *then*

$$\mathbf{P}\{\mu_0(n,M) = \alpha\} = \frac{1}{\sqrt{\mathbf{D}\mu_0(n,M)}}\varphi\left(\frac{\alpha - \mathbf{E}\mu_0(n,M)}{\sqrt{\mathbf{D}\mu_0(n,M)}}\right)\left(1 + O\left(\frac{1}{\sqrt{\mathbf{D}\mu_0(n,M)}}\right)\right). \quad (8)$$

This assertion for $M = N$ is proved in [1].

LEMMA 4. *Under the conditions of Corollary* 3

$$\mathbf{D}\mu_0(n, N_i) = o(\mathbf{D}\mu_0(n,M)), \qquad i = 1, \dots, l. \quad (9)$$

PROOF. Under the conditions of Theorem 1 $\mathbf{D}\mu_0(n, M) = O(Mpq)$. As follows from Lemma 1,

$$\mathbf{E}\mu_0(n,M) = Mp,$$
$$\mathbf{D}\mu_0(n,M) = \mathbf{E}\mu_0(n,M)(V_1 + MV_2),$$

where

$$V_1 = 1 - \prod_{i=1}^{n}(1 - m_i/(N-1)),$$
$$V_2 = \prod_{i=1}^{n}(1 - m_i/(N-1)) - \prod_{i=1}^{n}(1 - m_i/N).$$

Let $N \to \infty$, then, since $V_1 \to q$, we see that $MV_2/V_1 \to c$, where c is a positive constant. Therefore, under the conditions of Corollary 3, $N_iV_2/V_1 \to 0$ for $i = 1, \dots, l$. Thus, relations (9) are proved.

6. PROOFS

PROOF OF THEOREM 1. We use representation (5) and estimate separately the asymptotic behaviour of the distribution of the number of empty cells among M cells and the corresponding hypergeometric probability.

The first factor in (5) is estimated in Theorem 3.

Using the local Moivre–Laplace theorem for estimating the binomial coefficients, we obtain

$$
\begin{aligned}
&\binom{M}{\alpha}^{-1} \prod_{i=1}^{k} \binom{N_i}{\alpha_i} \\
&= \frac{\sqrt{2\pi Mpq} \prod_{i=1}^{k} \exp\{-(\alpha_i - N_i p)^2/(2N_i pq)\}}{(2\pi)^{k/2} \prod_{i=1}^{k} \sqrt{N_i pq} \exp\{-(\alpha - Mp)^2/(Mpq)\}} \\
&\quad \times \left(1 + O\left(\frac{1}{\sqrt{N(1,k)}}\right)\right) \\
&= \prod_{i=1}^{k} \sqrt{N_i} \exp\left\{ -\frac{1}{2} \sum_{i=1}^{k} \frac{u_i^2 \mathbf{D}\mu_0(n, N_i)}{N_i pq} - \frac{1}{2Mpq} \left(\sum_{i=1}^{k} u_i \sqrt{\mathbf{D}\mu_0(n, N_i)} \right)^2 \right\} \qquad (10) \\
&\quad \times \frac{\sqrt{M}}{(2\pi pq)^{(k-1)/2}} \left(1 + O\left(1/\sqrt{N(1,k)}\right)\right).
\end{aligned}
$$

Relation (1) follows from Theorem 3 and (10).

PROOF OF COROLLARIES 1 AND 2. The proofs of the corollaries follow directly from Theorem 1 by taking into account the fact that, as is proved in Lemma 4,

$$\mathbf{D}\mu_0(n, N_i) = o(\mathbf{D}\mu_0(n, M)),$$

for $i = 1, \dots, l$ in Corollary 1 and for $i = 1, \dots, k-1$ in Corollary 2.

Thus, we obtained that under the conditions of Corollary 1 the distributions of the first l components are asymptotically normal and independent of each other and of the other components of the vector $\bar{\mu}_0(n, N_1, \dots, N_k)$, and under the conditions of Corollary 2 all components of this vector are asymptotically normal and independent of each other.

PROOF OF COROLLARY 3. As it was noted, under the conditions of Corollary 1 the first l components of the vector $\bar{\mu}_0(n, N_1, \dots, N_k)$ are asymptotically normal and independent of each other and of the other components. Therefore to prove Corollary 3, we can estimate the probabilities $\mathbf{P}\{\mu_0(n, N_i) = \alpha_i\}$, $i = 1, \dots, l$ and, $\mathbf{P}\{\mu_0(n, N_i), i = l+1, k\}$ separately. Using Theorem 3, we obtain for $i = 1, \dots, l$,

$$
\begin{aligned}
\mathbf{P}\{\mu_0(n, N_i) = \alpha_i\} &= \frac{1}{\sqrt{\mathbf{D}\mu_0(n, N_i)}} \varphi\left(\frac{\alpha_i - \mathbf{E}\mu_0(n, N_i)}{\sqrt{\mathbf{D}\mu_0(n, N_i)}}\right) \left(1 + O\left(\frac{1}{\sqrt{\mathbf{D}\mu_0(n, N_i)}}\right)\right) \\
&= \frac{1}{\sqrt{2\pi \mathbf{D}\mu_0(n, N_i)}} \exp\left\{-\frac{u_i^2}{2M_i pq}\right\} \left(1 + O\left(\frac{1}{\sqrt{\mathbf{D}\mu_0(n, N_i)}}\right)\right). \qquad (11)
\end{aligned}
$$

The joint limit distribution of the components with numbers $l+1, \dots, k$ is given by

Theorem 1:

$$\mathbf{P}\{\mu_0(n, N_i) = \alpha_i, i = l+1, \ldots, k\} = \sqrt{\frac{M}{2\pi}}\left((2\pi pq)^{k-l-1}\mathbf{D}\mu_0(n, M)\prod_{i=1}^{k} N_i\right)^{-1/2}$$

$$\times \exp\left\{-\frac{1}{2}\sum_{i=l+1}^{k}\frac{u_i^2 M}{N_i} + \frac{1}{2}\left(\sum_{i=l+1}^{k} u_i\right)^2\left(\frac{Mpq}{\mathbf{D}\mu_0(n, M)} - 1\right)\right\}$$

$$\times\left(1 + O\left(\frac{1}{\sqrt{N(l+1, k)}}\right)\right). \quad (12)$$

The assertion of Corollary 3 follows from (11) and (12).

7. SUPPLEMENT

The results obtained can be used in analysis of the so-called inverse problem of allocating particles. We consider the effect of small groups of cells on the asymptotic behaviour of the waiting times.

Let $t_1, \ldots, t_k$ be positive integers which determine levels such that the content of the ith group attains the required level if the number of empty cells in the group becomes less than t_i, $i = 1, \ldots, k$.

Let $\nu^{(i)}$ be the waiting time equal to the number of allocated groups of particles until for the first time the contents of at least i groups attain the required levels, $1 \leq i \leq k$.

It can be shown (see [6]) that the moments of reaching the levels in the groups depend on the ratios t_i/N_i, $i = 1, \ldots, k$.

As an example we give here without proofs some features which demonstrate the effects of small groups on the waiting times. We set

$$A_i = \frac{\ln(1 - t_i/N_i)}{\ln(1 - m/N)},$$

$$B_i = t_i(1 - t_i/N_i) + \frac{(N_i - t_i)^2 m \ln(1 - t_i/N_i)}{N(N-m)\ln(1 - m/N)},$$

$$\delta_i = \lim_{N\to\infty}(t_i/N_i), \qquad i = 1, \ldots, k.$$

We consider the case of one small group only. Let it be the group with number one. This means that $N_1 = o(M)$ and the sizes of all other groups are of order M. We suppose that $m_1 = \ldots = m_n = m = o(N_1)$.

In such a case, if $0 < \delta_1 < \delta_2 \leq \delta_3 \leq \ldots \leq \delta_{k-1} < \delta_k$, then

$$\mathbf{P}\{\nu_*^{(1)} \leq x,\ \nu_*^{(2)} \leq y,\ \nu_*^{(k)} \leq z\} = \Phi(x)\,\Phi(y)\,\Phi(z) + o(1),$$

where $\Phi(x)$ is the distribution function of the standard normal distribution and

$$\nu_*^{(1)} = \frac{\nu^{(1)} - A_1}{B_1}, \quad \nu_*^{(2)} = \frac{\nu^{(2)} - A_2}{B_2}, \quad \nu_*^{(k)} = \frac{\nu^{(k)} - A_k}{B_k};$$

if $0 < \delta_2 \le \delta_1 < \delta_3 \le \ldots \le \delta_{k-1}$, then

$$\mathbf{P}\{\nu_*^{(1)} \le x,\ \nu_*^{(2)} \le y,\ \nu_*^{(k)} \le z\} = \Phi(x)\,\Phi(y)\,\Phi(z) + o(1),$$

where

$$\nu_*^{(1)} = \frac{\nu^{(1)} - A_2}{B_2}, \quad \nu_*^{(2)} = \frac{\nu^{(2)} - A_1}{B_1}, \quad \nu_*^{(k)} = \frac{\nu^{(k)} - A_k}{B_k};$$

and if $0 < \delta_2 \le \delta_3 \le \ldots \le \delta_{l-1} \le \delta_1 < \delta_{l+1} \le \ldots \le \delta_k$, then

$$\mathbf{P}\{\nu^{(1)} < x_1, \ldots, \nu^{(k)} < x_k\}$$
$$= \mathbf{P}\{\nu^{(1)} < x_1, \ldots, \nu^{(l-1)} < x_{l-1}\}\,\mathbf{P}\{\nu^{(l)} < x_l\}\,\mathbf{P}\{\nu^{(l+1)} < x_{l+1}, \ldots, \nu^{(k)} < x_k\} + o(1).$$

We restrict ourselves to the illustrative example; the study of the waiting times in this scheme of allocating particles will be done in another paper.

REFERENCES

1. V. A. Vatutin and V. G. Mikhailov, Limit theorems for the number of empty cells in equiprobable scheme for group allocation of particles. *Theory Probab. Appl.* (1982) **27**, 734–742.
2. G. I. Ivchenko and S. A. Khonov, On an asymptotic estimation for stratified finite populations. *Discrete Math. Appl.* (1991) **3**, 311–320.
3. N. Yu. Enatskaya and E. R. Khakimullin, The rate of convergence of the waiting time distribution in an equiprobable scheme of allocating particles in groups. In: *Probab. Problems of Discrete Math.* MIEM, Moscow, 1990, pp. 63–68 (in Russian).
4. N. Yu. Enatskaya and E. R. Khakimullin, Asymptotic behaviour of the joint distribution of waiting times in an equiprobable scheme of allocating particles in groups. In: *Probab. Processes Appl.*, MIEM, Moscow, 1991, pp. 91–96 (in Russian).
5. N. Yu. Enatskaya and E. R. Khakimullin, Asymptotic behaviour of waiting times in a scheme of particle allocation, *Probabilistic Methods in Discrete Mathematics, Proc. III Intern. Petrozavodsk Conf.* VSP/TVP, Utrecht/Moscow, 1993, pp. 155–185.

Probabilistic Methods in Discrete Mathematics, pp. 199–210
V. F. Kolchin et al. (Eds.)

ESTIMATES OF THE DEVIATION OF THE DISTRIBUTION OF r-INDEPENDENT RANDOM VARIABLES FROM THE NORMAL DISTRIBUTION

B. V. GLADKOV

Moscow

Abstract — We consider random variables $\xi_1, \dots, \xi_N$ such that any r of them are independent, $2 \le r < N$, r is an even number. We obtain upper bounds for the deviation

$$\Delta = \sup_x \left| \mathbf{P} \left\{ \beta^{-1} \sum_{k=1}^{N} \xi_k - \alpha \right\} - \Phi(x) \right|$$

where $\beta > 0$ and α are arbitrary real numbers, $\Phi(x)$ is the distribution function of the standard normal law. The existence of the moments of the random variables $\xi_1, \dots, \xi_N$ is not assumed and the properties of the m-dimensional distributions of this random sequence for $m \ge r + 1$ are not used.

In the case where the moments of these random variables exist, we obtain estimates of Δ for special values of α and β. Under some additional conditions we investigate the probabilities of large deviations of sums of r-independent random variables in the triangular array scheme of summation as $N, r \to \infty$.

1. Let random variables

$$\xi_1, \dots, \xi_N \tag{1}$$

with the distribution functions $V_1(x), \dots, V_N(x)$, respectively, be r-independent, that is, any r of these random variables be independent, $2 \le r < N$ (see [1, 2]). We assume that at least one of these random variables has non-degenerate distribution. There are no other assumptions on the distribution of the random variables (1) but the r-independence, so that they can be dependent. Since r-independent random variables are obviously $(r-1)$-independent, without loss of generality, we assume that r is even.

We set

$$\Delta = \sup_x \left| \mathbf{P} \left\{ \beta^{-1} \sum_{k=1}^{N} \xi_k - \alpha \right\} - \Phi(x) \right|, \tag{2}$$

where $\beta > 0$ and α are some real numbers, $\Phi(x)$ is the distribution function of the standard normal law.

In this paper we obtain estimates of the deviation Δ for arbitrary $\beta > 0$, α, r, and N (r is even, $2 \le r < N$). The existence of the moments of the random variables $\xi_1, \ldots, \xi_N$ is not assumed and the properties of the m-dimensional distributions of this random sequence for $m \ge r+1$ are not used.

In the case where the moments of these random variables exist, we obtain estimates of Δ for special values of α and β. Under some additional conditions we investigate the probabilities of large deviations of sums of r-independent random variables in the triangular array scheme of summation as $N, r \to \infty$.

2. We introduce some auxiliary random variables and notations. We consider the accompanying random variables

$$\tilde{\xi}_1, \ldots, \tilde{\xi}_N \tag{3}$$

which are independent and have the same distribution functions $V_1(x), \ldots, V_N(x)$ as the r-independent random variables (1).

Let ε be some positive number. Truncating the random variables (1) and (3) by the constant ε, we obtain the random variables $\{\xi_k^*, k = 1, \ldots, N\}$ and $\{\tilde{\xi}_k^*, k = 1, \ldots N\}$ such that

$$\xi_k^* = \begin{cases} \xi_k, & |\xi_k| < \varepsilon, \\ 0, & |\xi_k| \ge \varepsilon, \end{cases} \qquad \tilde{\xi}_k^* = \begin{cases} \tilde{\xi}_k, & |\tilde{\xi}_k| < \varepsilon, \\ 0, & |\tilde{\xi}_k| \le \varepsilon. \end{cases} \tag{4}$$

We set

$$\begin{aligned}
Q(\varepsilon) &= \sum \int_{|x|\ge\varepsilon} dV_k(x) = \sum \mathbf{P}\{|\xi_k| \ge \varepsilon\} = \sum \mathbf{P}\{|\tilde{\xi}_k| \ge \varepsilon\}, \\
E(\varepsilon) &= \sum \int_{|x|<\varepsilon} x\,dV_k(x) = \sum \mathbf{E}\xi_k^* = \sum \mathbf{E}\tilde{\xi}_k^*, \\
D(\varepsilon) &= \sum \left(\int_{|x|<\varepsilon} x^2 dV_k(x) - \left(\int_{|x|<\varepsilon} x\,dV_k(x) \right)^2 \right) = \sum \mathbf{D}\xi_k^* = \sum \mathbf{D}\tilde{\xi}_k^*, \\
H(\varepsilon) &= (\mathbf{D}(\varepsilon))^{-3/2} \sum \int_{|x|<\varepsilon} \left| x - \int_{|x|<\varepsilon} x\,dV_k(x) \right|^3 dV_k(x) \\
&= (\mathbf{D}(\varepsilon))^{-3/2} \sum \mathbf{E}|\xi_k^* - \mathbf{E}\xi_k^*|^3 = (\mathbf{D}(\varepsilon))^{-3/2} \sum \mathbf{E}|\tilde{\xi}_k^* - \mathbf{E}\tilde{\xi}_k^*|^3.
\end{aligned} \tag{5}$$

Here and later the symbol $\sum$ means the summation over k from 1 to N.

Let ε be such that $D(\varepsilon)$ and $H(\varepsilon)$ be positive. It is clear that such ε exists since at least one of the distributions $V_1(x), \ldots, V_N(x)$ is non-degenerate.

Centering and normalizing the random variables (4), we obtain

$$\{\eta_k^* = (\xi_k^* - \mathbf{E}\xi_k^*)/\sqrt{D(\varepsilon)}\}, \quad \{\tilde{\eta}_k^* = (\tilde{\xi}_k^* - \mathbf{E}\tilde{\xi}_k^*)/\sqrt{D(\varepsilon)}\}, \quad k = 1, \ldots, N. \tag{6}$$

It is obvious that the random variables $\{\eta_k^*\}$ are r-independent, have all moments, and the random variables $\{\tilde{\eta}_k^*\}$ are independent and for any $k = 1, \ldots, N$ the random variable η_k^* has the same distribution as $\tilde{\eta}_k^*$. In addition,

$$|\eta_k^*|, |\tilde{\eta}_k^*| < \frac{2\varepsilon}{\sqrt{D(\varepsilon)}}, \quad \mathbf{E}\eta_k^* = \mathbf{E}\tilde{\eta}_k^* = 0, \quad k = 1, \ldots, N,$$

$$\sum \mathbf{D}\eta_k^* = \sum \mathbf{D}\tilde{\eta}_k^* = 1, \qquad \sum \mathbf{E}|\eta_k^*|^3 = \sum \mathbf{E}|\tilde{\eta}_k^*|^3 = H(\varepsilon) \le \frac{2\varepsilon}{\sqrt{D(\varepsilon)}}. \tag{7}$$

3. We turn now to estimating the deviation Δ given by (2). First we obtain inequalities estimating Δ without assumption on the existence of the moments of the random variables (1) and (3). In [3] it is shown that without assumption on the existence of the moments of the independent random variables $\{\tilde{\xi}_k\}$ for any $\beta > 0$, α, and any $\varepsilon > 0$ such that $D(\varepsilon)$ and $H(\varepsilon)$ are positive, for the deviation

$$\tilde{\Delta} = \sup_x \left| \mathbf{P}\left\{\beta^{-1} \sum \xi_k - \alpha\right\} - \Phi(x)\right| \tag{8}$$

the estimate

$$\tilde{\Delta} \le Q(\varepsilon) + R_0(\alpha, \beta) + AH(\varepsilon) = \tilde{R}(\varepsilon) \tag{9}$$

holds, where

$$R_0(\alpha, \beta) = \frac{|\alpha\beta - E(\varepsilon)|}{\sqrt{2\pi D(\varepsilon)}} + \frac{1}{\sqrt{2\pi e}} |1 - D(\varepsilon)/\beta^2| \max(1, \beta^2/D(\varepsilon), \tag{10}$$

A is the absolute constant from the right-hand side of the Esseen inequality [4]

$$\tilde{\Delta}^* = \sup_x \left| \mathbf{P}\left\{\sum \tilde{\eta}_k^* < x\right\} - \Phi(x)\right| \le AH(\varepsilon), \tag{11}$$

in which $\sum \tilde{\eta}_k^*$ is the sum of bounded random variables (6), and $Q(\varepsilon)$, $E(\varepsilon)$, $D(\varepsilon)$, $H(\varepsilon)$ are determined in (5).

The inequalities similar to (9) which estimate from above the deviation Δ for the random variables $\{\xi_k\}$ with arbitrary distributions are given in the following theorem.

THEOREM 1. *Let $\xi_1, \ldots, \xi_N$ be r-independent random variables with arbitrary distribution functions $V_1(x), \ldots, V_N(x)$, respectively, at least one of which is non-degenerate, let r be an even number, $2 \le r < N$, and $\beta > 0$, α be arbitrary real numbers.*

Then for any $\varepsilon > 0$ such that $D(\varepsilon)$, $H(\varepsilon)$ are positive and any $T \in (0, 1/(4H(\varepsilon))]$ and $c > r/2$,

$$\Delta = \sup_x \left| \mathbf{P}\left\{\beta^{-1} \sum \xi_k - \alpha < x\right\} - \Phi(x)\right| < \tilde{R}(\varepsilon) + R_1^{(i)}(\varepsilon, c, T), \quad i = 1, 2, \tag{12}$$

where $\tilde{R}(\varepsilon)$ is the right-hand side of (9),

$$R_1^{(i)}(\varepsilon, c, T) = \left(\frac{2}{\pi r}\right)^{3/2} (K^{(i)} + M) + \frac{24}{\pi\sqrt{2\pi}}(T^{-1} - 4H(\varepsilon)), \qquad i = 1, 2, \tag{13}$$

$$K^{(1)} = K^{(1)}(\varepsilon, c, T) = \left(\frac{ecT}{r\sqrt{D(\varepsilon)}}\right)^r \sum \mathbf{E}|\xi_k^* - \mathbf{E}\xi_k^*|^r, \tag{14}$$

$$K^{(2)} = K^{(2)}(\varepsilon, c, T) = \left(\frac{ecT}{r}\exp\{1 - 2/r\} \ln \frac{2\varepsilon}{\sqrt{D(\varepsilon)}}\right)^r, \quad K^{(1)} < K^{(2)}, \tag{15}$$

$$M = M(c, T) = \left(\frac{\sqrt{c}T}{r}\exp\{1 + c/r\}\right)^r B(r/2, c - r/2), \tag{16}$$

the random variables $\{\xi_k^*, k = 1, \dots, N\}$ *are defined in* (4), *and*

$$B(u, v) = \int_0^1 t^{u-1}(1 - t)^{v-1} dt.$$

PROOF. Following [3], we obtain that for any $p, q \in \mathbf{R}$,

$$\begin{aligned}\Delta \le \sup_x &\left|\mathbf{P}\left\{\beta^{-1}\sum \xi_k - \alpha < x\right\} \mathbf{P}\left\{\beta^{-1}\sum \xi_k^* - \alpha < x\right\}\right| \\ &+ \sup_x \left|\mathbf{P}\left\{\beta^{-1}\sum \xi_k - \alpha <\right\} - \Phi(px + q)\right| \\ &+ \sup_x |\Phi(px + q) - \Phi(px)| + \sup_x |\Phi(px) - \Phi(x)|.\end{aligned} \tag{17}$$

We set

$$p = \frac{\beta}{\sqrt{D(\varepsilon)}} > 0, \qquad q = \frac{\alpha\beta - E(\varepsilon}{\sqrt{D(\varepsilon)}}.$$

Then

$$\begin{aligned}\mathbf{P}\left\{\beta^{-1}\sum \xi_k^* - \alpha < x\right\} &= \mathbf{P}\left\{\sum \eta_k^* < px + q\right\}, \\ \mathbf{P}\left\{\beta^{-1}\tilde{\xi}_k^* - \alpha < x\right\} &= \mathbf{P}\left\{\tilde{\eta}_k^* < px + q\right\}.\end{aligned} \tag{18}$$

Using Lemma 3 in [1] (see also [3]) for the estimation of the first term in the right-hand side of (17) and the inequalities from [3] for the estimation of the two last terms, and taking (10) and (18) into account, we obtain from (17)

$$\Delta \le Q(\varepsilon) + R_0(\alpha, \beta) + \sup_x \left|\mathbf{P}\left\{\sum \eta_k^* < x\right\} - \Phi(x)\right|. \tag{19}$$

Let $\psi(t)$ and $\tilde{\psi}(t)$ be the characteristic functions of the sums $\sum \eta_k^*$ and $\sum \tilde{\eta}_k^*$, respectively, and let $\varphi(t) = e^{-x^2/2}$ be the characteristic function of the standard normal distribution. Using the Esseen theorem [4] for the estimation of the last term in (19), after some transformations, we see that for any $T \in (0, 1/(4H(\varepsilon))]$

$$\begin{aligned}\Delta &= Q(\varepsilon) + R_0(\alpha,\beta) + \frac{2}{\pi}\int_0^T \frac{|\psi(t)-\varphi(t)|}{t}dt + \frac{24}{\pi\sqrt{2\pi}}T^{-1}\\ &\le Q(\varepsilon) + R_0(\alpha,\beta) + \frac{2}{\pi}\int_0^{4H(\varepsilon))^{-1}} \frac{|\tilde\psi(t)-\varphi(t)|}{t}dt + \frac{24}{\pi\sqrt{2\pi}}4H(\varepsilon)\\ &\quad + \frac{2}{\pi}\int_0^T \frac{|\psi(t)-\tilde\psi(t)|}{t}dt + \frac{24}{\pi\sqrt{2\pi}}(T^{-1}-4H(\varepsilon)). \end{aligned} \tag{20}$$

The sum of the third and the fourth terms in the right-hand side of (20), according to the Esseen inequality [4] (or [3]) is equal to $AH(\varepsilon)$, and together with the first and the second terms it gives the first term $\tilde R(\varepsilon)$ in the right-hand sides of (12).

It is clear that the random variables

$$\{\eta_k^*\} = \{(\xi_k^* - \mathbf{E}\xi_k^*)/\sqrt{d(\varepsilon)}, \qquad \{\tilde\eta_k^*\} = \{(\tilde\xi_k^* - \mathbf{E}\tilde\xi_k^*)/\sqrt{\mathbf{D}(\varepsilon)}\}$$

satisfy the conditions of Lemma 3 in [2] (see also Lemma 1 in [1]) and Theorem 2 in [5]). Therefore, for even r and any $t \in \mathbf{R}$ and $c > r/2$,

$$|\psi(t) - \tilde\psi(t)| < \rho^{(1)}(\varepsilon, c, t) < \rho^{(2)}(\varepsilon, c, t), \tag{21}$$

where

$$\begin{aligned}\rho^{(1)}(\varepsilon,c,t) &= \sqrt{\frac{2}{\pi r}}\Bigg(\left(\frac{ec|t|}{r\sqrt{D(\varepsilon)}}\right)^r \sum \mathbf{E}|\xi_k^* - \mathbf{E}\xi_k^*|^r\\ &\quad + r(\sqrt{c}|t|\exp\{1+c/r\}/r)^r B(r/2, c-r/2)\Bigg)\end{aligned} \tag{22}$$

$$\begin{aligned}\rho^{(2)}(\varepsilon,c,t) &= \sqrt{\frac{2}{\pi r}}((ec|t|\exp\{(1-2/r)\ln(2\varepsilon/\sqrt{\mathbf{D}(\varepsilon)})\}/r)^r\\ &\quad + r(\sqrt{c}|t|\exp\{1+c/r\}/r)^r B(r/2, c-r/2)).\end{aligned} \tag{23}$$

Formulae (22) and (23) are obtained from the expressions of $\rho^{(i)}(\varepsilon, c, t)$, $i = 1, 2$, given in Lemma 3 in [2] (formulae (9) and (10)), where we omit the index n and substitute

$$\sum \mathbf{E}|\eta_k^*|^r = (D(\varepsilon))^{-r/2} \sum \mathbf{E}|\xi_k^* - \mathbf{E}\xi_k^*|^r$$

for $A_r(\varepsilon_n)$, $\varepsilon/\sqrt{D(\varepsilon)}$ for ε_n, and, according to (7), $\sum \mathbf{D}\eta_k^* = 1$ for $D_n(\varepsilon_n)$.

Using (20)–(23) and taking into account (14)–(16), we see that the last term in the right-hand side of (20) is bounded by the first terms of (13), namely, by

$$\left(\frac{2}{\pi r}\right)^{3/2}(K^{(i)} + m), \qquad i = 1, 2.$$

This estimate and (20) imply (12).

The theorem is proved.

REMARK 1. For given α, β, r and N and given distribution functions $V_k(x)$, $k = 1, \ldots, N$, the optimal estimates of the deviation Δ is equal obviously to

$$\inf_{\varepsilon,c,T}(\tilde{R}(\varepsilon) + R_1^{(i)}(\varepsilon, c, T))$$

and if ε is fixed,

$$\tilde{R}(\varepsilon) + \inf_{c,T} R_1^{(i)}(\varepsilon, c, T),$$

where $\tilde{R}(\varepsilon)$ and $R_1^{(i)}(\varepsilon, c, T)$, $i = 1, 2$, are determined in (9) and (13), respectively, and the infimum is taken over the values of the parameters determined in Theorem 1. In reality, it is impossible to find the exact value of this infimum, therefore in what follows, along with the estimates of the general form (12) depending on the parameters ε, c, and T, we consider the estimates of Δ with fixed values of some of these parameters (for example, c and T) which do not coincide with the optimal values provided the infimum. The choice of the values of c and T is arbitrary to some extent and is determined by the convenience in obtaining the estimate of Δ possessing good properties for large r and N.

Taking Remark 1 into account, we obtain the following corollary to Theorem 1.

COROLLARY 1. *Under the conditions of Theorem 1*

$$\Delta = \sup_x \left| \mathbf{P}\left\{\beta^{-1}\sum \xi_k - \alpha < x\right\} - \Phi(x)\right| < \tilde{R}(\varepsilon) + R_{11}^{(2)}(\varepsilon), \tag{24}$$

where

$$R_{11}^{(2)}(\varepsilon = 6\left(\frac{2}{\pi}\right)^{3/2}(r^{-3/2}/2 + T_1^{-1} - 4H(\varepsilon)), \tag{25}$$

$$T_1^{-1} = \max\left(\frac{e}{2}(1 + 2/r)\exp\{(1 - 2/r)\ln(2\varepsilon/\sqrt{D(\varepsilon)}\}, \frac{1}{\sqrt{2r}}e^{3/2+2/r}, 4H(\varepsilon)\right), \tag{26}$$

where

$$R_{11}^{(2)} \geq R_1^{(2)}(\varepsilon, 1 + r/2, T_1).$$

If in addition $\varepsilon \leq \sqrt{D(\varepsilon)}/2$, *then in* (25) *we can take*

$$T_1^{-1} = \max\left(4\exp\{(1 - 2/r)\ln(2\varepsilon/\sqrt{D(\varepsilon)}\}, \frac{1}{\sqrt{2r}}e^{3/2+2/r}\right). \tag{27}$$

PROOF. Setting in (13), (15), and (16) $c = 1 + r/2$ and $T^{-1} = T_1^{-1}$ given by (26), and using the relations $B(r/2, 1) = 2/r$ and $1 + 2/r < e^{2/r}$, we obtain $K^{(2)} \leq 1$ and $M \leq 2$. Taking (12) into account, we obtain (24). Under the additional condition $\varepsilon \leq \sqrt{D(\varepsilon)}/2$, estimate (24) with T_1^{-1} given by (27) follows if we take (7) into account and compare (26) and (27).

4. In obtaining the estimates of the deviation Δ, the following theorem can be useful.

THEOREM 2. *Under the conditions of Theorem 1, for any $\varepsilon >$ such that $D(\varepsilon)$ and $H(\varepsilon)$ are positive and any $T \in (0, 1/(4h(\varepsilon)]$ and $c > r/2$,*

$$\Delta = \sup_x \left| \mathbf{P}\left\{\beta^{-1}\sum \xi_k - \alpha <\right\} - \Phi(x)\right| \tag{28}$$
$$< \tilde{\Delta} + 2Q(\varepsilon) + 2AH(\varepsilon) + R_1^{(i)}(\varepsilon, c, T), \qquad i = 1, 2.$$

where $\tilde{\Delta}$ is determined in (8).

PROOF. For any real p and q,

$$\begin{aligned}\Delta \le & \sup_x \left| \mathbf{P}\left\{\beta^{-1}\sum \xi_k - \alpha < x\right\} - \mathbf{P}\left\{\beta^{-1}\sum \xi_k^* - \alpha < x\right\}\right| \\ & + \sup_x \left| \mathbf{P}\left\{\beta^{-1}\sum \tilde{\xi}_k^* - \alpha < x\right\} - \mathbf{P}\left\{\beta^{-1}\sum \tilde{\xi}_k - \alpha < x\right\}\right| \\ & + \sup_x \left| \mathbf{P}\left\{\beta^{-1}\sum \xi_k^* - \alpha < x\right\} - \Phi(px+q)\right| \\ & + \sup_x \left| \mathbf{P}\left\{\beta^{-1}\sum \tilde{\xi}_k^* - \alpha < x\right\} - \Phi(px+q)\right| \\ & + \sup_x \left| \mathbf{P}\left\{\beta^{-1}\sum \tilde{\xi}_k - \alpha < x\right\} - \Phi(x)\right|. \end{aligned} \tag{29}$$

Setting $p = \beta\sqrt{D(\varepsilon)}$, $q = (\alpha\beta - E(\varepsilon)/\sqrt{D(\varepsilon)}$ and using the inequalities from Lemma 3 in [1], and (11), (18) and (13)–(16), we obtain (28) from (29).

The theorem is proved.

REMARK 2. Inequality (28) is useful in obtaining estimates of Δ if the random variables (1) and (3) have the absolute moment of order $2 + \delta$, where $0 \le \delta \le 1$. In this case, setting

$$\beta^2 = B = \sum \mathbf{D}\xi_k = \sum \mathbf{D}\tilde{\xi}_k,$$
$$\alpha = \frac{1}{\sqrt{B}}\sum \mathbf{E}\xi_k = \frac{1}{\sqrt{B}}\sum \mathbf{E}\tilde{\xi}_k,$$

in the estimation of $\tilde{\Delta}$, we can use the inequalities from [3] instead of (9).

5. Here we give estimates of Δ provided that the random variables $\{\xi_k\}$ and $\{\tilde{\xi}_k\}$ introduced in (1) and (3), respectively, have the rth moment. In this case there is no need to use the truncation of these random variables.

THEOREM 3. *Let the random variables (1) (and (3)) have the third moment if $r = 2$ and the rth moment if $r \geq 4$ (r is even), and let*

$$\beta^2 = B = \sum \mathbf{D}\xi_k = \sum \mathbf{D}\tilde{\xi}_k,$$
$$\alpha = \frac{1}{\sqrt{B}} \sum \mathbf{E}\xi_k = \frac{1}{\sqrt{B}} \sum \mathbf{E}\tilde{\xi}_k,$$
$$L = B^{-3/2} \sum \mathbf{E}|\xi_k - \mathbf{E}\xi_k|^3 = B^{-3/2} \sum \mathbf{E}|\tilde{\xi}_k - \mathbf{E}\tilde{\xi}|^3.$$

Then for any $T \in (0, 1/(4l)]$ and $c > r/2$,

$$\Delta = \sup_x \left| \mathbf{P}\left\{ \frac{1}{\sqrt{B}} \sum (\xi_k - \mathbf{E}\xi_k) < x \right\} - \Phi(x) \right| < AL + R_3(c, T), \tag{30}$$

where A is the absolute constant in the Esseen inequality for $\tilde{\Delta}$,

$$R_3(c, T) = \left(\frac{2}{\pi r}\right)^{3/2} \left(\frac{ecT}{r\sqrt{B}}\right)^r \sum \mathbf{E}|\xi_k - \mathbf{E}\xi_k|^r$$
$$+ r\left(\frac{2}{\pi r}\right)^{3/2} \left(\frac{\sqrt{c}T}{r} e^{1+c/r}\right)^r B(r/2, c - r/2) + \frac{24}{\pi\sqrt{2\pi}}(T^{-1} - 4L). \tag{31}$$

The proof of Theorem 3 repeats the derivation of the estimate of the last term in the right-hand side of (19) in the proof of Theorem 1.

Taking into account Remark 1, under some additional conditions, we obtain the following assertion in the same way as Corollary 1.

COROLLARY 2. *If under the conditions of Theorem 3 r-independent random variables ξ_k, $k = 1, \dots, N$, are identically distributed and*

$$\sigma_1^2 = \mathbf{D}\xi_1, \quad a_1 = \mathbf{E}\xi_1, \quad \rho_1 = \sigma_1^{-3}\mathbf{E}|\xi_1 - a_1|^3,$$

then

$$\Delta = \sup_x \left| \mathbf{P}\left\{ \frac{1}{\sigma_1\sqrt{n}} \sum (\xi_k - a_1) < x \right\} - \Phi(x) \right| < \frac{A\rho_1}{\sqrt{n}} + R_{31}, \tag{32}$$

where A is the absolute constant in the Berry–Esseen inequality [3],

$$R_{31} = 6\left(\frac{2}{\pi}\right)^{3/2} \left(\frac{1}{6} r^{-3/2}\left(2 + \frac{1}{r}\mathbf{E}|\xi_1 - a_1|^r\right) + T_1^{-1} - \frac{4\rho_1}{\sqrt{n}}\right), \tag{33}$$

$$T_1^{-1} = \max\left(\frac{1}{2\sigma_1}(1 + 2/r)e^{1+\ln r/r} N^{-(1-2/r)/2}, \frac{1}{\sqrt{2r}} e^{3/2+2/r}, \frac{4\rho_1}{\sqrt{n}}\right). \tag{34}$$

If in addition,

$$|\xi_k| \leq C, \qquad k = 1, \dots, N,$$

where $C \le \sigma_1\sqrt{n}/2$, *then in* (32) *we can take*

$$R_{31} = 6\,(2\pi)^{3/2}\left(\frac{1}{2}r^{-3/2} + T_1^{-1} - \frac{4\rho_1}{\sqrt{n}}\right), \tag{35}$$

where

$$T_1^{-1} = \max\left(4\left(\frac{2C}{\sigma_1\sqrt{n}}\right)^{1-2/r}, \frac{1}{\sqrt{2r}}e^{3/2+2/r}\right). \tag{36}$$

In both cases $R_{31} \ge R_3(1 + r/2, T_1)$.

The proof is similar to the proof of Corollary 1. In (30) and (31), we should set $c = r/2$ and $T^{-1} = T_1^{-1}$, where T_1^{-1} is given in (34) or (36), and take into account that under the conditions of the corollary $L = \rho_1/\sqrt{n}$ and $\rho_1 \le 2C/\sigma_1$ if $|\xi_k| \le C$, $k = 1, \dots, N$.

6. We begin with a remark.

REMARK 3. Using the estimates obtained here for the deviation

$$\Delta = \sup_x \left|\mathbf{P}\left\{\beta^{-1}\sum \xi_k - \alpha < x\right\} - \Phi(x)\right|$$

and the known estimates for

$$\tilde{\Delta} = \sup_x \left|\mathbf{P}\left\{\beta^{-1}\sum \tilde{\xi}_k - \alpha < x\right\} - \Phi(x)\right|,$$

we can estimate the difference

$$\sup_x \left|\mathbf{P}\left\{\beta^{-1}\sum \xi_k - \alpha < x\right\} - \Phi(x)\right| - \sup_x \left|\mathbf{P}\left\{\beta^{-1}\sum \tilde{\xi}_k - \alpha < x\right\} - \Phi(x)\right| \le \Delta + \tilde{\Delta}.$$

The estimates of Δ obtained above and determined by (24) and (32) are useful for large values of r and N only.

We explain this assertion. To this end, we consider a scheme of summation of random variables which form a triangular array. Instead of the random variables (1) and (3) we consider the sequence of series of random variables

$$\{\xi_{nk},\ k = 1, \dots, N_n, n = 1, 2, \dots\}$$

which are r_n-independent in the nth series (r_n is even), $n = 1, 2, \dots$, and the sequence of series of random variables $\{\tilde{\xi}_{nk}, k = 1, \dots, N_n, n = 1, 2, \dots\}$ which are independent in each series and for any n and k the random variables ξ_{nk} and $\tilde{\xi}_{nk}$ have the same distribution $V_{nk}(x)$.

Let the random variables $\{\xi_{nk}\}$ and $\{\tilde{\xi}_{nk}\}$ be uniformly infinitesimal and the sum $\sum \tilde{\xi}_{nk}$ be asymptotically normal with the parameters (a, σ) as $n, N_n \to \infty$. In this case according to Theorem 1 in [2], there exists a sequence of positive constants $\{\varepsilon_n\}$, $n = 1, 2, \dots$, such that

$$\varepsilon_n \to 0, \quad Q_n(\varepsilon_n) \to 0, \quad E_n(\varepsilon_n) \to a, \quad D_n(\varepsilon_n) \to \sigma^2$$

as $n, N_n \to \infty$, where $Q_n(\varepsilon_n)$, $E_n(\varepsilon_n)$, and $D_n(\varepsilon_n)$ are obtained from $Q(\varepsilon)$, $E(\varepsilon)$, and $D(\varepsilon)$ by substitution of ε_n for ε and $V_{nk}(x)$ for $V_k(x)$. By Theorem 1 for fixed n (and fixed r_n and N_n) for the deviation

$$\Delta_n = \sup_x \left| \mathbf{P}\left\{\beta^{-1} \sum \xi_{nk} - \alpha < x\right\} - \Phi(x)\right|$$

estimates (12) and (24) are valid provided that we insert

$$\alpha_n, \quad \beta_n, \quad Q_n(\varepsilon_n, \quad E_n(\varepsilon_n), \quad D_n(\varepsilon_n), \quad H_n(\varepsilon_n) \le \frac{2\varepsilon_n}{\sqrt{D_n(\varepsilon_n)}}$$

into the right-hand side of these estimates. If $n, N_n, r_n = 2r_n' \to \infty$ and $\alpha_n \to a/\sigma$, $\beta_n \to \sigma$, then, as it is easy to see, the right-hand side of (24) which estimates the deviation Δ_n tends to zero. In this sense estimate (24) is useful for large r and N.

Let the r_n-independent random variables $\{\xi_{nk}\}$ in the nth series satisfy the conditions of Theorem 3. If the random variables $\{\xi_{nk}\}$ in the nth series are identically distributed and, as $N, N_n, r_n = 2r_n' \to \infty$,

$$\sigma_{n1} \to \infty, \quad \rho_{n1} = \mathbf{E}|\xi_{n1} - \mathbf{E}\xi_{n1}|^3/\sigma_{n1}^3 = o(\sqrt{N_n}),$$

where $\sigma_{n1}^2 = \mathbf{D}\xi_{n1}$, and

$$\limsup_{n, r_n \to \infty} \frac{1}{r_n} \mathbf{E}|\xi_{n1} - \mathbf{E}\xi_{n1}|^{r_n} < \infty;$$

then it is obvious that the right-hand side of (32), which estimates Δ_n with R_{31} determined in (33), tends to zero.

Note that if all random variables $\{\xi_{nk},\ k = 1, \dots, N_n,\ n = 1, 2, \dots\}$ have the same distribution $V_1(x)$ whose all moments exist, then $\sigma_{n1} = \sigma_1$ and $\rho_{n1} = \rho_1$ are constants and the condition

$$\limsup_{n, r_n \to \infty} \frac{1}{r_n} \mathbf{E}|\xi_1 - \mathbf{E}\xi_1|^{r_n} < \infty$$

is a sufficient condition for the uniqueness of the problem of moments for the distribution $V_1(x)$, that is, under this condition the moments

$$\mathbf{E}\xi_1^s = \int_{-\infty}^{\infty} x^s dV_1(x), \qquad s = 1, 2, \dots,$$

uniquely determine the distribution $V_1(X)$ (see, for example [6]). If in addition, $|\xi_{nk}| \le C$, $k = 1, \dots, N_n$, $n = 1, 2, \dots$, then the right-hand side of (32), which estimates Δ_n with R_{31} determined in (35), tends to zero as $n, N_n r_n = 2r_n' \to \infty$. Thus, estimate (32) under the conditions given above is also useful for large r and N.

7. Using estimate (32) in the scheme of series, we can obtain theorems on large deviations in the tight zone of normal convergence. We present one of such theorems.

THEOREM 4. *Let the random variables* $\{\xi_{nk}, k = 1, \dots, N_n, n = 1, 2, \dots\}$ *be* r_n*-independent in the* n*th series (*r_n *are even), have the same distribution* $V_1(x)$ *and be bounded, that is,* $|\xi_{nk}| \le C$ *for all* n *and* k*, and let*

$$F_n(xc) = \mathbf{P}\left\{\frac{1}{\sigma_1\sqrt{N}}\sum(\xi_{mk} - a_1) < x\right\},$$

where $a_1 = \mathbf{E}\xi_{nk}$, $\sigma_1 = \mathbf{D}\xi_{nk}$.

If $n, N_n, r_n = 2r_n' \to \infty$, *then uniformly with respect to* x *such that*

$$0 \le x \le x_n = (1 - \lambda)\sqrt{\ln r_n},$$

where $\lambda \in (0, 1)$ *is fixed,*

$$\frac{1 - F_n(x)}{1 - \Phi(x)} \to 1, \qquad \frac{f_n(-x)}{\Phi(-x)} \to 1.$$

PROOF. For any $x \in [0, x_n]$, $x_n > 1$, that is, if

$$r_n > e^{1/(1-\lambda)^2},$$

we have

$$\Phi(-x) = 1 - \Phi(x) \ge 1 - \Phi(x_n) \ge \frac{1}{\sqrt{2\pi}}(1 - x_n^{-2})e^{-x^2/2},$$

and

$$\left|\frac{1 - F_n(x)}{1 - \Phi(x)} - 1\right|, \left|\frac{F_n(-x)}{\Phi(-x)} - 1\right| \le \Delta_n x_n \sqrt{2\pi}(1 - x_n^{-2})^{-1} e^{x_n^2/2}, \tag{37}$$

where by virtue of (32), (35), and (36) for the nth series,

$$\Delta_n = \sup_x |F_n(x) - \Phi(x)| < \frac{A\rho_1}{\sqrt{N_n}} + 6\left(\frac{2}{\pi}\right)^{3/2}\left(\frac{1}{2}r_n^{-3/2} + T_{n1}^{-1} - \frac{4\rho_1}{\sqrt{N_n}}\right) \tag{38}$$

and

$$T_{n1}^{-1} = \max\left(4\left(\frac{2C}{\sigma_1\sqrt{N_n}}\right)^{1-2/r}, \frac{1}{\sqrt{2r_n}}e^{3/2+2/r_n}\right), \tag{39}$$

A, σ_1 and ρ_1 do not depend on n. The assertion of the theorem follows from (37)–(39).

REFERENCES

1. B. V. Gladkov, Sums of random variables, any r of which are independent. *Math. Notes* (1982) **32**, 385–399.
2. B. V. Gladkov, Central limit theorem for the sums of random variables any r of which are independent. *Discrete Math. Appl.* (1991) **1**, 73–79.
3. V. V. Petrov, *Sums of Independent Random Variables.* Springer, New York, 1975.
4. M. Loève, *Probability Theory.* Van Nostrand, Princeton, 1960.
5. S. V. Nagaev and I. F. Pinelis, Some inequalities for the distributions of sums of independent random variables. *Theory Probab. Appl.* (1977) **22**, 248–256 (in Russian).
6. A. N. Shiryaev, *Probability.* Nauka, Moscow, 1980 (in Russian).

Probabilistic Methods in Discrete Mathematics, pp. 211–218
V. F. Kolchin et al. (Eds.)

ON THE NUMBER OF PERMUTATIONS OF n OBJECTS WITH GREATEST CYCLE LENGTH k

S. W. GOLOMB and P. GAAL

Los Angeles, the University of Southern California

Abstract — Let $L_{k,n}$ be the number of permutations on n distinct objects with greatest cycle length k. We observe each of the following.

For all $n \geq 1$,

$$L_{1,n} = 1, \qquad L_{n,n} = (n-1)!, \qquad \sum_{k=1}^{n} L_{k,n} = n!.$$

For $n/2 < k \leq n$,

$$L_{k,n} = \frac{n!}{k}.$$

For $n/3 < k \leq n/2$,

$$L_{k,n} = \frac{n!}{k}\left(1 - \frac{1}{2k} - \sum_{j=1}^{n-2k} \frac{1}{k+j}\right).$$

For $n/4 < k \leq n/3$,

$$L_{k,n} = \frac{n!}{k}\left(1 - \frac{1}{2k}\left(1 - \frac{1}{3k}\sum_{j=1}^{n-3k}\frac{1}{k+j}\right) - \sum_{j=1}^{n-2k}\frac{1}{k+j} + \sum_{j=1}^{n-3k-1}\sum_{l=1}^{n-3k-j}\frac{1}{2(k+j)(k+l)}\right).$$

For $n/5 < k \leq n/4$,

$$\begin{aligned} L_{k,n} = \frac{n!}{k}\Bigg(1 &- \sum_{j=1}^{n-2k}\frac{1}{k+j} + \sum_{j=1}^{n-3k-1}\sum_{l=1}^{n-3k-j}\frac{1}{2!(k+j)(k+l)} \\ &- \sum_{j=1}^{n-4k-2}\sum_{l=1}^{n-4k-j-1}\sum_{m=1}^{n-4k-j-l}\frac{1}{3!(k+j)(k+l)(k+m)} \\ &- \frac{1}{2k}\left(1 - \sum_{j=1}^{n-3k}\frac{1}{k+j} + \sum_{j=1}^{n-4k-1}\sum_{l=1}^{n-4k-j}\frac{1}{2!(k+j)(k+l)} - \frac{1}{3k}\left(1 - \sum_{j=1}^{n-4k}\frac{1}{k+j} - \frac{1}{4k}\right)\right)\Bigg), \end{aligned}$$

etc. For $1 < k \leq n$,

$$L_{k,n} = \sum_{j=1}^{\lfloor n/k \rfloor} \frac{n!}{j!\,k^j(n-kj)!} \sum_{t=1}^{\min(k-1,n-kj)} L_{t,n-kj}.$$

The expected length E_n of the longest cycle in a random permutation on n objects is given by

$$E_n = \frac{1}{n!}\sum_{k=1}^{n} kL_{k,n},$$

where for $n/2 < k \le n$, the contribution of the kth term is 1. Similarly, for $n/3 < k \le n/2$, the contribution is

$$1 - \frac{1}{2k} - \sum_{j=1}^{n-2k} \frac{1}{k+j},$$

etc. The relative expected length λ_n of the longest cycle of a random permutation on n objects is $\lambda_n = E_n/n$, which has a limit $\lambda = 0.62432965\ldots$ (Golomb's constant) as $n \to \infty$. The total contribution to λ_n of all k, $n/2 < k \le n$, is $(n - \lfloor n/2 \rfloor)/n\lambda_n \to 80.0859\%$ as $n \to \infty$. Similarly, the total contribution of all k, $n/3 < k \le n/2$, to λ is $0.11003365\ldots$, or 17.6243%, as $n \to \infty$, etc.

Table 1 shows $L_{k,n}$ in triangular form for all $n \in [1, 14]$ and all $k \in [1, n]$.

We consider the symmetric group S_n of all $n!$ permutations on n objects. As a convenient conceptual model, we regard each element σ of S_n as a deck of n cards (the cards numbered from 1 to n) in a particular permuted order, so that we have $n!$ decks, each containing the same n cards, but each deck having the cards in a different order. For each deck, we partition the n cards into cycles as follows. We deal out cards, starting from the top of the deck, in order, until the card numbered 1 appears. This completes the first cycle. If there are cards still remaining in the deck, there is a lowest-numbered remaining card, bearing the number a_2, with $1 < a_2 \le n$. We continue dealing cards until the card numbered a_2 appears, and this completes the second cycle. If any cards still remain in the deck, we deal these until the lowest remaining number a_3, with $a_2 < a_3 \le n$, appears, which completes the third cycle. We continue this process until all the cards are dealt. For example, in a deck of 8 cards, if the numbers in order were 6 3 4 1 2 8 5 7, the cycles would be (6341)(2)(85)(7). Since the number 1 occurs equally often in each position of the deck, the first cycle has a length uniformly distributed from 1 to n, with an expected length of $(n + 1)/2$.

If a permutation σ in S_n contains a cycle of length k, where $n/2 < k \le n$, this is necessarily the longest cycle in σ, since the number of remaining elements in σ is $n - k < n/2 < k$. The number of permutations in S_n containing a cycle of length k, $n/2 < k \le n$, is exactly $n!/k$, so this is also the number of permutations in S_n for which the greatest cycle length is k. To obtain the number $n!/k$, we observe that the k elements to be on the cycle of length k can be chosen in $\binom{n}{k}$ ways, and permuted in $(k - 1)!$ ways to form a cycle of length k; and then the remaining $n - k$ elements can occur in any of $(n - k)!$ permuted orders. Thus, the number of permutations in S_n for which the longest cycle has length $k > n/2$ is

$$\binom{n}{k}(k-1)!(n-k)! = \frac{n!}{k}.$$

We define $L_{k,n}$ to be the number of elements in S_n for which the greatest cycle length is k. Clearly this requires

$$\sum_{k=1}^{n} L_{k,n} = n!$$

since every element of S_n is counted in one and only one of the $L_{k,n}$'s. We have just shown that $L_{k,n} = n!/k$ for $n/2 < k < n$. If, however, $1 \le k \le n/2$, there are two reasons why $n!/k$ overestimates $L_{k,n}$. If $k < n/2$, there may be a cycle of length greater than k in the permutation. If $k \le n/2$, even if there is no cycle of length greater than k, the cycle of length k may not be unique. For example, if $n = 2k$, there could be two cycles, each of length $n/2$. The number of ways this can happen is

$$\frac{1}{2}\binom{2k}{k}\binom{k}{k}((k-1)!)^2 = \frac{(2k)!}{2k^2},$$

and since these cases have been counted twice in $n!/k = (2k)!/k$, they should be subtracted once, to obtain

$$L_{k,2k} = \frac{(2k)!}{k} - \frac{(2k)!}{2k^2} = \frac{(2k)!}{k}\left(1 - \frac{1}{2k}\right),$$

for all $k \ge 1$. Other cases where k is only slightly less than $n/2$ are also relatively easy to count. For $n/3 < k \le n/2$, the corrections to $n!/k$ for the number of permutations on n objects with greatest cycle length k are:

$-n!/(2k^2)$ for the possibility that two cycles of length k are tied for longest;

$-n!/(k(k+j))$ for the possibility that there is a longer cycle, of length $k+j$, for each j, $1 \le j \le n-2k$.

For $n/3 < k \le n/2$, there are at most two cycles with length no less than k. Hence,

$$L_{k,2k} = \frac{(2k)!}{k}\left(1 - \frac{1}{2k}\right),$$

and

$$L_{k,n} = \frac{n!}{k}\left(1 - \frac{1}{2k} - \sum_{j=1}^{n-2k}\frac{1}{k+j}\right), \qquad \frac{n}{3} < k < \frac{n}{2}.$$

Because there may be a tie for longest cycle in a particular permutation, the expressions in brackets do not represent the probability that a cycle of length k is longest, nor the probability that it is uniquely longest, but is the weighting factor required to count the longest cycle only once per permutation, even when there is a tie for longest. For this reason also, as we go to values of k where $n/(r+1) < k < n/r$ with increasing r, the usual inclusion-exclusion rules of counting do not apply, but must be specially modified for the possibility that as many as r cycles of length k may be tied for longest.

Specifically, we find that

$$L_{k,3k} = \frac{(3k)!}{k}\left(1 - \frac{1}{2k}\left(1 - \frac{1}{3k}\right) - \sum_{j=1}^{k}\frac{1}{k+j}\right), \qquad k \ge 1,$$

$$L_{k,3k+1} = \frac{(3k+1)!}{k}\left(1 - \frac{1}{2k}\left(1 - \frac{1}{3k}\right) - \frac{1}{k+1}\left(1 - \frac{1}{2k}\right) - \sum_{j=2}^{k+1}\frac{1}{k+j}\right), k > 1,$$

Table 1. The values of $L_{k,n}$ for $1 \leq n \leq 13$ and $1 \leq k \leq n$

1													
1	3	2											
1	9	8	6										
1	25	40	30	24									
1	75	200	180	144	120								
1	231	980	1260	1008	840	720							
1	763	5152	8820	8064	6720	5760	5040						
1	2619	28448	61236	72576	60480	51840	45360	40320					
1	9495	162080	461160	653184	604800	518400	453600	403200	362880				
1	35695	979000	3647160	5854464	6652800	5702400	4989600	4435200	3991680	3628800			
1	140151	6179360	30044520	56567808	73180800	68428800	59875200	53222400	47900160	43545600	39916800		
1	568503	40575392	256524840	579705984	803088000	889574400	778377600	691891200	622702080	566092800	518918400	479001600	
1	2390479	279199648	2301835536	6178588416	9427017600	11564467200	10897286400	9686476800	8717829120	7925299200	7264857600	6706022400	6227020800

and more generally,

$$L_{k,n} = \frac{n!}{k}\left(1 - \sum_{j=1}^{n-2k}\frac{1}{k+j} + \sum_{j=1}^{n-3k-1}\sum_{l=1}^{n-3k-j}\frac{1}{2(k+j)(k+l)} - \frac{1}{2k}\left(1 - \sum_{j=1}^{n-3k}\frac{1}{k+j} - \frac{1}{3k}\right)\right),$$

for $n/4 < k \le n/3$.

In the next region, with $n/5 < k \le n/4$, we have

$$\begin{aligned} L_{k,n} = \frac{n!}{k}\Bigg(1 &- \sum_{j=1}^{n-2k}\frac{1}{k+j} + \sum_{j=1}^{n-3k-1}\sum_{l=1}^{n-3k-j}\frac{1}{2!(k+j)(k+l)} \\ &- \sum_{j=1}^{n-4k-2}\sum_{l=1}^{n-4k-j-1}\sum_{m=1}^{n-4k-j-l}\frac{1}{3!(k+j)(k+l)(k+m)} \\ &- \frac{1}{2k}\Bigg(1 - \sum_{j=1}^{n-3k}\frac{1}{k+j} + \sum_{j=1}^{n-4k-1}\sum_{l=1}^{n-4k-j}\frac{1}{2!(k+j)(k+l)} \\ &\qquad - \frac{1}{3k}\left(1 - \sum_{j=1}^{n-4k}\frac{1}{k+j} - \frac{1}{4k}\right)\Bigg)\Bigg). \end{aligned}$$

For the specific case $n = 5k$, it is necessary to modify only the right-most term, $-1/(4k)$, replacing it by

$$-\frac{1}{4k}\left(1 - \frac{1}{5k}\right).$$

The general pattern, for $n/(r+1) < k \le n/r$, is readily deduced from the cases $r = 2, 3$, and 4, illustrated above.

Starting from the other direction, at $k = 1$, we clearly have $L_{1,n} = 1$, for all $n \ge 1$, since there is exactly one permutation which has each object in a cycle by itself.

To evaluate $L_{2,n}$, we count the permutations of order 2 in S_n. We find the number of such permutations with j 2-cycles for each j, $1 \le j \le \lfloor n/2 \rfloor$, to obtain

$$L_{2,n} = \sum_{j=1}^{\lfloor n/2 \rfloor}\frac{1}{j!}\binom{n}{2}\binom{n-2}{2}\cdots\binom{n-2j+2}{2}, \qquad n \ge 2.$$

Similarly,

$$L_{3,n} = \sum_{j=1}^{\lfloor n/3 \rfloor}\frac{2^j}{j!}\binom{n}{3}\binom{n-3}{3}\cdots\binom{n-3j+3}{3}(1 + L_{2,n-3j}), \qquad n \ge 3.$$

The general result is easily shown to be

$$\begin{aligned} L_{k,n} &= \sum_{j=1}^{\lfloor n/k \rfloor}\frac{\{(k-1)!\}^j}{j!}\binom{n}{k}\binom{n-k}{k}\cdots\binom{n-kj+k}{k}\sum_{t=1}^{\min(k-1,n-kj)} L_{t,n-kj} \\ &= \sum_{j=1}^{\lfloor n/k \rfloor}\frac{n!}{j!k^j(n-kj)!}\sum_{t=1}^{\min(k-1,n-kj)} L_{t,n-kj}, \qquad 1 < k \le n, \end{aligned}$$

which expresses $L_{k,n}$ recursively in terms of $L_{k',n'}$ where $k' < k$ and $n' < n$. Table 1 shows the numerical triangle for $L_{k,n}$ for all n, $1 \le n \le 14$, and all k, $1 \le k \le n$.

The expected length of the longest cycle, over the set of elements in S_n, is given by

$$E_n = \sum_{k=1}^{n} kL_{k,n},$$

and the relative expected length of the longest cycle is $\lambda_n = E_n/n$. This subject has been discussed in [1–4], where it is shown that

$$\lim_{n\to\infty} \lambda_n = \lambda$$

exists, with $\lambda = 0.62432965\ldots$ (D. Knuth [5] has referred to λ as Golomb's constant.) It is interesting that E_n/n is a decreasing sequence, while $E_n/(n + 1)$ is an increasing sequence, both converging to λ. (The sequence $E_n/(n + 1/2)$ appears to converge to λ considerably faster.)

The proof that $b_n = E_n/(n + 1)$ is an increasing sequence, first given in [1], is as follows. In our deck-of-cards model, suppose that an $(n + 1)$st card is introduced in a random position (there are $n + 1$ equally likely positions) in a random deck of n cards. The probability would be $E_n/(n + 1)$ that it would land in the previously longest cycle, thereby increasing the length of the longest cycle by 1, if longest cycles were always unique. Since longest cycles are not necessarily unique for $n \ge 2$, the actual probability that the new card will increase the expected length by 1 exceeds $E_n/(n + 1)$ for all $n \ge 2$. Thus,

$$E_{n+1} \ge E_n + E_n/(n + 1) = E_n(1 + 1/(n + 1)) = E_n((n + 2)/(n + 1))$$

for $n \ge 1$, with strict inequality for $n \ge 2$. Dividing by $n + 2$, we find that

$$b_{n+1} = \frac{E_{n+1}}{n+2} \ge \frac{E_n}{n+1} = b_n,$$

for $n \ge 1$, with strict inequality for $n \ge 2$.

Since $\{b_n\}$ is a monotonic increasing sequence for $n \ge 2$, and is clearly bounded from above by 1, it has a limit λ, by a well-known theorem of Weierstrass. This is the proof, given in [1], that the limit λ exists. In response to the research problem [2], several closed-form expressions for $\lambda = 0.62432965\ldots$ were found, all involving definite integrals of non-elementary functions (see [3] and [4]). Perhaps the simplest of these, due to N. Fine, is

$$\lambda = \int_0^1 e^{\mathrm{Li}(t)}\, dt$$

where

$$\mathrm{Li}(x) = \int_0^x \frac{dt}{\ln t}$$

is the integral logarithm function popularized by Gauss in his study of the distribution of prime numbers. Alternatively, as shown by L. Shepp,

$$\lambda = \int_0^\infty e^{-t} e^{-\mathrm{Ei}(t)}\, dt,$$

where

$$\mathrm{Ei}(t) = \int_t^\infty \frac{e^{-u}}{u}\, du.$$

The cycles of length k, for $n/2 < k \le n$, each contribute 1 to E_n, for a total contribution of $\lfloor (n+1)/2 \rfloor$ to E_n, and the contribution to the limit λ is $1/2 = 0.50000$, or $80.0859\ldots\%$ of $\lambda = 0.62432965\ldots$. Numerical calculation shows that the total contribution to λ of k with $n/3 < k \le n/2$ is $0.11003365\ldots$, or $17.6243\ldots\%$ of λ, which leaves less than 2.3% for all $k \le n/3$. Of this amount, k with $n/4 < k \le n/3$ contributes $0.01316960\ldots$ or $2.1094\ldots\%$ of λ, leaving only $0.1804\ldots\%$ as the total contribution of all cycles of length less than $n/4$ to the relative expected length of the longest cycle.

For $n = 6a$, the total contribution of k, $n/3 < k \le n/2$, to E_n is given by

$$\sum_{k=n/3+1}^{n/2} \left(1 - \frac{1}{2k} - \sum_{j=1}^{n-2k} \frac{1}{k+j}\right) = a - \frac{1}{2}\sum_{k=2a+1}^{3a} \frac{1}{k} - \sum_{j=1}^{a-1} \frac{j}{2a+1+j} - \sum_{j=1}^{a-1} \frac{j}{4a-j},$$

which can be used either numerically or analytically to obtain the limiting value $0.11003365\ldots$ mentioned above as the portion of λ which arises from this region of k-values. The analytic approach uses

$$\begin{aligned}
\frac{1}{n}&\left(a - \frac{1}{2}\sum_{k=2a+1}^{3a} \frac{1}{k} - \sum_{j=1}^{a-1} \frac{j}{2a+1+j} - \sum_{j=1}^{a-1} \frac{j}{4a-j}\right)\\
&\sim \frac{1}{6a}\left(a - \frac{1}{2}\int_{2a}^{3a} \frac{dt}{t} - \int_1^a \frac{t\,dt}{2a+1+t} - \int_1^a \frac{t\,dt}{4a-t}\right)\\
&\sim \frac{1}{6} - \frac{1}{12a}\ln\frac{3}{2} - \frac{1}{6a}\left(a - (2a+1)\ln\frac{3}{2}\right) - \frac{1}{6a}\left(4a\ln\frac{4}{3} - a\right)\\
&\sim \frac{1}{6} - \frac{1}{3}\ln\frac{2}{3} - \frac{2}{3}\ln\frac{4}{3} = \frac{1}{6} + \ln 3 - \frac{5}{3}\ln 2 = 0.1100336544\ldots
\end{aligned}$$

as $a \to \infty$, which agrees precisely with the numerical data for large n.

Similar, though ever more complicated expressions correspond to each r-region, $n/(r+1) < k \le n/r$.

In the collection of the $n!$ decks of n cards each, representing the permutations in S_n, there is a grand total of $n\,n!$ cards, divided up into cycles with lengths from 1 to n. Among the $n\,n!$ cards, there are exactly $n!$ cards in cycles of length k, for every k from 1 to n. Thus the total number of cycles, among all $n!$ permutations, is

$$n!\left(1 + \frac{1}{2} + \frac{1}{3} + \frac{1}{4} + \ldots + \frac{1}{n}\right),$$

so the average (or expected) number of cycles in a random permutation on n objects is

$$\sum_{k=1}^{n} \frac{1}{k} \sim \ln n + \gamma + o\left(\frac{1}{n}\right)$$

as $n \to \infty$, where $\gamma = 0.577\ldots$ is the Euler–Mascheroni constant. Thus, in a random permutation on 10^3 or 2^{10} objects, the expected number of cycles is only about 7.5. From our card model, in a random sequence of the numbers from 1 to n, this is also the expected number of terms in the sequence which are smaller than all subsequent terms.

Several other random processes have been shown to have distributions which mimic the distribution of cycle lengths in a random permutation. Dickman [6] studied the sizes of the prime factors of a random integer N on the range $[1, x]$. Among other things, he showed that if $\bar{p}_1$ is the expected largest prime factor of N, then

$$\lim_{x \to \infty} \frac{\log \bar{p}_1}{\log N} = \lambda = 0.62432965\ldots$$

This may be stated as follows. The expected fraction of the digits of N which come from its largest prime factor tends to the limit λ as $N \to \infty$.

Since a ratio of logarithms is invariant under a change of logarithmic base, it does not matter whether the digits are in base 2, or base 10, or any other base $b > 1$. The analysis of cycle lengths in this paper, based on $k \in (n/(r+1), n/r]$, should correspond directly to the analysis of the sizes of prime factors based on $p \in (N^{1/(r+1)}, N^{1/r}]$, or equivalently to $\log p \in ((\log N)/(r+1), (\log N)/r]$. (Here $\log p$ corresponds to k, and $\log N$ to n, in the analogy with cycles of a random permutation.)

Another model for this distribution arises in the study of the degrees of the irreducible factors of a random polynomial of degree n over GF(q). If $\bar{d}_1$ is the expected value of the highest degree of any irreducible factor of a polynomial of degree n over GF(q), then

$$\lim_{n \to \infty} \frac{\bar{d}_1}{n} = \lambda.$$

Another interesting result (see [1–3]) is the following. The probability of the longest cycle in a random permutation on n objects (n large) having a relative length $L/n > a$ is $\ln(1/a) = -\ln a$ for $1/2 \le a \le 1$. In particular, the probability that the longest cycle has relative length no less than 0.5 is $\ln 2 = 0.69315$; and it is an even bet that the relative length exceeds $1/\sqrt{e} = 0.60653\ldots$

REFERENCES

1. S. W. Golomb, *Cycles from Nonlinear Shift Registers*. Jet Propulsion Laboratory Progress Report 20–389, August, 1959.
2. S. W. Golomb, Random permutations, research problem No.11. *Bull. Amer. Math. Soc.* (1964) **70**, 747.
3. S. W. Golomb, *Shift Register Sequences*, Holden–Day, San Francisco, 1967.
4. S. W. Golomb, *Shift Register Sequences,* Revised Edition, Aegean Park Press, Laguna Hills, 1982.
5. D. E. Knuth, *The Art of Computer Programming, Vol. 2: Seminumerical Algorithms*. Addison–Wesley, Reading, MA, 1969.
6. K. Dickman, On the frequency of numbers containing prime factors of a certain relative magnitude. *Arkiv för Matematik, Astronomi och Fysik* (1930) **22A**, No. 10, 1-14.

Probabilistic Methods in Discrete Mathematics, pp. 219–230
V. F. Kolchin et al. (Eds.)

EXPLICIT BOUNDS FOR PROBABILITIES OF LARGE DEVIATIONS OF SUMS OF RANDOM VECTORS WITH A GIVEN GRAPH OF DEPENDENCIES

A. B. GORCHAKOV

Moscow

Abstract — For the mixed cumulants of order k of a sum of random vectors with a given graph of dependencies, we obtain an upper bounds of the form $k!\,H\Delta^{2-k}$, where H is a constant depending on the second moment of the sum and $\Delta \leq B\Gamma D^{-1/2}$ if the random vectors have bounded coordinates, the vertices of their graph of dependencies have the maximum degrees equal to Γ, and the minimum of the variances of sums of coordinates of the vectors is equal to D. With the use of these bounds we obtain new explicit estimates of the closeness of the distribution of a sum of random vectors with a small number of dependencies to the multidimensional normal distribution. These estimates are effective in the domain of large deviations.

1. INTRODUCTION

We consider a scheme of summation of dependent random vectors such that a sum of dependent random vectors with a small number of dependencies is a particular case of this scheme. This scheme is a generalization to the multidimensional case of the scheme considered in [8].

Let ρ be a function from $\mathbf{R}^d \times \mathbf{R}^d$ to the set of non-negative real numbers $\mathbf{R}_+$ such that $\rho(i,i) = 0$ for any $i \in \mathbf{R}^d$ and $\rho(i,j) = \rho(j,i)$ for any $i,j \in \mathbf{R}^d$.

Let $\alpha(p,r)$ be a function from $\mathbf{R}_+ \times \mathbf{N}$ to $[0,1]$ such that $\alpha(p_1,r) \leq \alpha(p_2,r)$ for any $p_1 \geq p_2$ and r, and $\alpha(p,r_1) \leq \alpha(p,r_2)$ for any $r_1 \leq r_2$ and p. Here and further $M \times N$ denotes the set $\{(m,n)\colon m \in M, n \in N\}$ and $\mathfrak{M}^k = \mathfrak{M} \times \ldots \times \mathfrak{M}$ is the direct product of k sets $\mathfrak{M}$.

Let ξ_i, $i \in \mathbf{R}^d$, be an s-dimensional non-homogeneous random field and V be some at most countable subset of $\mathbf{R}^d$. Random vectors ξ_i, $i \in V$, are called $\rho\alpha$-dependent random vectors if for any finite sets $V_1, V_2 \subset V$ the sets of random vectors $\{\xi_i,\ i \in V_1\}$ and $\{\xi_i,\ i \in V_2\}$ satisfy the strong mixing condition with the mixing coefficient $\alpha(p,r)$, where $r = |V_1|+|V_2|$, $|M|$ is the cardinality of the set M, $p = \min\{\rho(i,j) \mid i \in V_1, j \in V_2\}$, i.e., for any $A \in \sigma\{\xi_i, i \in V_1\}$ and $B \in \sigma\{\xi_i, i \in V_2\}$

$$|\mathbf{P}(AB) - \mathbf{P}(A)\,\mathbf{P}(B)| \leq \alpha(p,r),$$

where $\sigma\{\xi_i, i \in W\}$ is the σ-algebra of events generated by the random vectors ξ_i, $i \in W \subset V$. Varying the functions ρ and α, we can obtain various schemes of dependence of random vectors ξ_i, $i \in V$. It is obvious that any random vectors ξ_i, $i \in V$, are $\rho\alpha$-dependent random vectors with $\rho(i,j) = 0$ and $\alpha(p,r) = 1$. It is clear that this trivial case is not interesting. Let us consider some non-trivial examples of $\rho\alpha$-dependent random vectors.

Let ρ be the Euclidean distance, i.e.,

$$\rho(i,j) = \sqrt{(i_1 - j_1)^2 + \ldots + (i_d - j_d)^2}$$

for any $i = (i_1, \ldots, i_d)$ and $j = (j_1, \ldots, j_d)$ from $\mathbf{R}^d$. Then $\rho\alpha$-dependent random vectors ξ_i, $i \in V$, constitute a random field satisfying the well-known Rosenblatt strong mixing condition with the mixing coefficient $\alpha(p, r)$ (see bibliography in [2]).

Let $X_1, \ldots, X_n$ be a sequence of random variables satisfying the Rosenblatt strong mixing condition with the mixing coefficient $\alpha(p)$ and let

$$\xi_i, \quad i = (i_1, \ldots, i_d) \in V = \{(i_1, \ldots, i_d) \mid 1 \le i_l \le n,\ 1 \le l \le d\},$$

be random vectors such that for any $i = (i_1, \ldots, i_d) \in V$ the random vector ξ_i is measurable with respect to $\sigma(X_j, j \in \{i_1, \ldots, i_d\}$. Note that the random vector

$$\xi = \sum_{i \in V} \xi_i$$

is the well-known generalized U-statistics (see bibliography in [3]).

Let

$$\rho(i,j) = \min\{|i_k - j_l| \mid 1 \le k \le d,\ 1 \le l \le d\}, \quad i = (i_1, \ldots, i_d), \quad j = (j_1, \ldots, j_d),$$

and

$$\alpha(p, r) = \min\{1, (2dr - 1)\alpha(p)\}.$$

Then the random vectors ξ_i, $i \in V$, are $\rho\alpha$-dependent random vectors.

Let a finite set V, $V \subset \mathbf{R}^d$, be the vertex set of some graph G and let T be the set of edges of the graph G. The graph G is called the dependence graph of random vectors ξ_i, $i \in V$, if $(i, i) \in T$ for any $i \in V$ and sets of random vectors $\{\xi_i, i \in V_1\}$ and $\{\xi_i, i \in V_2\}$ are independent for any finite sets $V_1, V_2 \subset V$ such that $(i,j) \notin T$ for any $i \in V_1$ and $j \in V_2$. Random vectors ξ_i, $i \in V$, with a dependence graph G are called dependent random vectors with a small number of dependencies if $|T|$ is essentially less than $|V|(|V| - 1)/2$ (see bibliography in [4]). It is easy to see that the random vectors ξ_i, $i \in V$, with the dependence graph G are $\rho\alpha$-dependent random vectors with $\rho(i,j) = 0$ if $(i,j) \in T$ and $\rho(i,j) = 1$ if $(i,j) \notin T$, and $\alpha(0, r) = 1$ and $\alpha(1, r) = 0$ for any $r \in \mathbf{N}$. Sums of random vectors with a small number of dependencies are of frequent use in combinatorial analysis (the corresponding examples can be found in [4]).

Let $\xi = (\xi^{(1)}, \dots, \xi^{(s)})$ be a random vector such that $\mathbf{E}|\xi^{(j)}|^{k_0}$ exists for any $1 \le j \le s$ and some $k_0 \in \mathbf{N}$. The quantities

$$S_\xi(v) = \frac{d^k}{i^k dz_{v_1} \dots dz_{v_k}} \operatorname{Ln} \mathbf{E} \exp\left\{ i \sum_{j=1}^{s} z_j \xi^{(j)} \right\} \Bigg|_{z_1 = \dots = z_s = 0},$$

where $k = 1, \dots, k_0$, $v = (v_1, \dots, v_k)$, $1 \le v_l \le s$, $1 \le l \le k$, and the symbol Ln denotes the principal branch of the logarithm, are called the cumulants of the random vector ξ.

For the case $s = 1$ upper bounds for the cumulants of sums of $\rho\alpha$-dependent random vectors were obtained in [8]. For this case a brief review of the results and bibliography are also contained in [8]. For the case $s > 1$ upper bounds for the cumulants of sums of independent random vectors were obtained in [7]. Upper bounds for the cumulants of quadratic forms

$$\xi = \sum_{i,j=1}^{n} (a_{ij}^{(1)} \xi_i^{(1)} \xi_j^{(1)}, \dots, a_{ij}^{(s)} \xi_i^{(s)} \xi_j^{(s)}),$$

where $\xi_i = (\xi_i^{(1)}, \dots, \xi_i^{(s)})$, $i \in \mathbf{N}$, is a Gaussian random field, can be found in [7].

We consider $\rho\alpha$-dependent random vectors $\xi_i = (\xi_i^{(1)}, \dots, \xi_i^{(s)})$, $i \in V$, and

$$\xi = \sum_{i \in V} \xi_i. \tag{1}$$

In the next section we obtain upper bounds for the cumulants of the random vector $\xi = (\xi^{(1)}, \dots, \xi^{(s)})$. These estimates generalize the results of [8] to the multidimensional case.

2. UPPER BOUNDS FOR THE CUMULANTS

We introduce some notation. Let

$$L_p(m) = \max_{1 \le l \le s} \sup \left\{ \sum_{j \in V:\ \rho(i,j) \le p} \left(\mathbf{E}|\xi_j^{(l)}|^m \right)^{1/m} \ \Big| \ i \in V \right\},$$

$$C(m) = \max_{1 \le l \le s} \sum_{j \in V} \left(\mathbf{E}|\xi_j^{(l)}|^m \right)^{1/m},$$

$$D_p(m) = \max_{1 \le l \le s,\, 1 \le t \le s} \sum_{i,j \in V:\ \rho(i,j) \le p} \left(\mathbf{E}|\xi_i^{(t)} \xi_j^{(l)}|^m \right)^{1/m}.$$

Let $Q_p(k)$ be non-random values such that for any $1 \le l \le s$, $i, i_1, \dots, i_{k-1} \in V$,

$$\sum_{j \in V:\ \rho(i,j) \le p} \mathbf{E}(|\xi_j^{(l)}| \mid \sigma(\xi_i, \xi_{i_m}, m \in \{1, \dots, k-1\})) \le Q_p(k) \quad \text{a.s.},$$

where $\mathbf{E}(\xi \mid \mathcal{F})$ is the conditional mathematical expectation of a random variable ξ with respect to a σ-algebra $\mathcal{F}$. Here and further we do not assume that $L_p(m)$, $C(m)$, $D_p(m)$, and $Q_p(k)$ are necessarily finite.

We introduce two more values which are uniquely determined by $L_p(m)$, $C(m)$, $D_p(m)$, and $Q_p(k)$.

Let $C(k) < \infty$. Then $M_p(k)$ is the minimum of the values

$$C(k)(L_p(k))^{k-1}, \quad D_p(1)(Q_p(k))^{k-2}, \quad D_p(m)(L_p(m(k-2)/(m-1)))^{k-2}, \quad m = 2, 3, \ldots$$

Let $\mathbf{E}|\xi_j^{(l)}|^{(1+\delta)k} < \infty$ for some $\delta > 0$ and some $k = 3, 4, \ldots$. Then

$$R_p(\delta, k) = \sum_{m \in \rho(V):\ m > p} (L_m((1+\delta)k))^{k-1} C((1+\delta)k) \lambda(6(\tilde{\alpha}(m,k))^{\delta/(1+\delta)}, k),$$

where

$$\tilde{\alpha}(p,k) = \max\{\alpha(p,r)/r \mid 1 \le r \le k\},$$

$\rho(V)$ is the set of non-negative real numbers such that $m \in \rho(V)$ if and only if $m = \rho(i,j)$ for some $i, j \in V$,

$$\lambda(i,k) = \sum_{r=1}^{[k/2]} \frac{i^r k!}{r(k-2r)!} (3r+1)^{k-2r},$$

$[z]$ is the integer part of z.

The following theorem is valid.

THEOREM 1. *Let* $\mathbf{E}\xi_i^{(l)} = 0$ *for any* $i \in V$ *and* $1 \le l \le s$, *let* $C((1+\delta)k) < \infty$ *for some positive* δ *and let* $R_{p_0}(\delta, k) < \infty$ *for some* $p_0 \in \mathbf{N}$.

Then for any $\nu = (\nu_1, \ldots, \nu_k)$, $1 \le \nu_l \le s$, $1 \le l \le k$,

$$|S_\xi(\nu)| \le k^{k-2} \min\{2^{k-1} M_p(k) + R_p(\delta, k) \mid p \in \rho(V)\}. \tag{2}$$

The proof of this theorem can be carried out by replacing ξ_{i_l}, $1 \le l \le k$, by $\xi_{i_l}^{(\nu_l)}$, $1 \le l \le k$, in the proof of Theorem 1 in [8].

We apply this theorem to the case where ξ_i, $i \in V$, are random vectors with dependence graph G.

THEOREM 2. *Let* ξ_i, $i \in V$, *be random vectors with dependence graph* G *satisfying the conditions of Theorem* 1, $\mathbf{D}\xi^{(l)} = 1$ *for any* $1 \le l \le s$, *and* $Q_1(k) = Q < \infty$ *for any* $k \in \mathbf{N}$.

Then for any $k \in \mathbf{N}$ *and* $\nu = (\nu_1, \ldots, \nu_k)$

$$|S_\xi(\nu)| \le \frac{2e^2 k!}{\sqrt{\pi k^3}} D_0(1)(4eQ)^{k-2}. \tag{3}$$

Theorem 2 follows from Theorem 1 and Stirling's formula.

Note that if

$$B = \max_{1 \le l \le s} \sup\{\lim_{m \to \infty} (\mathbf{E}|\xi_i^{(l)}|^m)^{1/m} \mid i \in V\} < \infty,$$

then $Q \le B\Gamma$, where Γ is the maximum degree of the vertices of G.

In the next section we obtain a new explicit bound for the closeness of the distribution of a random vector $\xi = (\xi^{(1)}, \ldots, \xi^{(s)})$ satisfying condition (3) to the multidimensional normal distribution.

3. EXPLICIT BOUNDS FOR THE CLOSENESS TO THE MULTIDIMENSIONAL NORMAL DISTRIBUTION

We introduce some notation. Let F be the distribution of ξ, $\Phi(\mathcal{A}, U)$ be the s-dimensional normal distribution with the zero mean vector and the covariance matrix U, let

$$\int_M f(x)G(dx)$$

stand for the Lebesgue integral

$$\int \cdots \int_M f(x_1, \dots, x_s)G(dx_1, \dots, dx_s)$$

over a Borelian convex set M and

$$\int f(x)G(dx) = \int_{\mathbf{R}^s} f(x)G(dx);$$

let

$$\|x\| = \left(\sum_{i=1}^{s} x_i^2\right)^{1/2},$$

and

$$\|B\| = \sup_{\|x\|=1} \|xB\|,$$

where B is a non-singular matrix.

Note that if B is a positive semi-definite symmetric matrix and $\lambda(B)$ is the maximum eigenvalue of B, then $\|B\| = \lambda(B)$.

Further we assume that $\xi = (\xi^{(1)}, \dots, \xi^{(s)})$ is a random vector with the zero mean vector and the covariance matrix V, $\mathbf{D}\xi_l = 1$ for any $1 \le l \le s$, and for any $k \in \mathbf{N}$ and $\nu = (\nu_1, \dots, \nu_k)$,

$$|S_\xi(\nu)| \le k!\, H\Delta^{2-k}, \tag{4}$$

where H and $\Delta < 1$ are constants.

THEOREM 3. *For any measurable bounded function f*

$$\left|\int f(x)F(dx) - \int f(x)\Phi(dx, V)\right| \le C(s, \Delta, H, V) \sup\{|f(x) - f(y)| : x, y \in \mathbf{R}^s\},$$

where

$$C(s, \Delta, H, V) = \left(\frac{8(10+\sqrt{3})}{18\sqrt{2\pi}} + a_1(s)\right)\rho_4^{3/4}\Delta_0^{-1} + a_2(s, \rho_4)\Delta_0^{-2}(2\ln\Delta_0)^{s/2}$$
$$+ \frac{4}{3\sqrt{\pi}} s\Delta_0^{-2}(2\ln\Delta_0)^{-1/2} + \frac{32}{3}s^{5/2}\rho_4^{9/4}\pi^{-1}(\Delta_0^2 2\ln\Delta_0)^{-3/2}$$
$$+ \frac{32}{3}(2s)^{(s-2)/2}(\Gamma(s/2))^{-2}m + \frac{4}{3}2^5\pi^{-1/3}s^{4/3}\rho_4^{3/4}\Delta_0^{-1}\frac{\Gamma((s+1)/2)}{\Gamma(s/2)},$$

$$a_1(s) = \frac{2}{9\sqrt{2\pi}}(4e^{-3/2} + 1) + \frac{2}{3\pi}e^{-1/2}s^{1/2}(1 - s^{-1}) + \frac{2}{3}(2/\pi)^{3/2}(s^{3/2} - 3s^{1/2} + 2),$$

$$a_2(s, \rho_4) = \frac{4\pi}{3\Gamma(s/2)}s^{s/2}(s+2)2^{s+1}(0.14 + \rho_4/24)$$
$$+ 0.19(2.3)^s s^{s/2}(s+2)(s+4)\rho_4^{3/2}(\Gamma(s/2))^{-1},$$

$$m = (2\ln\Delta_0)^{s/2}\int_l^\infty \left(r^{s-1}e^{-0.26r^2} + r^{s-1}e^{-r^2/2} + \frac{1}{6}\rho_4^{3/4}\Delta_0^{-1}r^{s+2}e^{-r^2/2}\right)dr,$$

$$l = \frac{\Delta_0}{2\sqrt{\rho_4}},$$

$$\rho_4 = \|V^{-1}\|^2 s^2\left(24H\frac{\Delta^2}{[\Delta^2]} + 3\right).$$

PROOF. We consider the random vector

$$Z = \Delta_0^{-1}(Y_1, \ldots, Y_{\Delta_0^2}),$$

where $\Delta_0^2 = [\Delta^2]$ and Y_i, $i = 1, \ldots, \Delta_0^2$, are independent identically distributed random vectors such that for any $k \in \mathbf{N}$ and $v = (v_1, \ldots, v_k)$

$$S_{Y_1}(v) = S_\xi(v)\Delta_0^{k-2}.$$

It is easy to see that the distribution function of Z is equal to F. From inequality (4) it follows that

$$\mathbf{E}\|Y_1\|^4 \le \left(\frac{24H\Delta^2}{[\Delta^2]} + 3\right)s^2.$$

Now Theorem 3 follows from Theorem 13.2 in [5].

However the result of Theorem 3 is interesting only if the integral

$$\int f(x)\Phi(dx, V)$$

is essentially lesser than Δ_0^{-1}. The last inequality is not valid in many probabilistic problems of discrete mathematics. Therefore we suggest explicit non-uniform bounds of closeness of the distribution of the random vector $\xi = (\xi^{(1)}, \ldots, \xi^{(s)})$ to the multidimensional normal distribution which may be used for the estimation of the probabilities of large deviations of ξ.

Let

$$T(B) = \sum_{i=1}^{s} |b_{ij}|,$$

where $B = (b_{ij})$ is an $s \times s$ symmetric matrix;

$$R(B, x) = \frac{1}{2}(xB, x),$$

where $x \in \mathbf{R}^s$,

$$(a,b) = \sum_{i=1}^{s} a_i b_i,$$

$a = (a_1, \dots, a_s), b = (b_1, \dots, b_s) \in \mathbf{R}^s$;

$$m(B,r) = \max\{R(B,x) \mid x = (x_1, \dots, x_s)\colon |x_l| \le r, 1 \le l \le s\};$$

and

$$\delta_0(x) = \sqrt{s}\|x\|\Delta^{-1}.$$

Further we assume that $\delta_0(x) < 1$ and

$$C(x) = \frac{27}{4} Hs\delta_0(x)(2 - \delta_0(x))(1 - \delta_0(x))^{-3} \le 1.$$

Let $\varepsilon(x)$ denote the minimal root on the interval $(0, 1 - \delta_0(x))$ of the equation

$$T(V^{-1})H\sqrt{s}\delta_0(x)(3(1-\lambda) - 2\delta_0(x)) = \lambda(1 - \lambda - \delta_0(x))^2.$$

Note that $\varepsilon(x)$ exists and $\varepsilon(x) \le (1 - \delta_0(x))/3$. Let

$$\mathcal{M} = \{x\colon \delta(x) < 1,\ C_1(x) < 1,\ C(x) \le 1,\ c_i x_i \ge T(V^{-1})C_2(x),\ i = 1, \dots, s\},$$

where

$$C_1(x) = 2\|V^{-1}\|H(3\delta(x) - 3\delta^2(x) + \delta^3(x))(1 - \delta(x))^{-3},$$
$$\delta(x) = \delta_0(x)(1 - \varepsilon(x))^{-1},$$
$$C_2(x) = H\Delta(3\delta^2(x) - 2\delta^3(x))^3(1 - \delta(x))^{-2},$$

and b is the nearest vector of $\mathcal{M}$ to the vector aV^{-1}.

We consider the set

$$M_s(a,c) = \{x\colon c_i x_i \ge a_i,\ i = 1, \dots, s\}, \tag{5}$$

where $c = (c_1, \dots, c_s)$, $c_i \in \{-1, 1\}$, $i = 1, \dots, s$, $a = (a_1, \dots, a_s) \in \mathbf{R}^s$. The following theorem is valid.

THEOREM 4. *The following estimates hold:*

$$F(M_s(a,c)) \ge D^{-1}\Phi(M_s(a_0,c), V) - C_0 e^{-R(V^{-1},a)},$$
$$F(M_s(a,c)) \le D\Phi(M_s(a_1,c), V) + C_0 e^{-R(V^{-1},a)},$$

where

$$a_\varepsilon^{(i)} = (a_i + (-1)^\varepsilon C_3(b)\operatorname{sign} c_i)(1 + (-1)^\varepsilon C_1(b))^{-1/2}, \quad 1 \le i \le s,$$

$$a_\varepsilon = (a_\varepsilon^{(1)}, \ldots, a_\varepsilon^{(s)}),$$
$$C_3(b) = H\Delta(3\delta^2(b) - \delta^3(b))(1 - \delta(b))^{-3},$$
$$D = \left(\frac{1 + C_1(b)}{1 - C_1(b)}\right)^{s/2} \exp\{H\Delta^2\delta^3(b)(1 - \delta(b))^{-3}\},$$
$$C_0 = W \exp\{H\Delta^2\delta^3(b)(1 - \delta(b))^{-1} + m(V^{-1}, C_2(b)) + \sqrt{s}C_2(b)\|aV^{-1} - b\| + R(V, aV^{-1} - b)\},$$

W is obtained by replacing $\|V^{-1}\|$ *by* $\|V^{-1}\|(1 - C_1(b))^{-1}$, *H by* $H(1 - \delta(b))^{-3}$, *and* Δ *by* $\Delta(1 - \delta(b))^{-1}$ *in the formula of* $C(s, \Delta, H, V)$.

REMARK 1. Explicit bounds for $F(M)$ were obtained in [6] under the assumptions that $V = E$, where E is the identity matrix, and

$$\left| \sum_{\nu=(\nu_1 \ldots, \nu_s)} S_\xi(\nu) \prod_{i=1}^{k} h_{\nu_i} \right| \le (k!)^\gamma H\Delta^{2-k}, \qquad \gamma = 1,$$

for any $h \in \mathbf{R}^s$ such that $\|h\| = 1$. Note that the last condition is similar to inequality (4). The explicit bounds of [6] were generalized to the case $\gamma > 1$ in [7]. However the application of these explicit bounds meets some difficulties, because these bounds depend on the cumulants of ξ which are unknown in many problems. Moreover some additional difficulties in applications of the bounds given in [6, 7] appear in the case $V \neq E$. Note that in the case $V = E$ and $a \in \mathcal{M}$ the explicit bounds given in Theorem 4 are better than the explicit bounds which are obtained from the bounds of [6, 7] by replacing the cumulants of ξ by their upper estimates.

REMARK 2. For the case $a \notin \mathcal{M}$ the value of C_0 may be so large that the explicit bounds of Theorem 4 cannot be used for the estimation of the probabilities of large deviations of ξ. Nevertheless, we can obtain explicit bounds for $F(M_s(a, c))$ using the equalities

$$F(M_s(a, c)) = F_{s-1}^{(i)}(M_{s-1}(a_0(i), c_i') - F(M_s(a, c_0(i))), \qquad i = 1, \ldots, s,$$

where $F_{s-1}^{(i)}$ is the distribution of $\xi(i) = (\xi^{(1)}, \ldots, \xi^{(i-1)}, \xi^{(i+1)}, \ldots, \xi^{(s)})$,

$$c_i' = (c_1, \ldots, c_{i-1}, c_{i+1}, \ldots, c_s),$$
$$a_0(i) = (a_1, \ldots, a_{i-1}, a_{i+1}, \ldots, a_s),$$
$$c_0(i) = (c_1, \ldots, c_{i-1}, -c_i, c_{i+1}, \ldots, c_s),$$

and the explicit bounds for $F_{s-1}^{(i)}(M_{s-1}(a_0(i), c_i'))$ and $F(M_s(a, c_0(i)))$ obtained with the use of Theorem 4.

PROOF OF THEOREM 4. The proof of this theorem follows the scheme used in [1]. Let $f(h)$ be the generating function of the moments of ξ, i.e.,

$$f(h) = \int e^{(h,x)} F(dx),$$

and let $\psi(h) = \ln f(h)$.

It is well known that

$$\psi(h) = \sum_{k=1}^{\infty} \frac{1}{k!} \sum_{v=(v_1,\dots,v_k)} S_\xi(v) \prod_{j=1}^{k} h_{v_j} \tag{6}$$

Further we take $h = (h_1, \dots, h_s)$ equal to the root of the equation

$$\operatorname{grad} \psi(h) = bV.$$

It is easy to see that h exists and it is unique if $C_1(b) < 1$. Let

$$Q(dx) = e^{-\psi(h)+(h,x)} F(dx).$$

Further s and the vector c are fixed and therefore we omit them in notations. It is easy to see that

$$I(a) = \int_{M(a)} F(dx) = e^{\psi(h)} \int_{M(a)} e^{-(h,x)} Q(dx).$$

Let $A(h)$ be obtained from $I(a)$ by replacing $Q(dx)$ by $\Phi_Q(dx)$, where $\Phi_Q(dx)$ is the s-dimensional normal distribution such that the mean vector and the covariance matrix of $\Phi_Q(dx)$ are equal to the mean vector and the covariance matrix of $Q(dx)$. Let η be a random vector with the distribution Q. It is clear that

$$A(h) = \exp\{\psi(h) - (h, bV) + R(B(h), h)\}\Phi(M(a - bV + hB(h), B(h)) \tag{7}$$

where

$$B(h) = \left. \frac{\partial^2 \psi(h)}{\partial h_i \partial h_j} \right|_{1 \le i \le s, 1 \le j \le s}$$

It is easy to see that

$$|I(a) - A(h)| \le \exp\{\psi(h) - (h, a)\} R, \tag{8}$$

where

$$R = \left| \int_{M(0} e^{-(h,x)} Q_0(dx) - \int_{M(0} e^{-(h,x)} \Phi(dx, B(h)) \right|,$$

Q_0 is the distribution of $\eta - a$.

Further we obtain upper estimates of the values included in (7) and (8). It is clear that

$$\begin{aligned}
\psi(h) - (h, a) &= \psi(h) - R(V, h) + R(V^{-1}, a - hV) - R(V^{-1}, a), \\
R(V^{-1}, a - hV) &= R(V^{-1}, a + bV - bV - hV) \\
&\le R(V^{-1}, bV - hV) + \sqrt{s} \|bV - hV\| \|aV^{-1} - b\| + R(V, aV^{-1} - b).
\end{aligned}$$

From (6) it follows that

$$\psi(h) - (h, bV) + R(B(h), h) = \sum_{k=3}^{\infty} \frac{1}{k!} \left(1 - \frac{3k}{2} + \frac{k^2}{2}\right) \sum_{\nu=(\nu_1,\dots,\nu_k)} S_\xi(\nu) \prod_{j=1}^{k} h_{\nu_j},$$

$$\psi(h) - R(V, h) = \sum_{k=3}^{\infty} \frac{1}{k!} \sum_{\nu=(\nu_1,\dots,\nu_k)} S_\xi(\nu) \prod_{j=1}^{k} h_{\nu_j},$$

$$g_m = \sum_{k=3}^{\infty} \frac{1}{k!} \sum_{\nu=(\nu_1,\dots,\nu_k)} (k-2) S_\xi(\nu) \times \sum_{t=1}^{k} \left(\mathbf{I}(\nu_t = m) \prod_{j \neq t} h_{\nu_j}\right), \qquad m = 1, \dots, s,$$

where $hB(h) - bV = (g_1, \dots, g_s)$.

Using the last equalities and the well-known equality

$$\sum_{k=0}^{\infty} \frac{(k+r)!}{k!} x^k = \frac{r!}{(1-x)^{r+1}}, \qquad |x| < 1, \tag{9}$$

we obtain that for $\delta_0(h) < 1$

$$|\psi(h) - (h, bV) + R(B(h), h)| \le H\Delta^2 \delta_0^3(h)(1 - \delta_0(h))^{-3}, \tag{10}$$

$$|\psi(h) - R(V, h)| \le H\Delta^2 \delta^3(h)(1 - \delta_0(h))^{-1}, \tag{11}$$

$$\left|\frac{\partial^2 \psi(h)}{\partial h_i \partial h_j} - \nu_{ij}\right| \le 2H(3\delta_0(h) - 3\delta_0^2(h) + \delta_0^3(h))(1 - \delta_0^{-3}(h))^{-3}, \tag{12}$$

where $V = (\nu_{ij})_{1 \le i \le s,\, 1 \le j \le s}$,

$$\left|\frac{\partial \psi(h)}{\partial_i} - \sum_{i=1}^{s} \nu_{ij} h_j\right| \le H\Delta(3\delta_0^2(h) - 2\delta_0^3(h))(1 - \delta_0(h))^{-2}, \tag{13}$$

$$\|bV - hV\| \le H\sqrt{s}\Delta(3\delta_0^2(h) - 2\delta_0^3(h))(1 - \delta_0(h))^{-2}, \tag{14}$$

$$|g_m| \le H\Delta(3\delta_0^2(h) - \delta_0^3(h))(1 - \delta_0(h))^{-3}. \tag{15}$$

It follows from the inequality

$$\|h\| \le \|b\| + \|V^{-1}\| \|bV - hV\|$$

and the definition of $\varepsilon(b)$ that

$$\|h\| \le \|b\|(1 - \varepsilon(b))^{-1}$$

and $\delta_0(h) < 1$ for $C(b) \le 1$. Consequently, we can replace $\delta_0(h)$ by $\delta(b)$ in the right-hand sides of inequalities (10)-(15). Hence

$$\psi(h) - (h, bV) + R(B(h), h) \le H\Delta^2 \delta^3(h)(1 - \delta(b))^{-3}$$

and

$$\psi(h) - (h, a) \le H\Delta^2\delta^3(h)(1 - \delta(b))^{-1} + m(V^{-1}, C_2(b)) + \sqrt{s}C_2(b)\|aV^{-1} - b\| + R(V, aV^{-}b) - R(V^{-1}, a)$$

Now let us obtain an upper bound for R. From equality (9) it follows that for any $k \in \mathbf{N}$ and $v = (v_1, \ldots, v_k)$

$$|S_{\eta-a}(v)| \le k!\, H_1\Delta_1^{2-k},$$

where $H_1 = H(1 - \delta(b))^{-3}$ and $\Delta_1 = \Delta(1 - \delta(b))^{-1}$.

From the definition of $\mathcal{M}$ we obtain that $h_i = 0$ or $h_i = c_i$ for any $\le i \le s$. Consequently, $e^{-(h,x)} \le 1$ for any $x \in M(0)$. Now from Theorem 3 it follows that

$$R \le C(s, \Delta_1, H_1, B(h)).$$

Now we obtain an upper bound for $\|B(h)^{-1}\|$. Let $\mu(B)$ be the minimum of the eigenvalues of the positive definite symmetric matrix B. Obviously,

$$\mu(B(h)) \ge \mu(V)(1 - C_1(b))$$

or

$$\|B(h)^{-1}\| \le \|V^{-1}\|C_1(b)^{-1}. \tag{16}$$

Then we obtain that $R \le W$. Hence

$$|I(a) - A(h)| \le C_0e^{-R(V^{-1},a)}.$$

Finally, let us obtain an upper bound for $\Phi(M(a - bV + hB(h), B(h))$. Note that for any $x \in \mathbf{R}^s$

$$R(B(h)^{-1}, x) = R(VB(h)^{-1}, x\sqrt{V^{-1}}).$$

Let $\{\lambda_1, \ldots, \lambda_s\}$ be the set of the eigenvalues of $V^{-1}B(h)$. Then for any $1 \le i \le s$

$$1 - C_1(b) \le \lambda_i \le 1 + C_1(b)$$

and, consequently,

$$(1 + C_1(b))^{-1} \le \lambda_i^{-1} \le (1 - C_1(b))^{-1}.$$

Thus,

$$R(V^{-1}, x)(1 + C_1(b))^{-1} \le R(B(h)^{-1}, x) \le R(V^{-1}, x)(1 - C_1(b))^{-1}.$$

It follows from the three last inequalities that

$$D^{-1}\Phi(M(x/\sqrt{C_1(b)}, V^{-1}) \le \Phi(M(x), B(h)),$$
$$\Phi(M(x, B(h)) \le D\Phi(M(x/\sqrt{C_1(b)}, V^{-1}).$$

Now Theorem 4 follows from the upper estimates of $\psi(h) - (h, bV) + R(B(h), h)$, $\psi(h) - (h, a)$, R, $\Phi(M(x), B(h))$, and inequality (15).

Note that if $\delta^3(b)\Delta^2$ is essentially greater than one, the inequalities given in Theorem 4 may be used in the case where $\Phi(M(a), V)$ is essentially greater than Δ_0^{-1}, since $e^{-R(V^{-1},a)}$ is contained in the right-hand sides of the inequalities.

REFERENCES

1. B. von Bahr, Multidimensional integral limit theorems on large deviations. *Arkiv für Math.* (1967) **7**, 89–99.
2. I. G. Zhurbenko, *Analysis of Stationary Homogeneous Random Systems*. Moscow Univ. Press, Moscow, 1987 (in Russian).
3. V. S. Korolyuk and Yu. V. Borovskikh, *Theory of U-statistics*. Naukova Dumka, Kiev, 1989 (in Russian).
4. V. G. Mikhailov, Explicit bounds in limit theorems for sums of random indicators. *Surveys in Appl. and Industr. Math.* (1994) **1**, 580–617 (in Russian).
5. R. N. Bhattacharya and R. Rango Rao, *Normal Approximation and Asymptotic Expansions*. Wiley, New York, 1976.
6. L. I. Saulis, On the large deviations of random vectors for some classes of sets. I. *Liet. Mat. Rink.* (1983) **23**, No. 3, 142–164 (in Russian).
7. L. I. Saulis, Probabilities of large deviations for random vectors. *Theory Probab. Appl.* (1991) **36**, 482–491 (in Russian).
8. A. B. Gorchakov, Upper estimates for semi-invariants of the sum of multi-indexed random variables. *Discrete Math. Appl.* (1995) **5**, 317–331.

Probabilistic Methods in Discrete Mathematics, pp. 231–236
V. F. Kolchin et al. (Eds.)

COMPOSITION OF A TRUSTED COMPUTER SECURITY SYSTEM ON THE BASE OF UNTRUSTED ELEMENTS

A. A. GRUSHO

Moscow

Abstract — We discuss the problem of composing a trusted computer security system on the base of untruster elements. We suggest a rigorous mathematical model of the situation and prove that such a composition is possible if all the conditions on the model are fulfilled.

We deal with a composition of a trusted computer security system with the use of untrustworthy hardware and software. First of all we recall the methodology of composing a trusted computer security system described in the Orange book [1]. On the base of analysis of the risk, a security policy is worked out which determines some rules of treating sensitive information in the computer. By definition, the security of the sensitive information means the fulfilment of the rules worked out. If the security policy is fixed, we have to create a subsystem to support the policy. A security system is good if it gives a reliable support for the security policy, and it is bad if the security policy support is unreliable. Usually a security support subsystem is complicated. That is why security subsystems are built using a hierarchy structure. The rules of the security policy are located at the highest level, the next level is intended for a support to the highest level and so on. The lowest level consists of the hardware which provides the realization of the protection algorithms. The Orange book [1] suggests two classes of security policies, discretionary and multilevel. For the trust support of these policies in universal computers, it is necessary to construct a special architecture of the hardware and software. This is a reason way it seems impossible to construct a trusted universal computer system using untrustworthy hardware and software imported from abroad.

However, almost all universality of modern computers is used to create a flexible information technology which, in many particular cases, intended for solving a narrow class of tasks. Therefore, if an information technology is chosen, then we can restrict ourselves to a simpler security policy. A simple security policy needs a simpler support, and what is more important, if an information technology is fixed, it lets us introduce some abundance in the computing process which gives us a possibility to guarantee the security of the system in an untrustworthy environment and even to prove that the rules of the security policy chosen are fulfilled.

We illustrate the ideas described above by the following example. Let us consider an information technology of receiving enciphered and electronically signed messages through an untrustworthy network from untrustworthy correspondents. It is necessary to decipher a message with an individual correspondent key, to authenticate the sign, and to transfer the plain text and the sign confirmation to an operator. The communication is established through a server of a distant access (SDA) which is an untrustworthy computer with an imported operational system (OS). To realize an information technology, it is possible to use any number of similar untrustworthy computers. In order to obtain a trusted computing, we could produce hardware and software by ourselves, the restriction is that we can produce only a limited number of programs and cannot create a universal OS. We assume that there are two vulnerabilities in the information technology. They are a flaw of the key information and troubles (initiated from outside) with availability of the technology for a long period of time. The vulnerabilities determine the following security policy. The information at the reception point is classified into the three levels:

(1) the list of all keys and the programs for deciphering cipher texts, all digital signature control programs, trusted hardware and software;

(2) clear messages, the hardware and software of the operator;

(3) information in the communication hardware and software, information in the transmitting cipher messages.

The rules for treating information according to its classification (the security policy) are the following:

(1) the information flow should be organized in such a way that the malicious user at any lower level cannot read the information of the higher level; any subject at the higher level cannot write information from its level downwards to a malicious one, except the operator (O) and the source (S) who are trusted subjects and can write information downwards;

(2) the operator initiated some abundance in the information at his levels in order to detect technology defects. We assume that the detection of a technology defect gives us a possibility to restore the technology without any loss.

If the security policy is fulfilled, then the information about the keys cannot be cleared by a malicious person who can be the operator, or any single source desiring to get the keys from other sources, or someone who controls the channel environment.

The abundance in communication traffic gives us an ability to reveal a technology trouble and to avoid the second vulnerability.

Since we assume that the system is functioning in an unfriendly environment, we should describe the malicious capacities:

- we assume that a malicious subject does not use a leakage of information through side physical channels connected with radiations, acoustics, power supply, visual control of displays, intrusion into hardware, and hardware bugs;

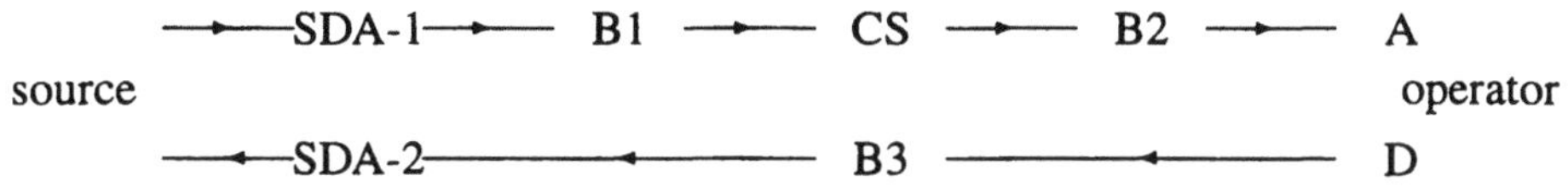

Figure 1.

- there is only one malicious person who can be the operator, any source, or someone who control channel environment;
- every source establishes a perfect protection of his own cryptographic facilities;
- the untrustworthy imported software may include bugs (Troyan horses) which do not oriented to specific applications (but if a bug is initiated and if there is any information in the channel to the computer, it can receive the application oriented software);
- to initiate a Troyan horse one may use some kind of a logic bomb, based on some kind of counter, or by sending a special signal through an information or control channel (resulting in appearance of some combination of digits in one of the registers).

If these assumptions are valid, then the following system architecture presented in Fig. 1 gives us a secure information technology.

In Fig. 1, SDA-1 and SDA-2 are servers of distant access which are untrustworthy and which establish communication with any of the sources.

By the symbol $\rightarrow$ we denote a one way information channel (like RS-232) which connects two sequential ports of two different devices (while it is possible to send control signals to the different direction).

B1, B2, and B3 are buffer devices each of which consists of two sequential ports (one for transmitting and another for receiving only). A similar idea was used in [2]. A transmission is possible from the buffer when there is the flag of the end of the message received. That is why it is impossible to modulate a control channel through the buffer. Besides there is some kind of coding in B1 which realizes a substitution of bits. It helps to prevent a malicious person to compose the text needed to activate a Troyan horse with the help of a special combination in some register. The buffer B3 realizes decoding fulfilled by the computer D. We assume that the buffers cannot contain bugs, since the operations carried out by these devices are simple and can be programmed by ourselves.

The cryptoserver (CS) gets messages from B1, decodes bits in such a way that there are no parts of the cipher text in the registers of the computer. Then according to an identifier the cryptoserver chooses the necessary key, deciphers the message, and check the signature. The result is sent to B2. The cryptoserver may be installed on the base of an untrustworthy computer, since the treatment of every message ends with wiping off all information in RAM when the result reaches B2. The next message treatment begins with the installation of the software from ROM.

The computer A is an untrustworthy computer which presents the result of work of the cryptoserver to the operator.

The computer D is a untrustworthy computer which collects the data base of the cipher texts. The operator can introduce distortions into the cipher messages. The operator chooses a source (not the same as the true message sender) and sends the message to this source through SDA-2. When such a message reaches the source, it should be returned back immediately. Knowing the result, the operator can detect the technology troubles. We use the buffer B3 to prevent an attempt of the untrustworthy SDA-2 to activate a possible Troyan horse in D through the control channel. Besides we should use a code for bits in SDA-2 for the transmission to B3, because it prevents the leakage of the data base if a logical bomb activates a Troyan horse in D. We assume that any Troyan horse is not oriented to the application in D, does not know that it is necessary to code the computer output but it sends everything through the existing information channel to anyone who gets it.

The system works to support the information technology and the security policy in the following way:

- a source communicates with SDA-1 through any accessible channel;
- SDA-1 sends a message through the B1 to CS;
- the cryptographic facility in CS treats the message and sends the result to A through B2;
- before sending the information to A the buffer B2 turn off the power in CS;
- the operator analyzes the message and the signature, makes a decision, and sends securely the cipher message to the data base of D;
- the operator uses *D* to introduce an abundance in the traffic which helps him to detect troubles in the information technology.

Under the assumption described above, we can prove that the architecture supports the security policy.

THEOREM 1. *If the assumptions on the malicious activity in the system defined above are fulfilled, then the information flows in the system satisfy rule* (1) *of the security policy.*

PROOF. There are only two channels which can be used for transmitting the information about keys and cryptographic facilities from CS, namely, the control channel to B1 and the information channel to B2.

Any information from B1 to SDA-1 can be transmitted only through the control channel. Since the channels from CD to B1 and from B1 to SDA-1 are independent of each other, there is no direct channel from CD to SDA-1. If a Troyan horse in CS is activated by a source or by a malicious person who controls the channel environment and SDA-1, then an application oriented software can be installed in CS. This means that the key information may be transmitted outside CD, but there is no channel from CD to SDA-1. There is only one way for it from CD, the way to the operator. Since we assume that there exists only one malicious person, the transmission of the key information to the operator does not mean a flaw.

Any counter is eliminated for the activation of a logical bomb in CD by the regular reinstallation of the CD software.

It is impossible to activate a Troyan horse in CD through the information channel with the use of the cipher text since each bit is coded in B1. Therefore it can be activated only by a plain text. This can be done only by a malicious source, but he cannot get the desirable information back.

If we have a malicious operator who has the information channel from CD to A, then it is sufficient to him to activate a Troyan horse to get the key information. However the operator cannot do it because he has no key to compose the special plain text and does not know the substitution used for coding in B1 to compose the special cipher text.

Each information flow downwards from A to D is originated by the operator who is a trusted subject at this level.

The operator does not control the covert channel from D to SDA-2 which can be activated by a logical bomb in D. But the coding of the D output and the buffer B3 eliminate this covert channel. If the logical bomb eliminates the data base in D, then, according to the security policy, we does not consider it as a loss.

By virtue of B3 there is no direct channel from SDA-2 to D. That prevent from the activation of a Troyan horse in D and from drawing in any application oriented software.

The downwards information flow from any source is defended by enciphering.

The theorem is proved.

The proof of item (2) of the security policy rules needs the detail consideration of the reaction of the system to the additional traffic which can be initiated by the operator. We show only that under the assumption of right functioning of the buffers, it is possible to detect the stop of the right functioning of only one object in the technology while the others are in order.

If a given source does not answer, then there is a trouble with that source or a wrong work of SDA-1 or SDA-2 which are under the control of a virus. The same answer after a repetition of the call confirms that the source is blocked.

If no answer is received for all control messages, then the trouble is in SDA-2.

If the whole traffic is stopped, then the trouble is either in SDA-1 or in CS. After SDA-1 is tested by the operator and the same result is obtained, we conclude that the trouble is in CS.

Troubles in A and D are detected by the operator visual control.

Thus, this paper gives an example of the analysis technique of an information technology which works with valuable information.

REFERENCES

1. Department of Defense, *Trusted Computer System Evaluation Criteria.* 1985.
2. S. Jajodia and B. Kogan, Integrating an object-oriented data model with multilevel security. In: *Proc. IEEE Computer Society Symposium on Research in Security and Privacy.* Oakland, 1990, pp. 76–85.
3. A. A. Grusho and E. E. Timonina, *Theoretical Foundations of Information Security.* Yachtsman, Moscow, 1996 (in Russian).

Probabilistic Methods in Discrete Mathematics, pp. 237–252
V. F. Kolchin et al. (Eds.)

DECOMPOSABLE STATISTICS AND WAITING TIME IN THE MARKOV–PÓLYA URN MODEL

A. V. IVANOV

Moscow

Abstract — We consider the Markov–Pólya urn model. The urn contains a given number of balls of each of N different colours. The balls are sequentially drawn from the urn one at a time, independently of each other and with equal probabilities to be drawn. After each draw the drawn ball is returned into the urn together with c balls of the same colour. The drawing stops when, for the first time, the frequences of k unspecified colours reach or exceed the corresponding (random) levels settled before the beginning of the trials. Limit distributions of the decomposable statistics

$$L_{Nk} = \sum_{j=1}^{N} g(\eta_j)$$

are studied, as $N \to \infty$ and $k = k(N)$, where g is a given function of integer argument and $\eta_1, \ldots, \eta_N$ are the frequences, the numbers of balls of the corresponding colours, at the stopping time.

1. INTRODUCTION

Let an urn contain initially n balls of N different colours A_j, $j = 1, \ldots, N$. The balls are randomly and equiprobably drawn from the urn one at a time. After each trial the colour of the drawn ball is fixed and the ball is returned into the urn together with c balls of the same colour. This scheme will be referred to as the Markov–Pólya urn model.

It is not difficult to see that if at the moment t, $t = 1, 2, \ldots$, k_j balls of colour A_j, $j = 1, \ldots, N$, have been drawn, then the probability to draw a ball of colour A_i, $i = 1, \ldots, N$, at the moment $t + 1$ is equal to

$$(n + ck_i) \Big/ \sum_{j=1}^{N} (n + ck_j). \tag{1}$$

Note that this scheme possesses a significant aftereffect property: if a ball of some colour is drawn, then the probability to draw a ball of the same colour in the next trail increases.

The scheme described may be thought as an allocation scheme of the following type. Consider N cells labelled $1, \ldots, N$. At the beginning all cells are empty. If at the moment t, $t = 1, 2, \ldots$, a ball of colour A_j, $j = 1, \ldots, N$, is drawn from the urn, then a particle is placed into the jth cell and the probability of hitting the ith cell at the moment $t+1$ is equal to (1), where $k_1, \dots, k_N$ are the contents of the corresponding cells after allocation of t particles.

Note that the sequence of vectors

$$\{\bar{\eta}(t) = (\eta_1(t), \dots, \eta_N(t)), \quad t = 0, 1, \dots\},$$

where $\eta_j(t)$ is the contents of the jth cell after t trials, $\eta_j(0) = 0$, $j = 1, \dots, N$, forms the N-dimensional Markov chain with transitions of the form

$$(k_1, \dots, k_i, \dots, k_N) \to (k_1, \dots, k_i + 1, \dots, k_N), \quad i = 1, \dots, N,$$

the probabilities of which are given by (1).

It is assumed that before the beginning of the trials some level ν_j for the jth cell is settled, where $\nu, \nu_1, \dots, \nu_N$ are independent non-negative integer-valued random variables. We say that the jth cell is filled at the moment t, if $\eta_j(t) \geq \nu_j$.

For an integer k, $1 \leq k \leq N$, we define a random variable $\nu(N,k)$, setting

$$\nu(N,k) = \min\left\{t:\ \sum_{j=1}^{N} \mathbf{I}(\eta_j(t) \geq \nu_j) = k\right\}, \tag{2}$$

which is called the waiting time or the stopping time (here and below $\mathbf{I}(A)$ is the indicator of the event A).

Let $\eta_j = \eta_j(\nu(N,k))$ be the content of the jth cell at the stopping time, $j = 1, \dots, N$. We consider the decomposable statistic (DS), i.e., the random variable of the form

$$L_{Nk} = \sum_{j=1}^{N} g(\eta_j), \tag{3}$$

where g is a given function of integer argument.

The DS L_{Nk} (3) is one of the most popular characteristics of the Markov–Pólya generalized urn model, where $c \in \mathbf{C} = \{-1, 0, 1, 2, \dots\}$. In particular, if $c = -1$, we deal with sampling without replacement and if $c = 0$, with sampling with replacement (or with the equiprobable allocation scheme). These special cases are well studied. For the first case the complete description of the class of limit distributions of DS L_{Nk} (3) as $N \to \infty$ and $k = k(N)$ is obtained in [6]. Similar results for equiprobable allocation scheme are obtained in [3]. That paper can be regarded as the origin for the further analysis of the limit distributions of DS (3) in the Markov–Pólya generalized urn model. Some limit theorems for DS's in the equiprobable allocation scheme are given in [7], where the idea of random levels is introduced.

Limit distributions of DS L_{Nk} (3) in the Markov–Pólya generalized urn model for $c \in \{1, 2, \dots\}$ are poorly known. The aim of this paper is to fill the gap.

Note that some special cases of the Markov–Pólya urn model with deterministic level ν were studied in [8].

This paper is constructed as follows. The method of analysis of the limit distributions of the DS (3) is described briefly in Section 2. This method is based on the ideas presented in [3]. The theorems which describe the limiting behaviour of L_{Nk} in three regions of variation of the parameter $k = k(N)$ are presented in Sections 3, 4, and 5.

2. THE METHOD OF INVESTIGATION

We introduce some notation. Let $\mathcal{L}(\xi)$ denote the distribution law of a random variable ξ; let $\overline{\mathrm{Bi}}(n, p)$ denote the negative binomial distribution with parameters n and p, the probabilities of which are given by

$$\bar{b}_r(\alpha, p) = \binom{\alpha + r - 1}{r} q^r p^\alpha, \quad r = 0, 1, \ldots, \quad p = 1 - q \in (0, 1), \quad \alpha > 0,$$

where

$$\binom{\alpha + r - 1}{r} = \frac{\alpha \ldots (\alpha + r - 1)}{r!};$$

let θ be a random variable having beta distribution with the density

$$\beta(x; k, N - k + 1) = N \binom{N-1}{k-1} x^{k-1} (1 - x)^{N-k}, \qquad x \in [0, 1].$$

Let

$$\mathrm{B}(p; k, N - k + 1) = \int_0^p \beta(x; k, N - k + 1)\, dx, \qquad p \in [0, 1],$$

be the incomplete beta function. It is known that $\mathrm{B}(p; k, N - k + 1) = \mathbf{P}\{Y(N, p) \geq k\}$, where $\mathcal{L}(Y(n, p)) = \mathrm{Bi}(n, p)$, and $\mathrm{Bi}(n, p)$ is the binomial distribution with parameters n and p, the probabilities of which are given by

$$b_r(n; p) = \binom{n}{r} p^r q^{n-r}, \quad r = 0, 1, \ldots, \quad p = 1 - q \in (0, 1).$$

Let further $\xi(t)$ be a random variable, which does not depend on ν and has the distribution $\overline{\mathrm{Bi}}(\alpha, \gamma(t))$, where $\alpha = n/c$ and $\gamma(t) = e^{-ct}$, $t > 0$, and let $\xi_j(t), j = 0, 1, 2$, be independent random variables with the distributions

$$\mathbf{P}\{\xi_0(t) = r + 1\} = (n + rc)\bar{b}_r(\alpha, \gamma(t))\mathbf{P}\{\nu = r + 1\}/h(t), \quad r = 0, 1, \ldots,$$

where

$$h(t) = \sum_{j=0}^{\infty} (n + jc)\bar{b}_r(\alpha, \gamma(t))\mathbf{P}\{\nu = j + 1\}$$

is the normalizing factor,

$$\mathcal{L}(\xi_1(t)) = \mathcal{L}(\xi(t) \mid \xi(t) \geq \nu),$$
$$\mathcal{L}(\xi_2(t)) = \mathcal{L}(\xi(t) \mid \xi(t) < \nu).$$

We introduce also the corresponding characteristic functions

$$f_j(\tau, t) = \mathbf{E} \exp\{i\tau g(\xi_j(t))\}, \quad j = 0, 1, 2.$$

Finally, let

$$H(t) = 1 - \bar{H}(t) = \mathbf{P}\{\xi(t) \geq \nu\}, \qquad t \geq 0.$$

We set

$$p = 1 - q = \mathbf{P}\{\nu = 0\} = H(0)$$

and suppose here and below that $p < 1$. Then for $p > 0$ the function $H(t)$ has a jump of size p at zero and for $t > 0$ the derivative

$$H'(t) = \sum_{j=1}^{\infty}(n + jc)\bar{b}_j(\alpha, \gamma(t))\mathbf{P}\{\nu = j + 1\}$$

exists and is equal to $h(t) > 0$. Thus, $H(t)$ increases strictly monotonically from p to 1 and therefore the inverse function

$$t(x) = H^{-1}(x), \quad x \in [p, 1],$$

is defined.

For asymptotic analysis of the distributions of the DS L_{Nk} we use the integral representation for its characteristic function (see [5], Theorem 2)

$$\mathbf{E}e^{i\tau L_{Nk}} = \int_0^1 (\mathbf{I}(x \leq p) + \mathbf{I}(x > p)f_{Nk}(\tau, t(x)))\beta(x; k, N - k + 1)\, dx, \tag{4}$$

where

$$f_{Nk}(\tau, t) = f_0(\tau, t)f_1^{k-1}(\tau, t)f_2^{N-k}(\tau, t).$$

Representation (4) can be written in the form

$$\mathbf{E}e^{i\tau L_{Nk}} = \mathrm{B}(p; k, N - k + 1) + \int_p^1 f_{Nk}(\tau, t(x))\beta(x; k, N - k + 1)\, dx. \tag{5}$$

Without loss of generality we suppose that here and below $g(0) = 0$.

As indicated in [5] representations (4)–(5) have the following probabilistic interpretation. Let $\{\xi_0(t(\theta)), \xi_{1j}(t(\theta)), j = 1, \ldots, k - 1, \xi_{2j}(t(\theta)), j = 1, \ldots, N - k\}$ be independent random variables for every value of $\theta \in (p; 1]$, where $\xi_{1j}(t)$ ($\xi_{2j}(t)$) are independent copies of the truncated random variables $\xi_1(t)$ ($\xi_2(t)$) respectively. We suppose that before the beginning of the trials an experiment is performed where the random variable θ is observed. Then if $\theta \leq p$ we put $L_{Nk} = 0$, otherwise (i.e., if $\theta \in (p, 1)$) the distribution of L_{Nk} is the same as the distribution of the sum

$$g(\xi_0(t(\theta)) + \sum_{j=1}^{k-1} g(\xi_{1j}(t(\theta))) + \sum_{j=1}^{N-k} g(\xi_{2j}(t(\theta))),$$

where all summands are conditionally independent, i.e., they are independent for every fixed θ.

Note that in the equiprobable allocations and in the sampling without replacement analogous constructions are generated by the Poisson (see [3]) and the binomial (see [6]) random variables respectively.

To obtain the limit distributions of the DS L_{Nk}, we use the method presented in [3]. We give a brief description of its main steps.

Three basic domains are distinguished:

- the left-hand domain, where k is a non-zero constant;
- the right-hand domain, where $s = N - k + 1$ is a non-zero constant;
- the middle domain of moderate values of k, where

$$k = \lambda N + o(\sqrt{N}), \qquad \lambda = 1 - \bar{\lambda} \in (0, 1).$$

In what follows we assume that the distribution of ν is fixed and does not depend on N. The parameters n and c are fixed (the cases where $n = n(N) \to \infty$ and $c = c(N) \to \infty$ as $N \to \infty$ will be discussed separately).

It is known that in the left-hand domain $\mathrm{B}(p; k, N - k + 1) \to 1$; in the right-hand domain $\mathrm{B}(p; k, N - k + 1) \to 0$; and in the middle domain

$$\mathrm{B}(p; k, N - k + 1) \to \begin{cases} 1 & \text{if } \lambda < p, \\ 1/2 & \text{if } \lambda = p, \\ 0 & \text{if } \lambda > p, \end{cases}$$

as $N \to \infty$ and $p \in (0, 1)$ is fixed.

Taking into account that $g(0) = 0$, we have that in the left-hand domain and for $\lambda < p$ in the middle domain

$$L_{Nk} \xrightarrow{\mathbf{P}} 0$$

as $N \to \infty$ and $p \neq 0$. We omit these trivial situations from further consideration.

Setting $x_N = a_N + x b_N$, we obtain under a suitable choice of the constants a_N and b_N

$$\beta_N(x) = \beta(x_N; k, N - k + 1)|b_N| \to \beta(x),$$

where the limit density $\beta(x)$ is either the density of extreme values or the normal density. In addition, in the left-hand domain

$$a_N = 0, \quad b_N = 1/N, \quad \beta(x) = \pi_{k-1}(x), \quad x > 0;$$

here and below

$$\pi_r(x) = \frac{x^r}{r!} e^{-x}, \qquad r = 0, 1, \ldots;$$

in the right-hand domain

$$a_N = 1, \quad b_N = -1/N, \quad \beta(x) = \pi_{s-1}(x), \quad x > 0;$$

in the middle domain

$$a_N = \lambda, \quad b_N = (\lambda\bar{\lambda}/N)^{1/2}, \quad \beta(x) = \varphi(x) = \frac{1}{\sqrt{2\pi}} e^{-x^2/2}, \quad x \in \mathbf{R}.$$

Setting $t_N = t(x_N)$ and performing in (4) the corresponding change of variables, we obtain for the three domains, respectively,

$$\int_0^\infty (\mathbf{I}(x \le Np) + \mathbf{I}(Np < x \le N) f_{Nk}(\tau, t_N))\beta_N(x)\, dx,$$

$$\int_0^\infty (\mathbf{I}(Nq \le x \le N) + \mathbf{I}(x < Nq) f_{Nk}(\tau, t_N))\beta_N(x)\, dx,$$

$$\int_{-\infty}^{\infty} \Bigg(\mathbf{I}\left(-\sqrt{\lambda N/\bar{\lambda}} \le x \le (p-\lambda)\sqrt{N/(\lambda\bar{\lambda})} \right) + \mathbf{I}\left((p-\lambda)\sqrt{N/(\lambda\bar{\lambda})} < x \le \sqrt{\bar{\lambda} N/\lambda} \right) f_{Nk}(\tau, t_N) \Bigg) \beta_N(x)\, dx. \tag{6}$$

We assume that for any x belonging to the corresponding region the limit

$$f(\tau, x) = \lim_{N\to\infty} f_{Nk}(\tau, t_N)$$

exists and use the following version of the majorant converges theorem [9]: if $\beta_N(x)$ and $\beta(x)$ are density functions, $|\psi_N(x)| \le 1$, $|\psi(x)| \le 1$, and for any fixed x

$$\beta_N(x) \to \beta(x), \qquad \psi_N(x) \to \psi(x),$$

as $N \to \infty$, then

$$\int_{-\infty}^{\infty} \psi_N(x)\beta_N(x)\, dx \to \int_{-\infty}^{\infty} \psi(x)\beta(x)\, dx.$$

Hence it follows that if the limit $f(\tau; x)$ exists, then the limiting characteristic function of the DS L_{Nk} has the form

$$\int_{-\infty}^{\infty} f(\tau, x)\beta(x)\, dx$$

except the case of $\lambda = p$ for the middle domain, where this limit equals

$$\frac{1}{2} + \int_0^\infty f(\tau, x)\varphi(x)\, dx.$$

Thus, to obtain the limit distributions of the DS L_{Nk}, it is needed to solve the following problems: to obtain both the conditions for existence and the form of the possible limits $f(\tau; x)$; to derive the limit characteristic functions; to invert these characteristic functions.

The following sections are devoted to the analytic calculations which realize the plan given above in the corresponding domains of the variation of the parameter k.

3. THE LIMIT DISTRIBUTIONS IN THE LEFT-HAND DOMAIN

We recall that in this case $p = \mathbf{P}\{\nu = 0\} = 0$, $g(0) = 0$. Let the integers m, l be determined by the conditions

$$m = \min\{r \geq 1 : \mathbf{P}\{\nu = r\} > 0\}, \qquad l = \min\{j \geq 1 : g(j) \neq 0\}$$

and $a = 1 - \bar{a} = \mathbf{P}\{\nu = m\}$.

In the left-hand domain, $x_N = x/N$, $x > 0$, therefore $t_N \to 0$ as $N \to \infty$ and from the representation

$$H(t) = \sum_{r-m}^{\infty} \bar{b}_r(\alpha, \gamma(t))\mathbf{P}\{\nu \leq r\} = a\binom{\alpha + m - 1}{m} c^m t^m (1 + o(1)),$$

which is valid for $t \to 0$ (here it is assumed that $n = n(N) \to \infty$, $c = c(N) \to \infty$ in such a way that $\alpha(N) = n(N)/c(N)$ is a constant and $n(N)t_N \to 0$ as $N \to \infty$), it follows that, as $N \to \infty$,

$$t_N = \left(x \Big/ \left(a\binom{\alpha + m - 1}{m} N\right)\right)^{1/m} (1 + o(1)).$$

Hence it follows that, as $N \to \infty$,

$$f_0(\tau, t_N) = \sum_{r=m-1}^{\infty} e^{i\tau g(r+1)}(n + rc)\bar{b}_r(\alpha, \gamma(t_N))\mathbf{P}\{\nu = r + 1\}/h(t_N) = e^{i\tau g(m)}(1 + o(1)),$$

$$\begin{aligned} f_1(\tau, t_N) &= \sum_{r=0}^{\infty} e^{i\tau g(r)}\bar{b}_r(\alpha, \gamma(t_N))\mathbf{P}\{\nu \leq r\}/H(t_N) \\ &= \frac{N}{x} e^{i\tau g(m)} a\binom{\alpha + m - 1}{m} c^m t_N^m (1 + o(1)) = e^{i\tau g(m)}(1 + o(1)). \end{aligned}$$

For the characteristic function $f_2(\tau, t_N)$ we have the representation

$$\begin{aligned} f_2(\tau, t_N) &= \sum_{r=0}^{\infty} e^{i\tau g(r)}\bar{b}_r(\alpha, \gamma(t_M))\mathbf{P}\{\nu > r\}/H(t_N) \\ &= 1 + \frac{1}{1 - x_N}\sum_{r=1}^{\infty}(e^{i\tau g(r)} - 1)\bar{b}_r(\alpha, \gamma(t_N))\mathbf{P}\{\nu > r\} \\ &= 1 + (e^{i\tau g(l)} - 1)\binom{\alpha + l - 1}{l}\mathbf{P}\{\nu > 1\}c^l t_N^l (1 + o(1)), \end{aligned}$$

which implies that if $l > m$, then

$$f_2^N(\tau, t_N) \to 1,$$

if $l = m$, then

$$f_2^N(\tau, t_N) \to \exp\{(e^{i\tau g(m)} - 1)\bar{a}x/a\},$$

and, finally, if $l < m$, then

$$f_2^N(N^{(l-m)/m}\tau, t_N) \to \exp\left\{ i\tau d_l \left(x \Big/ \left(a \binom{\alpha + m - 1}{m} \right) \right)^{l/m} \right\},$$

where

$$d_l = g(l) \binom{\alpha + l - 1}{l}.$$

Inverting the corresponding characteristic functions

$$\int_0^\infty f(\tau, x)\pi_{k-1}(x)\,dx,$$

we obtain the following result.

THEOREM 1. *Let $N \to \infty$, k be fixed, $k \geq 1$, and $n, c \to \infty$ in such a way that $\alpha = n/c$ is a constant and $nt \to 0$.*

Then for $l > m$ the DS $L_N k$ is asymptotically degenerated at zero; for $l = m$, the random variable $g^{-1}(m)L_{Nk} - k$ has in limit the negative binomial distribution $\overline{\mathrm{Bi}}(k, \alpha)$, i.e.,

$$\mathbf{P}\{g^{-1}(m)L_{Nk} - k = r\} \to \binom{k + r - 1}{r} \bar{a}^r a^k, \quad r = 0, 1, \ldots;$$

and for $l < m$

$$\mathbf{P}\left\{ N^{(l-m)/m} d_l^{-1} \binom{\alpha + m - 1}{m}^{l/m} L_{Nk} \leq x \right\} \to \Gamma_k(ax^{m/l}), \quad x > 0,$$

where

$$\Gamma_k(t) = \int_0^t \pi_{k-1}(y)\,dy.$$

REMARK 1. Comparing Theorem 1 with the corresponding results for sampling with and without replacement (see Theorem 2 in [3] and Theorem 1 in [6] respectively), we see that some difference between these three schemes exists only in normalization in the case $l > m$ and for fixed n.

Consider some examples to illustrate the application of Theorem 1.

Let $g(x) = x$. Then $L_{Nk} = \nu(N, k)$ and for the waiting time we obtain that for $m = 1$

$$\mathcal{L}(\nu(N, k) - k) \to \overline{\mathrm{Bi}}(k, a)$$

(for the deterministic level, i.e., in the case $a = \mathbf{P}\{\nu = 1\} = 1$ we have $\nu(N, k) \to k$ in probability); for $m > 1$

$$\mathbf{P}\left\{ N^{(l-m)/m} \alpha^{-1} \binom{\alpha + m - 1}{m}^{l/m} \nu(N, k) \leq x \right\} \to \Gamma_k(ax^m), \qquad x > 0.$$

As one more example we consider the random variables

$$\mu_r = \sum_{j=1}^{N} \mathbf{I}(\eta_j = r), \qquad r \geq 0.$$

Then for $r > m$ the random variables $\mu_r \to 0$ in probability; for $r = m$

$$\mathcal{L}(\mu_m - k) \to \overline{\mathrm{Bi}}(k, a)$$

(for $a = 1$ the random variable $\mu \to k$ in probability); and for $r < m$

$$\mathbf{P}\left\{ N^{(r-m)/m} d_r^{-1} \binom{\alpha + m - 1}{m}^{r/m} \mu_r \leq x \right\} \to \Gamma_k(ax^{m/r}), \quad x > 0.$$

In particular, the results of these examples for the case of deterministic level are contained in [8].

Finally, we note that $\mathcal{L}(N - \mu_0) \sim \mathcal{L}(\mu_1)$ as $N \to \infty$.

4. THE LIMIT DISTRIBUTIONS IN THE RIGHT-HAND DOMAIN

In this case the class of limit distributions of the DS L_{Nk} is described in terms of their characteristic functions

$$\int_0^\infty f(\tau, x)\pi_{s-1}(x)\, dx,$$

but here $x_N = 1 - x/N$, $x > 0$, and, consequently, $t_N \to \infty$ as $N \to \infty$ and therefore the asymptotic behaviour of the function $\bar{H}(t)$ as $t \to \infty$ plays the main role in the analysis of this case.

Here and below we assume that the distribution of the random level ν is such that

$$\sum_{r=0}^{\infty} r \binom{\alpha + r - 1}{r} \mathbf{P}\{\nu > r\} < \infty. \tag{7}$$

REMARK 2. In particular, condition (7) holds if the random level ν is bounded, i.e., if $\mathbf{P}\{\nu \leq m\} = 1$ for some $m \geq 0$.

We set

$$A_\nu = \sum_{r=0}^{\infty} \binom{\alpha + r - 1}{r} \mathbf{P}\{\nu > r\},$$

where the convergence of the series follows from (7). Then, as $t \to \infty$,

$$H(t) = e^{-nt} A_\nu (1 + o(1)).$$

Taking into account that $H(t) = x/N$, we obtain

$$t_N = \frac{1}{n}(\ln c_N - \ln x) + o(1), \tag{8}$$

where $c_N = NA_\nu$.

It follows from (8) that

$$N\bar{b}_r(\alpha, \gamma(t_N)) = \binom{\alpha + r - 1}{r} \frac{x}{A_\nu}(1 + o(1))$$

as $N \to \infty$.

Taking the aforesaid into account, we obtain

$$\begin{aligned} f_0(\tau, t_N) &= \sum_{r=0}^{\infty} e^{i\tau g(r+1)}(n + rc)\bar{b}_r(\alpha, \gamma(t_N))\mathbf{P}\{\nu = r + 1\}/h(t_N) \\ &\sim \sum_{r=0}^{\infty} e^{i\tau g(r+1)}(n + rc)\binom{\alpha + r - 1}{r}\mathbf{P}\{\nu = r + 1\}/B_0 = \mathbf{E}e^{i\tau g(\xi^0)}, \end{aligned}$$

where

$$\mathbf{P}\{\xi^0 = r + 1\} = (n + rc)\binom{\alpha + r - 1}{r}\mathbf{P}\{\nu = r + 1\}/B_0, \qquad r = 0, 1, \ldots$$

and

$$B_0 = \sum_{r=0}^{\infty}(n + rc)\binom{\alpha + r - 1}{r}\mathbf{P}\{\nu = r + 1\}$$

is the normalizing factor,

$$\begin{aligned} f_2(\tau, t_N) &= \sum_{r=0}^{\infty} e^{i\tau g(r)}\bar{b}_r(\alpha, \gamma(t_N))\mathbf{P}\{\nu > r\}/\bar{H}(t_N) \\ &\sim \frac{1}{A_\nu}\sum_{r=0}^{\infty} e^{i\tau g(r)}\binom{\alpha + r - 1}{r}\mathbf{P}\{\nu > r\} = \mathbf{E}e^{i\tau g(\xi^1)}, \end{aligned}$$

where

$$\mathbf{P}\{\xi^1 = r\} = \binom{\alpha + r - 1}{r}\mathbf{P}\{\nu > r\}/A_\nu, \qquad r = 0, 1, \ldots$$

For the characteristic function $f_1(\tau, t_N)$ we obtain

$$
\begin{aligned}
f_1(\tau, t_N) &= \sum_{r=0}^{\infty} e^{i\tau g(r)} \bar{b}_r(\alpha, \gamma(t_N)) \mathbf{P}\{\nu > r\} / \bar{H}(t_N) \\
&= 1 + \frac{1}{x_N}\left(\mathbf{E}e^{i\tau g(\xi(t_N))} - 1 - \sum_{r=0}^{\infty}(e^{i\tau g(r)} - 1)\bar{b}_r(\alpha, \gamma(t_N))\mathbf{P}\{\nu > r\}\right) \\
&= 1 + \left(\mathbf{E}e^{i\tau g(\xi(t_N))} - 1\right)(1 + o(1)) \\
&\quad - \left(\frac{x}{NA_\nu}\sum_{r=0}^{\infty}(e^{i\tau g(r)} - 1)\binom{\alpha + r - 1}{r}\mathbf{P}\{\nu > r\}\right).
\end{aligned}
$$

Let $g = g_N$ be such that the limit

$$\lim_{N\to\infty} N(\mathbf{E}e^{i\tau g_N(\xi(t_N))} - 1) = a_1(\tau, x), \tag{9}$$

exists and for each $r \geq 0$ the limit

$$g^*(r) = \lim_{N\to\infty} g_N(r) \tag{10}$$

exists.

If (9)–(10) holds, then

$$f_1^N(\tau, t_N) \to \exp\{a_1(\tau, x) + a_2(\tau, x)\},$$

where

$$a_2(\tau, x) = -\frac{x}{A_\nu}\sum_{r=0}^{\infty}(e^{i\tau g^*(r)} - 1)\binom{\alpha + r - 1}{r}\mathbf{P}\{\nu > r\}.$$

Thus, the following theorem is proved.

THEOREM 2. *Let $N \to \infty$ while $s = N - k + 1 \geq 1$ be a constant, and let n and c be fixed. If g_N satisfies conditions* (9)–(10), *then*

$$\mathbf{E}e^{i\tau L_{Nk}} = \mathbf{E}e^{i\tau g^*(\xi^0)}(\mathbf{E}e^{i\tau g^*(\xi^1)})^{s-1}\int_0^\infty e^{a(\tau,x)}\pi_{s-1}(x)\,dx, \tag{11}$$

where $a(\tau, x) = a_1(\tau, x) + a_2(\tau, x)$.

This theorem lets us derive many special results in the cases where we can find explicit forms of the limits $a_1(\tau, x)$ and of $a_2(\tau; x)$.

As an important example let us consider the waiting time $\nu(N, k)$. Setting $\varepsilon_N = N^{-(\alpha+1)/\alpha}$ and $g_N(r) = \varepsilon_N r$, we see that in (9)

$$a_1(\tau, x) = i\tau A_\nu^{1/\alpha}.$$

Note that in this case $f_0(\tau, t_N) \to 1, f_2(\tau, t_N) \to 1$, and $a_2(\tau, x) = 0$. Thus, by virtue of Theorem 2,

$$\mathbf{E}\exp\{i\tau\varepsilon_N \nu(N,k)\} \to \int_0^\infty e^{a_1(\tau,x)}\pi_{s-1}(x)\,dx.$$

Inverting the limiting characteristic function, we obtain the explicit form of the limiting distribution function. Since the waiting time is one of the most important characteristics in applications, we formulate this result as a theorem.

THEOREM 3. *Let $N \to \infty$ while $s = N - k + 1 \geq 1$ be a constant, and let n and c be fixed. Then the waiting time $\nu(N, k)$ satisfies the relation*

$$\mathbf{P}\{d_N \nu(N,k) \leq x\} \to \Gamma_s(x^{-\alpha}), \qquad x > 0,$$

where $d_N = N^{-(\alpha+1)/\alpha} A_\nu^{-1/\alpha}\alpha^{-1}$.

Now we consider indicator characteristics. Let $g(x) = \mathbf{I}(x = r)$, $r \geq 1$. Then $L_{Nk} = \mu_r$ and $g^*(x) = \mathbf{I}(x = r)$. Carrying out the corresponding calculations, we obtain

$$\mathbf{E}e^{i\tau g(\xi^0)} = 1 + (e^{i\tau} - 1)B_0, \qquad \mathbf{E}e^{i\tau g(\xi^1)} = 1 + (e^{i\tau} - 1)B_1,$$

where $B_1 = \mathbf{P}\{\xi^2 = r\}$.

For the function $a(\tau; x)$ we find that

$$a(\tau, x) = x(e^{i\tau} - 1)B_2,$$

where

$$B_2 = \binom{\alpha + r - 1}{r}\mathbf{P}\{\nu \leq r\}/A_\nu.$$

Hence it follows that

$$\mathbf{E}e^{i\tau\mu_r} \to (1 + (e^{i\tau} - 1)B_0)(1 + (e^{i\tau} - 1)B_1)^{s-1}\int_0^\infty \exp\{x(e^{i\tau} - 1)B_2\}\pi_{s-1}(x)\,dx.$$

Inverting the limiting characteristic function, we obtain the following result.

COROLLARY 1. *Let $N \to \infty$ while $s = N - k + 1 \geq 1$ be a constant, and let n and c be fixed. Then*

$$\mathcal{L}(\mu_r) \to \mathcal{L}(\mathbf{I}(r \geq 1)\zeta_0 + \zeta_1 + \zeta_2).$$

where $\zeta_j, j = 0, 1, 2$, are independent random variables with the distributions

$$\mathcal{L}(\zeta_0) = \mathrm{Bi}(1, B_0), \quad \mathcal{L}(\zeta_1) = \mathrm{Bi}(s-1, B_1), \quad \mathcal{L}(\zeta_2) = \overline{\mathrm{Bi}}(s, B_2/(1 + B_2)).$$

5. THE LIMIT DISTRIBUTIONS FOR THE MIDDLE DOMAIN

As pointed out in Section 2, for $k = \lambda N + o(\sqrt{N})$, $\lambda = 1 - \bar{\lambda} \in (0, 1)$, the asymptotic behaviour of the DS L_{Nk} differs essentially for the cases $\lambda > p$ and $\lambda = p > 0$. For the first case, asymptotic normality of the DS holds.

THEOREM 4. *Let $N, k \to \infty$ in such a way that $k = \lambda N + o(\sqrt{N})$, $\lambda = 1 - \bar{\lambda} \in (0, 1)$, $\lambda > p$, and let the function g be such that for all $t > 0$ the variance $\mathbf{D}g(\xi(t))$ exists.*
Then

$$\mathcal{L}(L_N k) \sim \mathcal{N}(N\mu_\lambda, N\sigma_\lambda^2),$$

where

$$\begin{aligned}
\mu_\lambda = \mathbf{E}g(\xi_\lambda), \quad \xi_\lambda = \xi(t(\lambda)), \quad \sigma_\lambda^2 = \mathbf{D}\tilde{g}_c(\xi_\lambda),\\
\tilde{g}(\xi) = g(\xi) - d_\lambda(g)\,\mathbf{I}(\xi \geq \nu),\\
d_\lambda(g) = t'(\lambda)c\delta(\lambda)\operatorname{cov}(g(\xi_\lambda), \xi_\lambda),\\
\delta(\lambda) = \gamma(t(\lambda))/(1 - \gamma(t(\lambda))).
\end{aligned}$$

PROOF. In this case $x_N = \lambda + x\sqrt{\lambda\bar{\lambda}/N}$ and for all $x \in R$

$$t_N(\lambda) = t(\lambda) + xt'(\lambda)\sqrt{\lambda\bar{\lambda}/N} + O(1/N), \qquad t'(\lambda) = 1/H'(t(\lambda)).$$

Taking this into account, we find that

$$\mathbf{E}g(\xi_j(t_n)) = \mu_{j\lambda} + \frac{xc_j}{\sqrt{N}} + O(1/N), \qquad j = 1, 2,$$

where

$$\begin{aligned}
\mu_{j\lambda} &= \mathbf{E}g(\xi_{j\lambda}),\\
c_1 &= (t'(\lambda)\mathbf{E}g(\xi_{1\lambda})(\delta(\lambda)c\xi_{1\lambda} - n) - \mu_{1\lambda}/\lambda)\sqrt{\lambda\bar{\lambda}},\\
c_2 &= (t'(\lambda)\mathbf{E}g(\xi_{2\lambda})(\delta(\lambda)c\xi_{2\lambda} - n) + \mu_{2\lambda}/\bar{\lambda})\sqrt{\lambda\bar{\lambda}},
\end{aligned}$$

and we also see that $\mathbf{D}g(\xi_j(t_N)) \to \mathbf{D}g(\xi_{j\lambda}) = \sigma_{j\lambda}^2, j = 1, 2$.

By virtue of the central limit theorem, for $j = 1, 2$,

$$\begin{aligned}
&e^{-i\tau\sqrt{N}\mu_{j\lambda}} f_j^N(\tau/\sqrt{N}, t_N)\\
&\quad = (\mathbf{E}\exp\{i\tau(g(\xi_j(t_N)) - \mathbf{E}g(\xi_j(t_N))/\sqrt{N}\})^N \exp\{i\tau\sqrt{N}(\mathbf{E}g(\xi_j(t_N)) - \mu_{j\lambda})\}\\
&\qquad \to \exp\{-\tau^2\sigma_j^2/2 + i\tau c_j x\}.
\end{aligned}$$

In its turn, this allows us to conclude that

$$\begin{aligned}
&\mathbf{E}\exp\{i\tau(L_{Nk} - N(\lambda\mu_{1\lambda} + \bar{\lambda}\mu_{2\lambda})/\sqrt{N})\}\\
&\qquad \to \int_{-\infty}^{\infty} \exp\{-\tau^2(\lambda\sigma_1^2 + \bar{\lambda}\sigma_2^2)/2 + i\tau x(\lambda c_1 + \bar{\lambda}c_2)\}\varphi(x)\,dx = e^{-\tau^2\sigma_\lambda^2/2},
\end{aligned}$$

where

$$\sigma_\lambda^2 = \lambda\sigma_1^2 + \bar{\lambda}\sigma_2^2 + (\lambda c_1 + \lambda c_2)^2.$$

It remains to make sure that $\lambda\mu_{1\lambda} + \bar{\lambda}\mu_{2\lambda} = \mu_\lambda$ and σ_λ^2 has the form indicated in the statement of the theorem.

The first equality follows from the representations

$$\mu_{1\lambda} = \frac{1}{\lambda}\mathbf{E}g(\xi_\lambda)\,\mathbf{I}(\xi_\lambda \geq \nu), \qquad \mu_{2\lambda} = \frac{1}{\bar{\lambda}}\mathbf{E}g(\xi_\lambda)\,\mathbf{I}(\xi_\lambda < \nu).$$

Further, taking into account that

$$\lambda c_1 + \bar{\lambda}c_2 = (d_\lambda(g) + mu_{2\lambda} - mu_{1\lambda})\sqrt{\lambda\bar{\lambda}},$$

we find that

$$\sigma_\lambda^2 = \mathbf{D}g(\xi_\lambda) + \lambda\bar{\lambda}d_\lambda^2(g) - 2g_\lambda(g)\lambda\bar{\lambda}(mu_{1\lambda} - mu_{2\lambda}).$$

But

$$\lambda\bar{\lambda}(mu_{1\lambda} - mu_{2\lambda}) = \operatorname{cov}(g_(\xi_\lambda), \mathbf{I}(\xi_\lambda \geq \nu),$$

hence

$$\sigma_\lambda^2 = \mathbf{D}(g(\xi_\lambda) - d_\lambda(g)\,\mathbf{I}(\xi_\lambda \geq \nu) = \mathbf{D}\tilde{g}(\xi_\lambda).$$

The theorem is proved.

Finally, the boundary effects (for $\lambda = p > 0$) are described as follows. Let the parameters m, l and a be defined as in Section 3. In addition we set

$$h = pq\binom{\alpha + 2m - 1}{2m}\left(a\binom{\alpha + m - 1}{m}\right)^{-2},$$

$$v_l = g(l)\binom{\alpha + l - 1}{l}\left(\frac{a}{\sqrt{pq}}\binom{\alpha + m - 1}{m}\right)^{-l/m}.$$

Let also $E(x)$ be the distribution function of the random variable degenerated at zero and $F(x)$ be the distribution function of the absolute value of the standard normal random variable. Then the following assertion takes place.

THEOREM 5. *Let $p = \mathbf{P}\{\nu = 0\} > 0$ and $N, k \to \infty$ in such a way that $k = pN + o(\sqrt{N})$.*

Then for $l > 2m$, the DS L_Nk is asymptotically degenerated at zero; for $l = 2m$, the random variable $g^{-1}(2m)L_Nk$ has the discrete limit distribution $\{p_r, r = 0, 1, \dots\}$, where

$$p_0 = \frac{1}{2}(1 + (1 + 2h)^{-1/2}),$$

$$p_r = \frac{(2r)!(1 + 2h)^{-1/2}}{(r!)^2 2^{2r+1}}\left(\frac{2h}{1 + 2h}\right)^r, \qquad r \geq 1;$$

and for $l < 2m$

$$\mathbf{P}\{N^{(1-2m)/(2m)}v_l{}^{-1}L_{Nk} \leq x\} \to \frac{1}{2}(E(x) + F(x^{m/l})), \qquad x > 0.$$

PROOF. The proof repeats the arguments of Section 3, therefore we mark only the basic moments. Here $x_N = p + x\sqrt{pq/N}$, $x > 0$, and $t_N \to 0$ as $N \to \infty$, therefore, from the representation

$$H(t) = p + a\binom{\alpha + m - 1}{m} c^m t^m (1 + o(1)), \qquad t \to 0,$$

we find that

$$t_N = \frac{1}{c}\left(\frac{a}{x\sqrt{pq/N}}\binom{\alpha + m - 1}{m}\right)^{-1/m}(1 + o(1)).$$

Hence it follows that

$$\begin{aligned} f_0(\tau, t_N) &= e^{i\tau g(m)}(1 + o(1)), \\ f_1(\tau, t_N) &= 1 + \frac{1}{x_N}\sum_{r=l}^{\infty}(e^{i\tau g(r)} - 1)\bar{b}_r(\alpha, \gamma(t_N))\mathbf{P}\{v \le r\} \\ &= 1 + (e^{i\tau g(l)} - 1)\mathbf{P}\{v \le l\}c^l t_N^l \frac{1}{p}(1 + o(1)), \\ f_2(\tau, t_N) &= 1 + (e^{i\tau g(l)} - 1)\mathbf{P}\{v > l\}c^l t_N^l \frac{1}{q}(1 + o(1)), \end{aligned}$$

and we obtain the following description of the limiting behaviour of the characteristic function $f_{Nk}(\tau, t_N)$:
if $l > 2m$, then $f_{Nk}(\tau, t_N) \to 1$;
if $l = 2m$, then $f_{Nk}(\tau, t_N) \to \exp\{(e^{i\tau g(2m)} - 1)hx_2\}$;
if $l < 2m$, then $f_2^N(N^{(l-2m)/(2m)}\tau, t_N) \to \exp\{i\tau v_1 x^{l/m}\}$.

Finally, the inversion of the characteristic function

$$\frac{1}{2} + \int_0^\infty f(\tau, x)\varphi(x)\, dx$$

in each of these three cases leads us to the limit distributions given in the statement of the theorem.

Let us emphasize that the result has no analogue for the schemes with deterministic level v, and in connection with the waiting time $v(N; k)$ only the first case with $l = 1$ is appropriate, where the following assertion is valid.

COROLLARY 2. *Let $p > 0$ and $N, k \to \infty$ in such a way that $k = pN + o(\sqrt{N})$. Then*

$$\mathbf{P}\{N^{(1-2m)/(2m)}{v_1}^{-1}v(N, k) \le x\} \to \frac{1}{2}(E(x) + F(x^m)), \qquad x > 0.$$

REFERENCES

1. W. Feller, *An Introduction to Probability Theory and its Applications, Vol.* 1. Wiley, New York, 1957.
2. G. I. Ivchenko and Yu. I. Medvedev, On Markov–Pólya urn scheme from 1917 to nowadays. *Survey Appl. and Industrial Math.* (1996) **3**, 484–511 (in Russian).
3. G. I. Ivchenko, Decomposable statistics in an inverse allocation problem. *Discrete Math. Appl.* (1991) **1**, 81–96.
4. G. I. Ivchenko, The waiting time and related statistics in the multinomial scheme: a survey. *Discrete Math. Appl.* (1993) **3**, 451–482.
5. G. I. Ivchenko and A. V. Ivanov, Decomposable statistics in inverse urn problems. *Discrete Math. Appl.* (1995) **5**, 159–172.
6. G. I. Ivchenko and A. V. Ivanov, Decomposable statistics and stopping times in sampling without replacement. *Discrete Math. Appl.* (1997) **7**, 47–64.
7. V. A. Ivanov, Limit theorems in allocation scheme with random levels. *Math. Notes* (1982) **31**, 619–631 (in Russian).
8. L. Holst and J. Husler, Sequential urn schemes and birth processes. *Adv. Appl. Probab.* (1985) **17**, 257–279.
9. J. W. Pratt, On interchanging limits and integrals. *Ann. Math. Statist.* (1960) **31**,74–77.

Probabilistic Methods in Discrete Mathematics, pp. 253–265
V. F. Kolchin et al. (Eds.)

THE EXACT AND ASYMPTOTIC MAXIMUM LIKELIHOOD ESTIMATES OF THE STRUCTURE OF A STRATIFIED POPULATION

G. I. IVCHENKO and S. A. KHONOV

Moscow

Abstract — We give an exact form of the integer-valued vectors $\bar{N} = (N_1, \ldots, N_k)$ which provides the maximum of the function

$$L(\bar{N}, \bar{l}) = \prod_{j=1}^{k} \binom{N_j}{l_j}$$

for arbitrary positive integers $l_1, \ldots, l_k$, $k \geq 1$ and given $N^0 = N_1 + \ldots + N_k$, and the asymptotic maximum likelihood estimate of $\bar{N}$ as $N^0 \to \infty$. Applications of these results in statistics are discussed.

1. INTRODUCTION

In sampling survey of finite populations the key elements are the structure of the examined population, the type of the sampling mechanism for obtaining relevant statistical data, and the object of estimation. These three elements determine mainly the methods of investigation and the results obtained in the end.

1.1. Population structure

In general, the elements of an examined finite population U are partitioned by some feature (sort, income, capacity, etc.) into a known number k of disjoint classes or strata $U_1, \ldots, U_k$,

$$\bigcup_{j=1}^{k} U_j = U, \quad U_i \cap U_j = \varnothing, \quad i \neq j,$$

Then U is referred to as a stratified population with k strata, the case $k = 1$ corresponds to an unstratified or homogeneous population. Let $N^0 = |U|$ be the total population size and $N_1, \ldots, N_k$ be the corresponding strata sizes. The vector $\bar{N} = (N_1, \ldots, N_k)$ is called the structure of the population U (for a homogeneous population the structure and the size coincide).

1.2. Sampling plan

In applications for obtaining statistical data (statistical information on the examined population U) the classical schemes of sampling with or without replacement are often used. We use a more general sampling procedure called the general sampling scheme (GSS) where s independent random samples without replacements of sizes $m_1, \ldots, m_s$ are sequentially selected from the population U so that the ith sample is such that any of the $\binom{N^0}{m_i}$ distinct combinations of the population elements are drawn with equal probabilities (each of these samples is returned to the population before the next sampling). Thus the number of outcomes of such a statistical experiment is $\prod_{i=1}^{s}\binom{N^0}{m_i}$ and all are equiprobable. The sampling plan (SP) corresponding to the GSS with the parameters $m_1, \ldots, m_s$ and the total number of trials $n = m_1 + \ldots + m_s$ is denoted by $\Pi_n(m_1, \ldots, m_s)$. If all the partial samples have the same size m, the SP is denoted by $\Pi_n(s, m)$, where $n = sm$. Then $\Pi_n(n, 1) = \Pi_n$ is the plan of a simple sampling with replacement and $\Pi_n(1, n) = \bar{\Pi}_n$, $n \leq N^0$, is the plan of simple sampling without replacement of size n. In what follows we use the notation

$$m = \max_{1 \leq i \leq s} m_i$$

for the size of the maximum partial sample and suppose that $N^0 \geq m$. Let us note that a GSS is classified as a sampling procedure with fixed total number of observations. The data obtained by a GSS are represented in the form

$$\bar{\mu} = (\mu_{jr}, \quad j = 1, \ldots, k, \quad r = 1, \ldots, s),$$

here μ_{jr} is the number of elements of the strata U_j sampled exactly r times each (thus, in this case we do not take into account the order of appearance of elements and any other information, but the marks which show whether the elements belong to the corresponding strata).

1.3. Object of estimation

In the problems under consideration, the structure $\bar{N}$ of the population U is an unknown parameter with some a prior information given by a certain set $\mathcal{N} = \{\bar{N}\}$ of its possible values. We assume that the set U is not empty, that is, $N^0 = N_1 + \ldots + N_k \geq m > 0$. The structure of the parametric set $\mathcal{N}$ may be different but it is usually determined by linear constraints, for example

$$\mathcal{N} = \{\bar{N} = (N_1, \ldots, N_k) : N_1 + \ldots + N_k = N^0\},$$

where N^0 is a given size of the population

$$U = \bigcup_{j=1}^{k} U_j;$$

this case is considered in the paper and is typical for the problems of quality control.

1.4. The method of estimation

We point out some common features of the methods of investigation of the model discussed. In the GSS, the sufficient statistic for estimating the parameter $\bar{N}$ is the vector

$$\bar{\eta} = (\eta_1, \ldots, \eta_k),$$

where

$$\eta_j = \sum_{r=1}^{s} \mu_{jr}$$

is the number of distinct elements drawn from the stratum U_j, $j = 1, \ldots, k$, in the whole sampling process. In some cases, depending on the structure of the parametric set $\mathcal{N}$, as in the case considered in this paper, the sufficient statistic is not complete. In such cases, where there is no unbiased estimate with minimal variance (UEMV), one has to find some substitute for such estimates, moreover, even if an UEMV exists, as a rule, it is of an extremely complicated form [4, 5].

This paper is devoted to an unresolved problem for the model described above. Namely, we consider the estimation of the parameter $\bar{N}$ by the maximum likelihood method which, in this case, is preferable from various points of view.

2. THE EXACT ESTIMATE OF THE MAXIMUM LIKELIHOOD

The exact form of the maximum likelihood estimate (MLE) $\hat{N} = (\hat{N}_1, \ldots, \hat{N}_k)$ of the parameter $\bar{N}$ can be determined in this case as the solution of the following discrete problem on conditional extremum:

$$\hat{N} = \arg\max_{\bar{N} \in \mathcal{N}} L(\bar{N}, \bar{l}), \qquad \mathcal{N} = \{\bar{N} : N_1 + \ldots + N_k = N^0\}, \tag{1}$$

where $\bar{l} = (l_1, \ldots, l_k)$ is the observed value of the random vector $\bar{\eta} = (\eta_1, \ldots, \eta_k)$.

The main feature of this problem is the fact that here we deal with functions of multidimensional arguments and there are no general methods developed for solving such extremal problems. Only for the simplest case $k = 2$, where the kernel L is a function of one argument N_1, the solution of the problem (1) is simple and is known. In this case,

$$\hat{N}_1 = [(N^0 + 1)l_1/l],$$

where $[a]$ is the integral part of a number a and $l = l_1 + \ldots + l_k$. If $(N^0 + 1)l_1/l$ is an integer, then $\hat{N}_1 - 1$ is the MLE of the parameter N_1 as well. For the general case $k \geq 2$, the exact form of the MLE (the solution of the problem (1)) is given here for the first time. We also discuss the question of uniqueness of the MLE $\hat{N}$ and give a simple algorithm of calculating this estimate.

In what follows the key role is played by the function

$$F(\lambda) = \sum_{j=1}^{k} [\lambda l_j], \qquad \lambda \geq 0. \tag{2}$$

The properties of this function are discussed in detail later. Here we point out only some of these properties which are needed to formulate and prove Theorem 1.

It is clear that $F(\lambda)$ is a continuous from the right integer-valued step-function whose points of jumps are divisible by the inverse values of the parameters $l_1, \dots, l_k$. The inverse function

$$F^{-1}(N) = \min\{\lambda : F(\lambda) \geq N\}, \quad N = 1, 2, \dots,$$

is defined uniquely and the range of this function coincides with the points of jumps of $F(\lambda)$. We introduce the set

$$J(\lambda) = \{j : \lambda l_j \text{ is an integer}\},$$

that is, the set of the numbers of those terms in (2) which have a jump at the point λ. Then the value of the jump of the function $F(\lambda)$ at the point λ is equal to

$$\Delta F(\lambda) = F(\lambda) - F(\lambda - 0) = |J(\lambda)|.$$

Let $\lambda^0 = F^{-1}(N^0)$ and $J^0 = J(\lambda^0)$, that is, λ^0 is the point of jump of $F(\lambda)$ determined by the condition

$$F(\lambda^0 - 0) < N^0 \leq F(\lambda^0),$$

and $|J^0|$ is the value of this jump. Note that $\lambda^0 \geq 1$ (since $N^0 \geq l$) and

$$\lambda^0 = 1 \iff N^0 = l,$$

and $|J^0| \leq k$. Finally, we set $\delta^0 = F(\lambda^0) - N^0$, and therefore, $0 \leq \delta^0 < |J^0|$.

Using this notation, we formulate the main result of this section.

THEOREM 1. *The solution of problem* (1) *is of the form*

$$\hat{N} = ([\lambda^0 l_j] - I_D), \qquad j = 1, \dots, k, \tag{3}$$

where D is an arbitrary subset of size δ^0 of the set J^0 and I_D is the indicator of this subset.

We make some remarks before we turn to the proof of this assertion. By definition, the set J^0 is non-empty, but the parameter δ^0 can be zero and then $D = \varnothing$. In this case only, that is, if $F(\lambda^0) = N^0$, the MLE $\hat{N}$ is determined uniquely and $\hat{N} = ([\lambda^0 l_j], j = 1, \dots, k)$. In other cases, where $\delta^0 > 0$, there is $\binom{|J^0|}{\delta^0}$ distinct variants of choice of the subset $D \subset J^0$ and, consequently, of the corresponding versions of the MLE. Thus, in the problem under consideration the MLE is defined, in general, non-uniquely (note that such a situation is usual for models with discrete parameters). However, in this case, the non-uniqueness is of a special character, namely, the versions of the MLE can differ only by the term I_D. Finally, the assumption that all the observed values $l_1, \dots, l_k$ are positive does not restrict the generality of the problem, since otherwise we can pass to the vector $\bar{N}$ of lesser dimension (note that in such a case the function $L(\bar{N}, \bar{l})$ depends in fact on $l_j > 0$). In terms of the MLE it means that we set $\hat{N}_j = 0$ if $l_j = 0$.

PROOF OF THEOREM 1. By definition, the MLE $\hat{N}$ is such a point of the hyperplane $\mathcal{N}$ (see (1)), where the inequality $L(\bar{N}+\bar{\delta},\bar{l}) \le L(\bar{N},\bar{l})$ holds for any non-zero integer-valued vector $\bar{\delta} = (\delta_1, \ldots, \delta_k)$ such that $\delta_1 + \ldots + \delta_k = 0$ and $N_j + \delta_j \ge l_j$ for all $j = 1, \ldots, k$. Setting

$$J_- = \{j\colon \delta_j < 0\}, \qquad J_+ = \{j\colon \delta_j > 0\},$$

we have

$$\begin{aligned}
&\frac{L(\bar{N}+\bar{\delta},\bar{l})}{L(\bar{N},\bar{l})}\\
&= \prod_{j\in J_-} \frac{(N_j+\delta_j+1-l_j)\ldots(N_j-l_j)}{(N_j+\delta_j+1)\ldots N_j} \prod_{j\in J_+} \frac{(N_j+1)\ldots(N_j+\delta_j)}{(N_j+1-l_j)\ldots(N_j+\delta_j-l_j)}\\
&= \prod_{j\in J_-}\left(1-\frac{l_j}{N_j+\delta_j+1}\right)\cdots\left(1-\frac{l_j}{N_j}\right)\prod_{j\in J_+}\left(1-\frac{l_j}{N_j+1}\right)^{-1}\cdots\left(1-\frac{l_j}{N_j+\delta_j}\right)^{-1}\\
&\le \prod_{j\in J_-}\left(1-\frac{l_j}{N_j}\right)^{|\delta_j|}\prod_{j\in J_+}\left(1-\frac{l_j}{N_j+1}\right)^{-\delta_j}.
\end{aligned}$$

In particular, if all non-zero δ_j are equal to ± 1, then the last inequality above is equality, and the ratio under consideration is no greater than one if and only if for all $i \ne j$

$$\frac{l_j}{N_j+1} \le \frac{l_i}{N_i}. \tag{4}$$

Since

$$\sum_{j\in J_-} |\delta_j| = \sum_{j\in J_+} \delta_j,$$

the point $\bar{N}$ which satisfies (4) is the point of global maximum of the function $L(\bar{N},\bar{l})$. Thus, any solution of inequality (4) which belongs to the hyperplane $\mathcal{N}$ is an MLE $\hat{N}$.

Let us check that any point given in (3) is an MLE, that is, it satisfies (4) and belongs to $\mathcal{N}$. For indices $j \in J^0$, the values $\lambda^0 l_j$ are integers, therefore in (3),

$$\hat{N}_j = \begin{cases} \lambda^0 l_j - 1, & j \in D,\\ \lambda^0 l_j, & j \in J^0 \setminus D.\end{cases}$$

If $j \notin J^0$, then

$$\lambda^0 l_j - 1 < \hat{N}_j = [\lambda^0 l_j] < \lambda^0 l_j.$$

Thus, in any case (for any j),

$$\lambda^0 l_j - 1 \le \hat{N}_j \le \lambda^0 l_j$$

or

$$\hat{N}_j \le \lambda^0 l_j \le \hat{N}_j + 1.$$

Hence it follows that for all j

$$\frac{\hat{N}_j}{l_j} \le \lambda^0 \le \frac{\hat{N}_j + 1}{l_j},$$

and consequently, $\hat{N} = (\hat{N}_1, \dots, \hat{N}_k)$ satisfies (4). Finally, the belonging of the point $\hat{N}$ to the hyperplane $\mathcal{N}$ follows from the obvious equalities

$$\sum_{j=1}^{k} \hat{N}_j = F(\lambda^0) - \sum_{j=1}^{k} I_D(j) = N^0 + \delta^0 - |D| = N^0.$$

Let us prove now that any solution of the system (4) which belongs to $\mathcal{N}$ can be represented in the form (3). Let a point $\bar{N} \in \mathcal{N}$ satisfies (4). We set

$$\lambda' = \max_j \frac{N_j}{l_j}, \qquad E = \{j\colon N_j/l_j = \lambda'\}$$

and prove that $\lambda' = \lambda^0$.

First, it is clear that $N_j \le \lambda' l_j$ for any j, that is, $N_j \le [\lambda' l_j]$, and consequently, $N^0 \le F(\lambda')$. Secondly, $N_j = \lambda' l_j$ for any $j \in E$, and by virtue of (4) $\lambda' l_j \le N_j + 1$ for all j. Hence, for any $\lambda < \lambda'$ we have $[\lambda l_j] \le \lambda_j < N_j$ for $j \in E$ and $[\lambda l_j] \le N_j$ for all j. Since the set E is not empty, hence it follows that $F(\lambda) < N^0$. Thus, for all $\lambda < \lambda'$,

$$F(\lambda) < N^0 \le F(\lambda'),$$

that is, $\lambda' = F^{-1}(N^0) = \lambda^0$.

Let

$$D = \{j\colon \lambda^0 l_j = N_j + 1\}.$$

It is clear that $D \subset J^0$ and $[\lambda^0 l_j] \le N_j$ for $j \notin D$. Taking into account the inequality $N_j \le [\lambda' l_j]$ obtained above, we see that $N_j = [\lambda^0 l_j]$. Thus, we have the representation of form (3)

$$N_j = \begin{cases} \lambda^0 l_j - 1, & j \in D, \\ [\lambda^0 l_j], & j \notin D, \end{cases}$$

that is,

$$N_j = [\lambda^0 l_j] - I_D(j), \qquad j = 1, \dots, k.$$

It follows from this representation that $N^0 = F(\lambda^0) - |D|$, that is, $|D| = F(\lambda^0) - N^0 = \delta^0$.

The theorem is proved.

To use the result (3) in practice, we need a simple algorithm to compute the parameter $\lambda^0 = F^{-1}(N^0)$, that is, we need an algorithm to invert the function F defined in (2). In what follows we give a detail analysis of this function and obtain its constructive description which lets us suggest an effective and convenient for realization algorithm.

We introduce two auxiliary functions, pre and next. The first of them for any $\lambda \ge 0$ calculates the point of jump where the function F becomes equal to $F(\lambda)$ for the first time, more precisely,

$$\mathrm{pre}(\lambda) = \min\{\lambda'\colon F(\lambda') = F(\lambda)\}.$$

The second function gives the next after λ point of jump of the function F, more precisely,

$$\operatorname{next}(\lambda) = \min\{\lambda' : F(\lambda') > F(\lambda)\}.$$

Thus, these functions together determine the half-interval where the function F is a constant and where the point λ lies, in other words,

$$F(\lambda') = F(\lambda) \Longleftrightarrow \lambda' \in [\operatorname{pre}(\lambda), \operatorname{next}(\lambda)).$$

To calculate these functions, we can use the formulae

$$\operatorname{pre}(\lambda) = \max_j \frac{[\lambda l_j]}{l_j},$$
$$\operatorname{next}(\lambda) = \min_j \frac{[\lambda l_j] + 1}{l_j}. \tag{5}$$

Indeed, since $[\lambda l_j] \le \lambda l_j < [\lambda l_j] + 1$, we have

$$\frac{[\lambda l_j]}{l_j} \le \operatorname{pre}(\lambda) \le \lambda < \operatorname{next}(\lambda) \le \frac{[\lambda l_j] + 1}{l_j}. \tag{6}$$

Hence it follows that, first, $[\lambda l_j] \le \operatorname{pre}(\lambda) l_j \le \lambda l_j$, that is, $[\operatorname{pre}(\lambda) l_j] = [\lambda l_j]$, and consequently, $F(\operatorname{pre}(\lambda)) = F(\lambda)$.

Secondly, for any $\lambda' \in [\lambda, \operatorname{next}(\lambda))$,

$$\lambda l_j \le \lambda' l_j < [\lambda l_j] + 1,$$

that is, $[\lambda' l_j] = [\lambda l_j]$, and consequently, $F(\lambda') = F(\lambda)$. Thus, the function F is a constant on the half-interval $[\operatorname{pre}(\lambda), \operatorname{next}(\lambda))$ and is equal to $F(\lambda)$. At the same time, $\operatorname{pre}(\lambda)$ (for $\lambda \ge (\max_j l_j)^{-1}$) and $\operatorname{next}(\lambda)$ (for any $\lambda \ge 0$) are points of jumps, since these values are divided by the inverse values of some l_j. Note that the value $|J(\operatorname{next}(\lambda))|$ which is equal to the size of the jump at the point $\operatorname{next}(\lambda)$ can be calculated by the formula

$$|J(\operatorname{next}(\lambda))| = |\{j : ([\lambda l_j] + 1)/l_j = \operatorname{next}(\lambda)\}|, \tag{7}$$

and similarly for the jump at the point $\operatorname{pre}(\lambda)$

$$|J(\operatorname{pre}(\lambda))| = |\{j : [\lambda l_j]/l_j = \operatorname{pre}(\lambda)\}|. \tag{8}$$

Indeed, by definition, $J(\operatorname{next}(\lambda))$ is the set of those j for which $\operatorname{next}(\lambda) l_j$ are integers, that is, the set contains j given in (7). For the remaining j by virtue of (6)

$$\lambda l_j < \operatorname{next}(\lambda) l_j < [\lambda l_j] + 1,$$

and $\operatorname{next}(\lambda) l_j$ is not an integer. Thus, $J(\operatorname{next}(\lambda))$ includes those and only those j which are given by (7). Equality (8) can be proved in a similar way.

The role of the function next is not exhausted by the applications described above. This function has the following important property. Since it calculates the point of jump which immediately follows λ, we can use this function to calculate sequentially all the point of jump on any interval $\lambda > \lambda_0 \geq 0$. Namely, this is the sequence

$$\{\lambda_t = \text{next}(\lambda_{t-1}) = \text{next}_t(\lambda_0), \quad t = 1, 2, \dots\}$$

where next_t is the tth iteration of the function next, that is, $\text{next}_t(\lambda) = \text{next}(\text{next}_{t-1}(\lambda))$ and $\text{next}_0(\lambda) = \lambda$. The sequence $\{|J(\lambda_t)|, t = 1, 2, \dots\}$ (see (7)) is the sequence of the corresponding sizes of jumps. Thus, the function $F(\lambda)$ for $\lambda \geq \lambda_0$ can be constructively described by the double sequence $\{(\lambda_t, |J(\lambda_t)|), t = 1, 2, \dots\}$, since

$$F(\lambda) = F(\lambda_0) + \sum_{i=1}^{t} |J(\lambda_i)|, \quad \lambda \in [\lambda_t, \lambda_{t+1}), \quad t = 0, 1, 2, \dots \tag{9}$$

(here we assume that the empty sum for $t = 0$ is equal to zero). In particular, for $\lambda_0 = 0$ formula (9) gives a constructive description of the function F on the half-line $\lambda \geq 0$.

Finally, representation (9) lets us suggest the following simple algorithm calculating the inverse function $F^{-1}(N)$ for any integer $N \geq F(\lambda_0)$ (if $\lambda = 0$, then this algorithm calculates the inverse function for all $N = 1, 2, \dots$):
for $N > F(\lambda_0)$

$$F^{-1}(N) = \lambda_{\tau(N)}, \tag{10}$$

where

$$\tau(N) = \min\{t\colon F(\lambda_t) \geq N\},$$

if $N = F(\lambda_0)$, then $F^{-1}(N) = p(\lambda_0)$.

REMARK 1. If it is required to calculate the value of the inverse function F^{-1} at only one given point N^0 (as in formulae (3)), then the general algorithm (10) can be simplified if we take into account that for $\lambda_0 = (N^0 + 1)/l$ we have

$$N^0 + 1 - k < F(\lambda_0) = \sum_{j=1}^{k} [(N^0 + 1)l_j/l] \leq N^0 + 1.$$

Here

$$F(\lambda_0) = N^0 + 1 \iff J(\lambda_0) = \{1, 2, \dots, k\},$$

that is, the point λ_0 is a point of jump and $\Delta F(\lambda_0) = |J(\lambda_0)| = k$. Consequently, in this case, $F^{-1}(N^0) = \lambda_0$. If $F(\lambda_0) \leq N^0$, then the general form of the algorithm (10) have to be used and in this case the inequality $F(\lambda_0) < N^0$ is possible for $k > 2$ only, and the number of steps of the algorithm is no greater than $N^0 - F(\lambda_0) \leq k - 2$ (we assume that the tth step of the algorithm is the calculation of $\lambda_t = \text{next}(\lambda_{t-1})$ and the comparing $F(\lambda_t)$ and N^0, $t = 1, 2, \dots$).

Note also that if $N^0 = l$, then $\lambda^0 = F^{-1}(N^0) = 1$, and obviously, $\hat{N} = \bar{l}$. Therefore, in what follows we assume that $N^0 > l$.

Using the simplified algorithm (according to Remark 1), we obtain that if $F(\lambda_0) = N^0 + 1$, then

$$\lambda^0 = \lambda_0 = \frac{N^0 + 1}{l}, \qquad \delta^0 = F(\lambda_0) - N^0 = 1,$$

and we have k versions of MLE:

$$\hat{N} = \lambda_0 \bar{l} - \bar{e}_i, \qquad i = 1, \dots, k,$$

where $\bar{e}_i$ is the ith unit vector, $i = 1, \dots, k$.

If $F(\lambda_0) = N^0$, then according to (10), $\lambda^0 = \mathrm{pre}(\lambda_0)$, $\delta^0 = 0$, and by formula (3), the MLE is determined uniquely and

$$\hat{N} = ([p(\lambda_0)l_j] = [\lambda_0 l_j], \quad j = 1, \dots, k).$$

If $k = 2$, then these two cases exhaust all possibilities and we have the well-known result [1].

If $k > 2$, then the case $F(\lambda_0) < N^0$ is possible and according to (10), $\lambda^0 = \lambda_{\tau^0}$, where $\tau^0 = \min\{t\colon F(\lambda_t) \geq N^0\}$. Thus, invoking (3), we obtain a constructive algorithm of calculating all possible MLE with the number of steps no greater than $k - 2$.

3. ASYMPTOTIC RESULTS

Asymptotic approach to the problem under consideration was used in many papers (see, for example, [3–6]). In this section, we give some results for the case of large samples (as $n \to \infty$) under assumption that the total size of the population and the sizes of the strata are also large. More precisely, we assume that there exist constants c_1 and c_2 such that, as $n, N^0, N_1, \dots, N_k \to \infty$,

$$\begin{aligned} N_j &= \beta_j N^0, \qquad & 0 < c_1 \leq \beta_j \leq c_2 < 1, \\ n/N &= \alpha, \qquad & 0 < c_1 \leq \alpha \leq c_2 < \infty. \end{aligned} \tag{11}$$

Since the model under consideration has many parameters (multidimensional parameter $\bar{N} = (N_1, \dots, N_k)$ which have to be estimated, arbitrary parameters $m_1, \dots, m_s$ of the GSS), meaningful results can be obtained under the corresponding conditions of regularity on these parameters. As such conditions, the following requirements are assumed in [3]:

$$\begin{aligned} \mathbf{E}_N \bar{\eta} &= N^0 \bar{\beta} a(1 + O(1/N)), \\ \mathbf{D}_N \bar{\eta} &= N^0 B(1 + O(1/N)), \\ \mathbf{D}_N \eta &= N b(1 + O(1/N)), \qquad N = N^0, \end{aligned} \tag{12}$$

where

$$a = a(\alpha) = \lim_{N\to\infty} a_N = \lim_{N\to\infty}\left(1 - \prod_{i=1}^{s}\left(1 - \frac{m_i}{N}\right)\right)$$

$$B = \|b_{ij}\|, \quad b_{ij} = \beta_i(\beta_j b + (\delta_{ij} - \beta_j)a(1-a)), \quad i,j = 1,\dots,k,$$

δ_{ij} is the Kronecker symbol,

$$|B| = (a(1-a))^{k-1} b\beta_1 \dots \beta_k,$$
$$b = \lim_{N\to\infty}(1 - a_N)(a_N - (N-1)(a_{N-1} - a_N)), \qquad i,j = 1,\dots,k.$$

Note that there were no direct conditions on the parameters $s, m_1, \dots, m_s$.

We consider the following two variants of conditions on the parameters of the GSS which from our point of view are of a significant interest in applications:

$m \geq 1$ is fixed and $s \to \infty$ in such a way that $n = m_1 + \dots + m_s \to \infty$,

$s \geq 2$ is fixed and $m_i \to \infty$ in such a way that there exists $\delta > 0$ such that $\rho_i = m_i/n \in [\delta, 1-\delta]$, $i = 1, \dots, s$.

If the asymptotic approach is used, then it is natural to consider as the estimated parameter the vector $\bar\beta = (\beta_1, \dots, \beta_k)$, which determines completely the structure of the population U, and below we construct the MLE of $\bar\beta$ under the conditions described above. Note that these conditions are typical for the theory of large samples in such problems and let us develop sufficiently complete (regular) asymptotic theory of the corresponding statistical inference. We begin with the theorem on the asymptotic normality of the sufficient statistic for the model under consideration.

THEOREM 2. *Let $N, n \to \infty$ and conditions* (11) *are fulfilled. Then uniformly with respect to the parameters $\bar\beta$, $\alpha = n/N$ (in the case where the total size of the population is known) and $\bar t = t_1, \dots, t_k$, $|t_j| \leq c < \infty$, $j = 1, \dots, k$,*

$$\mathbf{P}_N\{\bar\eta = \bar l\} = (2\pi N)^{-k/2}|B|^{-1}\exp\{\bar t' B^{-1}\bar t/2\}(1 + O(1+\sqrt{N})), \tag{13}$$

where

$$\bar l = N^0\bar\beta a + \bar t\sqrt{N^0},$$

$$B^{-1} = \|b^{ij}\|, \qquad b^{ij} = \frac{1}{b} = \frac{\delta_{ij} - \beta_j}{a(1-a)\beta_j}, \qquad i,j = 1,\dots,k,$$

$\bar t$ is a column vector, $\bar t'$ is a row vector.

This theorem is proved in [5]. In the model considered here the parameters $\bar\beta$ and α determine completely the structure of the population and (13) gives the likelihood function for these parameters. Note that in [3] a more general case is considered where the total size N of the population may be unknown. The following theorem gives us the exact form of the MLE of the parameter $\bar\beta$.

THEOREM 3. *If $n, N \to \infty$ and the conditions (12) are fulfilled, then the asymptotic MLE of the parameter $\bar\beta$ is equal to*

$$\hat\beta_n = \bar\eta/\eta \tag{14}$$

PROOF. The MLE of $\bar\beta$ is the solution of the problem of finding the conditional maximum of the asymptotic distribution (13) under the condition $\beta_1 + \ldots + \beta_k = 1$. By some not complicated transformations we obtain the following system of equations with respect to the unknowns $\beta_1, \ldots, \beta_k$:

$$\frac{d}{d\beta_j}\left(\sum_{j=1}^{k}\frac{(\eta_j/(Na)-\beta_j)^2}{\beta_j} + \frac{1-a}{Na}\ln\beta_j\right) + \varkappa + o(1) = 0, \quad j = 1, \ldots, k, \quad \sum_{j=1}^{k}\beta_j = 1, \tag{15}$$

where $\varkappa$ does not depend on $\beta_1, \ldots, \beta_k$. Hence it follows that

$$\beta_j = \frac{\eta_j}{Na\sqrt{1+\varkappa}} + o(1), \qquad j = 1, \ldots, k, \qquad \frac{\eta}{Na\sqrt{1+\varkappa}} = 1 + o(1)$$

or

$$\beta_j = \frac{\eta_j}{\eta} + o(1), \qquad j = 1, \ldots, k.$$

At the point $\hat\beta = \bar\eta/\eta$,

$$\frac{d^2}{d\beta_j^2}\left(\sum_{j=1}^{k}\frac{(\eta_j/(Na)-\beta_j)^2}{\beta_j} + \frac{1-a}{Na}\ln\beta_j\right)\Bigg|_{\beta=\hat\beta} = -\frac{Na}{\eta_j} < 0.$$

Thus, the likelihood function takes the maximum value at the point $\hat\beta_n$. Note that the expression for $\hat\beta_n$ does not contain the parameter α, therefore we can maximize the likelihood function separately with respect to $\bar\beta$ and α (if the total size of the population is unknown).

It is interesting to note that the consistent asymptotically normal estimates of the parameter $\bar\beta$ are of the same form (see [5]).

THEOREM 4. *For both variants of conditions on the parameters of the GSS given above, if conditions (11) and (12) are fulfilled, then the statistic $\hat\beta_n = \bar\eta/\eta$ is a consistent asymptotically normal estimate of $\bar\beta$ and*

$$\mathcal{L}(\hat\beta_n) \sim \mathcal{N}(\bar\beta, \Sigma(\alpha, \hat\beta)/n),$$

where

$$\Sigma(\alpha, \hat\beta) = \Psi(\alpha)\|\beta_i(\delta_{ij} - \beta_j)\|, \qquad i,j = 1, \ldots, k,$$

$$\Psi(\alpha) = \frac{1-\alpha G(\alpha)}{G(\alpha)}, \qquad G(\alpha) = \frac{a(\alpha)}{\alpha}.$$

Moreover, in the formulae for the second moments in these asymptotic relations, the unknown parameters $\bar\beta$, and α if N is unknown, may be replaced by their estimates $\hat\alpha_n$ and $\hat\beta_n$.

This theorem is proved in [5]. Note that in the case considered here α is known, since we assume that $N = N^0$ is known.

Applying this theorem to constructing confidence intervals and regions in the standard way (see [5]), we conclude that in our case the asymptotic $(1-\delta)$-confidence intervals for α and β_j are of the form

$$\left(\hat{\alpha}_n \pm t_{\delta/2}\sigma(\hat{\alpha}_n)/\sqrt{n}\right)$$

and

$$\left(\hat{\beta}_{jn} \pm t_{\delta/2}\sqrt{\Psi(\hat{\alpha}_n)\hat{\beta}_{jn}(1-\hat{\beta}_{jn})/n}\right), \tag{16}$$

and the asymptotic γ-confidence region for the vector $\bar{\beta}$ is of the form

$$\mathcal{G}_{\gamma n} = \left\{\bar{\beta}:\ \sum_{j=1}^{k}\frac{(\hat{\beta}_{jn}-\beta_j)^2}{\hat{\beta}_{jn}} < \frac{\chi^2_{\gamma,k-1}\Psi(\hat{\alpha}_n)}{n},\quad \sum_{j=1}^{k}\beta_j = 1\right\}, \tag{17}$$

where $\chi^2_{\gamma,k-1}$ is the γ-quantile of the χ^2-distribution with $k-1$ degrees of freedom.

We can write (17) is the equivalent form

$$\mathcal{G}_{\gamma n} = \left\{\bar{\beta}^0:\ |\bar{\lambda}'(\hat{\beta}^0_n - \bar{\beta}^0)| < \sqrt{\chi^2_{\gamma,k-1}\Psi(\hat{\alpha}_n)(\bar{\lambda}'H_n\bar{\lambda})/n}\quad \forall\bar{\lambda}\right\}, \tag{18}$$

where

$$\bar{\beta}^0 = (\beta_1,\ldots,\beta_{k-1}),$$
$$H_n = H(\hat{\beta}^0_n),\quad H(\bar{\beta}^0) = \|\beta_i(\delta_{ij}-\beta_j)\|,\quad i,j = 1,\ldots,k-1.$$

It follows from (17) that for all linear combinations $\{\bar{\lambda}'\bar{\beta}^0\}$ the system of confidence intervals

$$\left(\bar{\lambda}'\hat{\beta}^0_n \pm t_{\delta/2}\sqrt{\Psi(\hat{\lambda}_n)(\bar{\lambda}'H_n\bar{\lambda})/n}\right) \tag{19}$$

has asymptotically (under the conditions of Theorem 4) the confidence level γ. In other words, (19) is the system of simultaneous asymptotic confidence intervals for arbitrary linear combinations $\bar{\lambda}'\bar{\beta}^0$. If we estimate a finite number of such linear combinations, then the system of the confidence intervals from (19) corresponding to those combinations has the asymptotic confidence level no less than γ. The asymptotic $(1-\delta)$-confidence interval for a single linear combination $\bar{\lambda}'\bar{\beta}^0 = \lambda_1\beta_1 + \ldots + \lambda_{k-1}\beta_{k-1}$ is of the form

$$\left(\bar{\lambda}'\hat{\beta}^0_n \pm t_{\delta/2}\sqrt{\Psi(\hat{\alpha}_n)(\bar{\lambda}'H_n\bar{\lambda})/n}\right). \tag{20}$$

In particular, it follows from the results above that the system of the confidence intervals for the parameters $\{\beta_j\}$ (or a part of them) is of the form

$$\left\{ \left(\hat{\beta}_{jn} \pm \sqrt{\chi^2_{\gamma,k-1} \Psi(\hat{\alpha}_n) \hat{\beta}_{jn} (1 - \hat{\beta}_{jn}/n)} \right), \quad j = 1, \ldots, k \right\}, \tag{21}$$

and their confidence level is asymptotically no less than γ. The similar system for the differences $\{\beta_i - \beta_j, j < i\}$ is of the form

$$\left\{ \left(\hat{\beta}_{in} - \hat{\beta}_{jn} \pm \sqrt{\chi^2_{\gamma,k-1} \Psi(\hat{\alpha}_n)(\hat{\beta}_{in} + \hat{\beta}_{jn} - (\hat{\beta}_{in} - \hat{\beta}_{jn})^2)/n} \right), \quad j < i \right\}. \tag{22}$$

If the total size of the population $N = N^0$ is known, as in the situation considered in this paper, we have to set $\hat{\alpha}_n = \alpha$ in (16)–(22).

REFERENCES

1. Yu. K. Belyaev, *Probabilistic Methods of Sampling Inspection.* Nauka, Moscow, 1975 (in Russian).
2. G. I. Ivchenko and S. A. Khonov, The maximum likelihood estimation for a stratified finite population. *Math. Methods Statist.* (1994) **3**, 346–361.
3. S. A. Khonov, The asymptotic maximum likelihood estimates of parameters of a stratified population. In: *Probab. Problems of Discrete Math.* MIEM, Moscow, 1990, pp. 127–129 (in Russian).
4. G. I. Ivchenko and S. A. Khonov, Statistical estimation of the structure of a finite population. *Discrete Math. Appl.* (1996) **6**, 1–39.
5. G. I. Ivchenko and S. A. Khonov, On the asymptotic estimation for stratified finite populations. *Discrete Math. Appl.* **1**, 311–320.

Probabilistic Methods in Discrete Mathematics, pp. 267–274
V. F. Kolchin et al. (Eds.)

DISCRETE DISTRIBUTIONS IN CONTROL PROBLEMS

V. A. KASHTANOV

Moscow

Abstract — We study the structure of the distribution functions which provide the extremum of additive accumulative functionals defined on the trajectories of controlled semi-Markov processes with finite number of states.

1. FORMULATION OF THE PROBLEM

We consider the discrete distributions which appear in the problems of control of a semi-Markov process $\zeta(t)$ with a finite set of states $E = \{1, 2, \dots, N\}$. The controlled semi-Markov process $\zeta(t)$ is defined by a Markov chain (ξ_n, θ_n, u_n), $n \geq 1$, where

$$\xi_n \in E, \quad \theta_n \in R^* = [0, \infty), \quad u_n \in U_i,$$

U_i are some spaces of control parameters with σ-algebras A_i of their subsets, $i = 1, \dots, N$. The Markov chain is determined by the transition characteristics (semi-Markov kernel)

$$\begin{aligned}\mathbf{P}\{\xi_n = j, \theta_n < t, u_n \in B \mid \xi_{n-1} = i, \theta_{n-1} = \tau, u_{n-1} = u\} \\ = \mathbf{P}\{\xi_n = j, \theta_n < t, u_n \in B \mid \xi_{n-1} = i\} = \bar{Q}_{ij}(t, B),\end{aligned}$$

where $t \in R^*$, $B \in A_i$, $n \geq 1$, $i, j \in E$, and by some initial distribution

$$\mathbf{P}\{\xi_0 = i\} = p_i, \quad i \in E.$$

We introduce the counting process

$$\nu(t) = \inf\left(n \colon \sum_{k=0}^{n} \theta_k > t\right), \qquad \theta_0 = 0.$$

Then the controlled semi-Markov process $\zeta(t)$ can be defined as the pair

$$\zeta(t) = (\xi(t), u(t)), \quad \xi(t) = \xi_{\nu(t)-1}, \quad u(t) = u_{\nu(t)}.$$

First of all we note that the future of the homogeneous Markov chain which determines the semi-Markov process depends on the first component only. Secondly, the components of the semi-Markov process $\zeta(t)$ are dependent random processes whose piecewise-constant realizations have coinciding points of discontinuity.

The semi-Markov kernel $\bar{Q}_{ij}(t, B)$ defines a family of probability measures on the measurable spaces (U_i, A_i)

$$G_i(B) = \sum_{j \in E} \bar{Q}_{ij}(\infty, B), \qquad i \in E.$$

Since for any $j \in E$ and $t \in R^*$,

$$G_i(B) > \bar{Q}_{ij}(t, B),$$

there exist functions $Q_{ij}(t, B)$ such that

$$\bar{Q}_{ij}(t, B) = \int_B Q_{ij}(t, u) G_i(du).$$

The function $Q_{ij}(t, u)$ can be regarded as the conditional probability

$$Q_{ij}(t, u) = \mathbf{P}\{\xi_n = j, \theta_n < t \mid \xi_{n-1} = i, u_n = u\}.$$

Thus, a controlled semi-Markov process can be determined by a family of matrices $(Q_{ij}(t, u))$, a set of probability measures $G_i(B)$, and an initial distribution p_i, $i, j = 1, 2, \ldots, N$.

If we have to classify the model described above, we can say that we use the class of Markov homogeneous randomized control strategies which are determined by the probability measures $G_i(B)$, $i = 1, 2, \ldots, N$.

In what follows we use the notation

$$\bar{G} = (G_1, G_2, \ldots, G_N), \quad \bar{G}_i = (G_1, G_2, \ldots, G_{i-1}, G_{i+1}, \ldots, G_N), \quad i \in E,$$

$p_{ij} = \bar{Q}_{ij}(\infty, U_i)$, P denotes the $N \times N$ matrix with elements p_{ij}, and $\bar{Q}_{ij}(t, U_i) = p_{ij} F_{ij}(t)$ for $p_{ij} > 0$, $i, j = 1, 2, \ldots, N$.

According to [1], the additive accumulative functional $S(t)$ is the mathematical expectation of the accumulated income for the time interval $(0, t)$ and is determined by the functions $R_{ij}(t)$ which are the conditional mathematical expectations of the accumulated incomes for intervals $(\tau, \tau + t)$ provided that at the moment of time τ the process $\xi(t)$ transits to the state i, stays in this state a time t, and transits to the state j at the moment of time $\tau + t$.

It is known [1] that if the embedded Markov chain with the transition matrix P is ergodic, for any $j \in E$ there exists $i \in E$ such that F_{ij} determines a non-lattice distribution, and

$$\int_0^\infty t \, dF_{ij}(t) < \infty, \qquad i, j \in E,$$

then, as $t \to \infty$,

$$S(t) = It + o(t)$$

where

$$I = \sum_{i=1}^{N} \rho_i \pi_i \left(\sum_{i=1}^{N} m_i \pi_i \right)^{-1}, \tag{1}$$

$$\rho_i = \sum_{j=1}^{N} \int_0^{\infty} R_{ij}(t)\, d\bar{Q}_{ij}(t),$$

$$m_{ij} = \int_0^{\infty} t\, d\bar{Q}_{ij}(t),$$

$$m_i = \sum_{j=1}^{N} m_{ij}, \quad i,j = 1, \dots, N,$$

the row vector $\bar{\pi} = (\pi_1, \pi_2, \dots, \pi_N)$ is the stationary distribution of the states of the embedded Markov chain with the transition matrix

$$P = (p_{ij}) = (\bar{Q}_{ij}(\infty, U_i))$$

which satisfies the relations

$$\bar{\pi} = \bar{\pi} P, \qquad \sum_{i=1}^{N} \pi_i = 1.$$

We divide the set E into two disjoint subsets E_0 and E_1, $E = E_0 \cup E_1$, F each trajectory of the semi-Markov process $\xi(t)$ we put into correspondei η of the first occurrence of the trajectory in the subset E_1. We set

$$S_i(t) = \mathbf{E}(\min(t, \eta) \mid \xi(0) = i).$$

Then, as $t \to \infty$,

$$S_i(t) \to S_i = \mathbf{E}(\eta \mid \xi(0) = i).$$

The functional $S_i(t)$ can be considered as a particular case of the functional $S(t)$ for a slightly modified semi-Markov process where E_1 consists of the absorbing states of the process and $R_{ij}(t) = t$ for $i \in E_0$ and $R_{ij}(t) = 0$ for $i \in E_1$. We assume that $E_0 = (1, 2, \dots, n)$, $0 < n < N$. Then the asymptotic expansion, as $t \to \infty$, is of the form

$$S_i(t) = S_i + o(1), \quad S_i = \mathbf{E}(\eta \mid \xi(0) = i), \quad i \in E_0.$$

The following two problems arise for the controlled semi-Markov process $\zeta(t)$. First we have to clarify how the functionals I and S_i depend on the probability measures $G_i(B)$, $i = 1, 2, \dots, N$, which determine a homogeneous Markov randomized control strategy, and secondly, we have to find the extrema of these functionals and determine the structure of the distributions $G_i(B)$, $i = 1, 2, \dots, N$, which provide these extrema.

2. THE MAIN RESULTS

The answer to the first question is given by the following theorem [2, 3].

THEOREM 1. *If for a controlled semi-Markov process with finite set of states the embedded Markov chain is ergodic, for any $j \in E$ there exists $i \in E$ such that $F_{ij}(t)$ determines a non-lattice distribution, and $m_{ij} < \infty$, $i, j \in E$, then the additive accumulative functional I is a linear-fractional functional with respect to $G_i(B)$, that is,*

$$I(\bar{G}) = \frac{I_1(\bar{G})}{I_2(\bar{G})}, \tag{3}$$

where

$$I_i(\bar{G}) = \int_{U^{(N)}} A_i(u_1, \ldots, u_N) G_1(du_1) \ldots G_N(du_N), \qquad i = 1, 2,$$

$U^{(N)} = U_1 \times \ldots \times U_N$ is a direct product of the control spaces.

PROOF. Under the conditions of the theorem, equality (1) is valid and system (2) has a normalized solution. Note that

$$\rho_i = \rho_i(G_i), \qquad i \in E, \tag{4}$$

$$m_i = m_i(G_i), \qquad i \in E, \tag{5}$$

$$p_{ij} = p_{ij}(G_i), \qquad i, j \in E, \tag{6}$$

are linear functionals depending on the function G_i, $i \in E$. Taking this remark into account, we obtain from (2) that the functionals $\pi_i = \pi_i(\bar{G})$, $i \in E$, can be represented in the form

$$\pi_i(\bar{G}) = J(\bar{G}) J_i(\bar{G}_i), \qquad i \in E, \tag{7}$$

where $J_k(\bar{G}_k)$ stands for linear functionals of G_s, $k \neq s$, $k, s \in E$. The assertion of the theorem follows from representations (4)-(7) and equality (1).

THEOREM 2. *If a semi-Markov process has a finite number of states and the set E_0 of absorbing states of its embedded Markov chain is not empty, then the functionals $\bar{S} = (S_1, S_2, \ldots, S_n)$ are linear-fractional with respect to the distribution functions G_i, $i = 1, 2, \ldots, n$.*

PROOF. By the formula of the total mathematical expectation, for $i \in E_0$,

$$S_i = m_i + \sum_{j=1}^{N} p_{ij} S_j,$$

where

$$m_i = \sum_{j=1}^{N} \int_0^{\infty} t \, d\bar{Q}_{ij}(t)$$

is the mathematical expectation of the duration of continuous stay in the state i of the semi-Markov process. Note that $m_i = m_i(G_i)$ is a linear functional of G_i. We can rewrite this system of equations in the matrix form as

$$\bar{S} = \bar{m} + \bar{S}\bar{P},$$

where

$$\bar{m} = (m_1, m_2, \ldots, m_n), \qquad \bar{P} = (p_{ij}), \quad i,j \in E_0.$$

If a chain is absorbing, then the inverse matrix $(I - \bar{P})^{-1}$ exists, and the solution of the system can be represented as a ratio of two determinants each of which is a linear functional of the distribution functions $G_1, G_2, \ldots, G_n$, since all the elements of the ith row of these determinants are linear functionals of G_i, $i \in E_0$.

The theorem is thus proved.

We denote by W the domain of definition of the functional $I(\bar{G})$ under consideration. Then the denominator in (3) is not equal to zero for $\bar{G} \in W$.

Suppose we have to find the extremum of the linear-fractional functional $I(\bar{G})$ over a set V of admissible distributions, $V \subset W$. We denote by W_1 and W_2 the sets of admissible distributions such that $I_2(\bar{G}) > 0$ and $I_2(\bar{G}) < 0$ respectively. We prove a lemma which reduces the problem to the investigation of linear functionals.

LEMMA 1. *If the maximum of the linear-fractional functional $I(\bar{G})$ over the set V exists, then*

$$\{\bar{\Phi}\colon I(\bar{\Phi}) = \max_{G \in V} I(\bar{G}) = c\} = \{\bar{\Phi}\colon J(\bar{\Phi}) = 0\} = H_1 \cup H_2,$$

where

$$H_1 = \{\bar{\Phi}\colon J(\bar{\Phi}) = \max_{G \in V_1} J(\bar{G}) = 0\}, \qquad V_1 = V \cap W_1,$$

$$H_2 = \{\bar{\Phi}\colon J(\bar{\Phi}) = \min_{G \in V_2} J(\bar{G}) = 0\}, \qquad V_2 = V \cap W_2,$$

$$J(\bar{G}) = C(u_1, \ldots, u_n) G_1(du_1) \ldots G_n(du_n),$$

$$C(u_1, \ldots, u_n) = A_1(u_1, \ldots, u_n) - cA_2(u_1, \ldots, u_n),$$

where we assume that $H_1 = 0$ and $H_2 = 0$ if, respectively,

$$\max_{G \in V_1} J(\bar{G}) < 0, \qquad \min_{G \in V_2} J(\bar{G}) > 0).$$

PROOF. By the condition of the lemma $\{\bar{\Phi}\colon I(\bar{\Phi}) = c\} \neq \varnothing$. Therefore, if $I(\bar{\Phi}) = c$, then $J(\bar{\Phi}) = 0$, and vice versa, since $I_2(\bar{\Phi}) \neq 0$ for $\bar{\Phi} \in V$. The first part of the assertion is thus proved.

Note that at least one of the sets V_1 and V_2 contains a distribution $\bar{G}$ such that $J(\bar{G}) = 0$, since

$$\{\bar{\Phi}\colon J(\bar{\Phi}) = 0\} \subset V$$

and is not empty. For any $\bar{G} \in V$, we have $I(\bar{G}) \leq c$ and if $\bar{G} \in V_1$, then $J(\bar{G}) \leq 0$, and if $\bar{G} \in V_2$, then $J(\bar{G}) > 0$. Thus, at least one of the equalities

$$\max_{\bar{G}\in V_1} J(\bar{G}) = 0, \qquad \min_{\bar{G}\in V_2} J(\bar{G}) = 0$$

is valid, that is, we proved the inclusion

$$\{\bar{\Phi}: J(\bar{\Phi}) = 0\} \subset H_1 \cup H_2.$$

The inverse inclusion is obvious.

REMARK 1. Lemma 1 gives us no possibility to find the maximum directly, since the constant c is unknown. However, if among the extremal distribution functions of the linear functional there are functions of a special class, then we can search the extremum of the linear-fractional functional over this class of distributions only. For example, the following assertion [2] can be easily proved using Lemma 1.

THEOREM 3. *If the maximum of the linear-fractional functional* (1) *exists and* $A_2(u_1, \dots, u_N)$ *is positive for* $(u_1, \dots, u_N) \in U^{(N)}$, *then*

$$\max_{\bar{G}\in W} I(\bar{G}) = \max_{(u_1,\dots,u_N)\in U^{(N)}} \frac{A_1(u_1, \dots, u_N)}{A_2(u_1, \dots, u_N)},$$

that is, the optimal distribution function $G_i(x)$ *can be found in the class of degenerate distribution functions.*

In the sequel, we restrict the investigations to the case $N = 1$ and the linear functional

$$J(\bar{G}) = \int_R A(x)\, dG(x), \tag{8}$$

where $R = (-\infty, \infty)$.

Let $G_1(x)$ and $G_2(x)$ be non-decreasing functions, $G_1(x) < G_2(x)$ for $x \in R$, and

$$Z = \{G\colon G_1(x) \leq G(x) \leq G_2(x)\},$$
$$L_G = \{x\colon G(x) = G_1(x)\} \cup \{x\colon G(x) = G_2(x)\}.$$

THEOREM 4. *Let the maximum* $\max_{G\in Z} J(G) = c$ *exist and* $L = \{G\colon J(G) = c\}$. *Let* M *be a subset of the set* Z *such that for each* $G \in Z$ *there is no* $F \in Z$ *such that* L_G *is a proper subset of* L_F.

Then for any $G \in M$ *there exists* $G^* \in M$ *such that* $L_{G^*} = L_G$ *and in any interval which contains no point of* L_{G^*} *the function* $G^*(x)$ *has no more than one point of increase.*

A proof of this theorem can be found in [4]. As a comment, we note that one can look for the extremal function among the functions whose values either coincide with one of the boundary, or are constant, or have a jump in the corresponding domains.

To distinguish between the domains mentioned above, we prove the following theorem.

THEOREM 5. *If maximum (8) over the set Z exists, the function $A(x)$ has a finite or countable number of intervals of monotonicity $T_i = [t_i, t_{i+1})$, $-\infty < i < \infty$, then on each interval of monotonicity T_i, the extremal function either coincides with the boundaries $G_i(x)$, $i = 1, 2$, on some connected subset of T_i, or is a constant, or has one jump at one of the ends of the interval T_i.*

PROOF. The function $A(x)$ generates an alternating measure. Without loss of generality, we assume that $A(x)$ is continuous at some point t_{i_0}. Then

$$\int_R A(x)\, dG(x) = A(t_{i_0}) - \int_R G(x)\, dA(x) + \int_R (1 - G(x))\, dA(x). \tag{9}$$

Let

$$Z_\pi = \{G\colon G(t_{i_0}) = \pi\}, \qquad 0 \le \pi \le 1.$$

Then for distinct π, the sets Z_π are disjoint,

$$Z = \bigcup_\pi (ZZ_\pi)$$

and

$$\max_{G \in Z} J(G) = \max_{0 \le \pi \le 1} \max_{G \in (ZZ_\pi)} J(G).$$

Therefore the analysis of structure of the extremal function of functional (9) for a fixed π can be reduced to the independent search of the minimum or maximum of the functional

$$\int_{-\infty}^{t_{i_0}} G(x)\, dA(x)$$

or

$$\int_{t_{i_0}}^{\infty} (1 - G(x))\, dA(x)$$

respectively. Let $G^* \in L$ and $G^*(t_i) = \pi_i$. Then by virtue of the equality

$$\int_{-\infty}^{t_{i_0}} G(x)\, dA(x) = \sum_{-\infty < i \le i_0} \int_{T_i} G(x)\, dA(x)$$

the function G^* provides the minimum to the ith term over the set of functions with fixed values at the ends of the interval T_i. If the function $A(x)$ increases (decreases) on T_i, then the extremal function is a minorant (majorant) on this interval for all admissible functions. It is clear that the majorizing and minorizing functions either coincide with one of the boundaries on some connected subset of the set T_i or have no more than one jump at the ends of the interval T_i.

Theorem 5 indicates the domains of the arguments where the extremal function coincides with the boundaries $G_i(x)$, $i = 1, 2$. These domains are the intervals of monotonicity of the function $A(x)$.

In conclusion, we give three corollaries to Theorems 3 and 4 in [4].

COROLLARY 1. *Let $-\infty < t_1 < t_2 < \ldots < t_m < \infty$, C_i, $i = 1, \ldots, m$, be closed sets on the real line, and*

$$Z = \bigcap_{i=1}^{m} \{G: G(t_i) \in C_i\}.$$

Then maximum (8) over the set Z, if it exists, is attained on the step-functions with no more than one jump on each of the intervals $(-\infty, t_1), (t_1, t_2), \ldots, (t_m, \infty)$.

COROLLARY 2. *Let $\bar{C}_i$ be a boundary of the set C_i, $i = 1, 2, \ldots, m$. Then under the conditions of Corollary 1, maximum (8) is attained on some step-function G with no more than one jump on any interval $[t_i, t_j]$, $0 < i < j < m + 1$, such that $G(t_k) < \bar{C}_k$ for all k, $i < k < j$.*

COROLLARY 3. *Let $C_i = \{\pi_i\}$ consist of one point, $i = 1, \ldots, m$, and*

$$0 = \pi_0 \leq \pi_1 \leq \pi_2 \leq \ldots \leq \pi_m \leq \pi_{m+1} = 1.$$

Then maximum (8), if it exists, is attained on a step-function G with exactly one jump of the value $\pi_{i+1} - \pi_i$ on the interval (t_i, t_{i+1}), $i = 0, 1, \ldots, m$, where $t_0 = -\infty$, $t_{m+1} = \infty$.

REFERENCES

1. W. S. Jewell, Markov-renewal programming. *Oper. Res.* (1963) **11**, 938–971.
2. B. V. Gnedenko, *Problems of Mathematical Reliability Theory.* Radio i Svyaz, Moscow, 1983 (in Russian).
3. V. A. Kashtanov, On a class of optimal discrete control stratagies for a semi-Markov process. *Proc. MIEM* (1975) **44**, 67–76 (in Russian).
4. E. Yu. Barzilovich, V. A. Kashtanov, and I. N. Kovalenko, On the minimax criteria in reliability problems. *Izv. Soviet Acad. Sci., Ser. Techn. Cybern.* (1971) No. 3, 87–98 (in Russian).

Probabilistic Methods in Discrete Mathematics, pp. 275–277
V. F. Kolchin et al. (Eds.)

A RELATION BETWEEN THE UNIFORM AND COMPOSED MEAN PROBABILISTIC METRICS

I. V. KHARLAMOV

Moscow

Abstract — We study two distance measures between random variables ξ and η with distribution functions $F_\xi(x)$ and $F_\eta(x)$

$$\rho(\xi,\eta) = \sup_{x\in\mathbf{R}} |F_\xi(x) - F_\eta(x)|,$$
$$\tau_1(\xi,\eta) = \mathbf{E}|\xi - \eta|.$$

It is proved that if ξ has a density function bounded by a constant C_0, then

$$\rho(\xi,\eta) \le 2\sqrt{C_0\tau_1(\xi,\eta)}.$$

Let ξ and η be random variables on the same probabilistic space, $F_\xi(x)$ and $F_\eta(x)$ be their distribution functions, respectively, and let

$$\rho(\xi,\eta) = \sup_{x\in\mathbf{R}} |F_\xi(x) - F_\eta(x)| \tag{1}$$

be the uniform metric and

$$\tau_1(\xi,\eta) = \mathbf{E}|\xi - \eta| \tag{2}$$

be the composed mean metric.

We prove the following theorem.

THEOREM. *If a random variable ξ has an absolutely continuous distribution with density function $p_\xi(x)$ and*

$$C_0 = \sup_{x\in\mathbf{R}} p_\xi(x) < \infty,$$

then for an arbitrary random variable η

$$\rho(\xi,\eta) \le 2\sqrt{C_0\tau_1(\xi,\eta)}.$$

The proof of this theorem is based on the following lemma.

LEMMA. *For any random variables* ξ *and* η, *and for an arbitrary* $\delta > 0$,

$$\rho(\xi, \eta) \leq Q(\xi, \delta) + \mathbf{P}\{|\xi - \eta| \geq \delta\},$$

where $\rho(\cdot, \cdot)$ *is defined in* (1) *and*

$$Q(\xi, \delta) = \sup_{x \in \mathbf{R}} \mathbf{P}\{x \leq \xi \leq x + \delta\}$$

is the concentration function of the random variable ξ.

PROOF OF THE LEMMA. By definition,

$$|F_\xi(x) - F_\eta(x)| = |\mathbf{P}\{\xi < x\} - \mathbf{P}\{\eta < x\}|.$$

Two cases are possible, either $\mathbf{P}\{\xi < x\} < \mathbf{P}\{\eta < x\}$, or $\mathbf{P}\{\xi < x\} \geq \mathbf{P}\{\eta < x\}$.

In the first case,

$$\begin{aligned} |F_\xi(x) - F_\eta(x)| &= \mathbf{P}\{\eta < x\} - \mathbf{P}\{\xi < x\} \\ &= \mathbf{P}\{\eta < x, \xi - \eta < \delta\} + \mathbf{P}\{\eta < x, \xi - \eta \geq \delta\} - \mathbf{P}\{\xi < x\} \\ &\leq \mathbf{P}\{\xi < x + \delta\} + \mathbf{P}\{|\xi - \eta| \geq \delta\} - \mathbf{P}\{\xi < x\} \\ &= \mathbf{P}\{x \leq \xi < x + \delta\} + \mathbf{P}\{|\xi - \eta| \geq \delta\} \\ &\leq Q(\xi, \delta) + \mathbf{P}\{|\xi - \eta| \geq \delta\}. \end{aligned}$$

In the second case,

$$\begin{aligned} |F_\xi(x) - F_\eta(x)| &= \mathbf{P}\{\eta \geq x\} - \mathbf{P}\{\xi \geq x\} \\ &= \mathbf{P}\{\eta \geq x, \xi - \eta < \delta\} + \mathbf{P}\{\eta \geq x, \xi - \eta \geq \delta\} - \mathbf{P}\{\xi \geq x\} \\ &\leq \mathbf{P}\{\xi \geq x - \delta\} + \mathbf{P}\{|\xi - \eta| \geq \delta\} - \mathbf{P}\{\xi \geq x\} \\ &= \mathbf{P}\{x - \delta \leq \xi < x\} + \mathbf{P}\{|\xi - \eta| \geq \delta\} \\ &\leq Q(\xi, \delta) + \mathbf{P}\{|\xi - \eta| \geq \delta\}. \end{aligned}$$

The lemma is proved.

PROOF OF THE THEOREM. Under the assumptions of the theorem,

$$Q(\xi, \delta) = \sup_{x \in \mathbf{R}} \int_x^{x+\delta} p_\xi(u)\, du \leq \delta C_0.$$

By the Markov inequality for an arbitrary $\delta > 0$,

$$\mathbf{P}\{|\xi - \eta| \geq \delta\} \leq \frac{\mathbf{E}|\xi - \eta|}{\delta} = \frac{\tau_1(\xi, \eta)}{\delta}.$$

Choosing the value of δ equal to $\delta_0 = \sqrt{\tau_1(\xi, \eta)/C_0}$ and applying the lemma, we obtain

$$\rho(\xi, \eta) \leq Q(\xi, \delta) + \mathbf{P}\{|\xi - \eta| \geq \delta_0\} \leq C_0\delta_0 + \frac{\tau_1(\xi, \eta)}{\delta_0} = 2\sqrt{C_0\tau_1(\xi, \eta)},$$

and the theorem is proved.

Note that a weaker inequality can be derived from the known relations between probabilistic metrics. Indeed, for the mean metric

$$\varkappa_1(\xi, \eta) = \int_{\mathbf{R}} |F_\xi(x) - F_\eta(x)|\, dx$$

and the Lévy metric

$$L(\xi, \eta) = \inf\{\varepsilon\colon F_\xi(x) \le F_\eta(x+\varepsilon) + \varepsilon,\ F_\eta(x) \le F_\xi(x+\varepsilon) + \varepsilon,\ x \in \mathbf{R}\}$$

under the theorem assumptions we have (see [1], pp. 107, 114, 57)

$$\begin{aligned} \rho(\xi, \eta) &\le (1 + C_0) L(\xi, \eta), \\ L^2(\xi, \eta) &\le \varkappa_1(\xi, \eta), \\ \varkappa_1(\xi, \eta) &\le \tau_1(\xi, \eta). \end{aligned}$$

Combining these inequalities, we obtain

$$\rho(\xi, \eta) \le (1 + C_0)\sqrt{\tau_1(\xi, \eta)},$$

which is weaker than the result proved above, especially for large and small values of the constant C_0.

The author wishes to express his thanks to B. V. Ryazanov for many helpful suggestions.

REFERENCES

1. V. M. Zolotarev, *Modern Theory of Summation of Random Variables*, Nauka, Moscow, 1986 (in Russian).

Probabilistic Methods in Discrete Mathematics, pp. 279–284
V. F. Kolchin et al. (Eds.)

ON THE MONTE-CARLO ESTIMATION OF THE DISTRIBUTION FUNCTION OF A FUNCTIONAL OF THE WIENER PROCESS

I. V. KHARLAMOV

Moscow

Abstract — A class of integral-type functionals of the Wiener process is studied. The Monte-Carlo approach to estimation of the distribution of such functionals is developed. The proximity between the studied functional and another simple functional is estimated.

This work is devoted to the estimation of the distribution of a random variable

$$V_\varepsilon = \int_0^1 t^{-\varepsilon} \exp\left\{-\frac{w^2(t)}{2t}\right\} dt \tag{1}$$

where ε is a real parameter $0 < \varepsilon < 1$, $w(t)$ is the standard Wiener random process.

Such a variable with $\varepsilon = 1/2$ appears, for example, in the random trajectory detection problem [1], which is connected with the study of a potential possibility of detection of a random trajectory against a noise background [2].

For practical realization of the statistical tests studied in [1] we need to calculate the distributions of V_ε and $U + kV_\varepsilon$, where U is the standard normal random variable, independent of V_ε, for various values of the constant k.

Since the analytical form of the distribution of variable (1) is unknown [1], we suggest to apply the Monte-Carlo method [3]. The peculiarity of our case is that only a functional depending on a finite number of points of a trajectory of the Wiener process $w(t)$ can be computed in practice.

Consider the functional of the form

$$J_\varepsilon(n) = \frac{1}{n}\sum_{k=1}^{n}\left(\left(\frac{k}{n}\right)^{-\varepsilon} \exp\left\{-\frac{w^2(k/n)}{2k/n}\right\}\right). \tag{2}$$

For evaluation of the accuracy of the Monte-Carlo method, we need to estimate the value of $\rho(V_\varepsilon, J_\varepsilon(n))$, where

$$\rho(\xi, \eta) = \sup_{x\in\mathbf{R}} |F_\xi(x) - F_\eta(x)|$$

is the uniform distance between the distributions of the random variables ξ and η.

In the case of $U + kV_\varepsilon$ we can use the following result.

PROPOSITION 1. *For arbitrary random variables ξ, η, and a random variable ζ, independent of ξ and η, for any non-zero k the inequality*

$$\rho(\zeta + k\xi, \zeta + k\eta) \le \rho(\xi, \eta)$$

holds.

PROOF. Indeed,

$$\begin{aligned}
\rho(\zeta + k\xi, \zeta + k\eta) &= \sup_{x\in\mathbf{R}} \left| \int_{\mathbf{R}} F_\xi((x-y)/k)\,dF_\zeta(y) - \int_{\mathbf{R}} F_\eta((x-y)/k)\,dF_\zeta(y) \right| \\
&\le \sup_{x\in\mathbf{R}} \int_{\mathbf{R}} \left|F_\xi((x-y)/k) - F_\eta((x-y)/k)\right|\,dF_\zeta(y) \\
&\le \int_{\mathbf{R}} \sup_{x\in\mathbf{R}} \left|F_\xi((x-y)/k) - F_\eta((x-y)/k)\right|\,dF_\zeta(y) \\
&= \int_{\mathbf{R}} \rho(\xi,\eta)\,dF_\zeta(y) = \rho(\xi,\eta).
\end{aligned}$$

The proposition is thus proved.

The value $\rho(V_\varepsilon, J_\varepsilon(n))$ can be estimated with the use of the inequality (see [4])

$$\rho(V_\varepsilon, J_\varepsilon) \le 2\sqrt{C_0(\varepsilon)\tau_1(V_\varepsilon, J_\varepsilon(n))},$$

where

$$C_0(\varepsilon) = \sup_{x\in\mathbf{R}} p_{V_\varepsilon}(x), \qquad \tau_1(V_\varepsilon, J_\varepsilon(n)) = \mathbf{E}\left|V_\varepsilon - J_\varepsilon(n)\right|,$$

and then by applying the following theorem.

THEOREM 1. *For V_ε and J_ε defined by* (1) *and* (2)

$$\tau_1(V_\varepsilon, J_\varepsilon(n)) \le g_\varepsilon(n),$$

where

$$g_\varepsilon(n) = \begin{cases} \dfrac{4}{\pi(1-2\varepsilon)\sqrt{n}}(1+\varphi_1(n)), & 0<\varepsilon<1/2, \\ \dfrac{2\ln n}{\pi\sqrt{n}}(1+\varphi_2(n)), & \varepsilon = 1/2, \\ \left(\dfrac{4}{\pi(2\varepsilon-1)} + \dfrac{1}{1-\varepsilon} + \dfrac{3}{2}(2+\sqrt{2})\right)\dfrac{1}{n^{1-\varepsilon}}, & 1/2<\varepsilon<1, \end{cases}$$

$$\varphi_1(n) = \frac{\pi(1-2\varepsilon)}{4n^{1/2-\varepsilon}}\left(\frac{1}{1-\varepsilon} + 1 + \varepsilon + \frac{1+\varepsilon}{\varepsilon 2^\varepsilon} + \frac{1}{2\varepsilon}\right),$$

$$\varphi_2(n) = \frac{\pi}{4\ln n}(9+3\sqrt{2}).$$

PROOF. We set

$$w(k/n) = \frac{1}{\sqrt{n}} \sum_{j=1}^{k} \xi_j = \frac{1}{\sqrt{n}} S_k,$$

where

$$\xi_j = \sqrt{n}(w(k/n) - w((k-1)/n)), \qquad j = 1, \ldots, k,$$

are independent standard normal random variables. Then

$$V_\varepsilon = \sum_{k=1}^{n} \int_{(k-1)/n}^{k/n} t^{-\varepsilon} \exp\left\{-\frac{w^2(t)}{2t}\right\} dt,$$

$$J_\varepsilon(n) = \frac{1}{n} \sum_{k=1}^{n} (k/n)^{-\varepsilon} \exp\{-S_k^2/(2k)\} = \sum_{k=1}^{n} \int_{k/n}^{(k-1)/n} \left(\frac{k}{n}\right)^{-\varepsilon} \exp\left\{-\frac{S_k^2}{2k}\right\} dt.$$

Separating the first item and moving the absolute value under the integral symbol, we obtain

$$|V_\varepsilon - J_\varepsilon(n)| \leq \int_0^{1/n} H(t)\, dt + \sum_{k=2}^{n} \int_{k/n}^{(k-1)/n} (T_k(t) + R_k(t) + Q_k(t))\, dt, \tag{3}$$

where

$$\begin{aligned} H(t) &= |t^{-\varepsilon} \exp\{-w^2(t)/(2t)\} - (1/n)^{-\varepsilon} \exp\{-S_1^2/2t\}|, \\ T_k(t) &= |t^{-\varepsilon} \exp\{-w^2(t)/(2t)\} - t^{-\varepsilon} \exp\{-S_k^2/(2nt)\}|, \\ R_k(t) &= |t^{-\varepsilon} \exp\{-S_k^2/(2nt)\} - (k/n)^{-\varepsilon} \exp\{-S_k^2/(2nt)\}|, \\ Q_k(t) &= |(k/n)^{-\varepsilon} \exp\{-S_k^2/(2nt)\} - (k/n)^{-\varepsilon} \exp\{-S_k^2/(2k)\}|. \end{aligned}$$

We estimate each of these values. For $H(t)$ we have

$$\begin{aligned} \int_0^{1/n} H(t)\, dt &= \int_0^{1/n} (t/n)^{-\varepsilon} | \exp\{-w^2(t)/(2t)\} - (1/n)^{-\varepsilon} \exp\{-S_1^2/2\} t^{-\varepsilon}|\, dt \\ &\leq \int_0^{1/n} t^{-\varepsilon} (1/n)^{-\varepsilon} (1/n)^{\varepsilon}\, dt = \left.\frac{t^{1-\varepsilon}}{1-\varepsilon}\right|_0^{1/n} = \frac{1}{(1-\varepsilon) n^{1-\varepsilon}}. \end{aligned} \tag{4}$$

Using the inequality

$$|e^{-a} - e^{-b}| \leq |a - b|, \qquad a, b > 0, \tag{5}$$

for $T_k(t)$ we have

$$T_k(t) = t^{-\varepsilon} | \exp\{-w^2(t)/(2t)\} - \exp\{-S_k^2/(2nt)\}| \leq t^{-\varepsilon} \frac{1}{2t} |w^2(t) - S_k^2/n|.$$

Note that $S_k^2/n = w(k/n)$. Let $\xi = w(k/n) - w(t)$, then $w(t)$ and ξ are independent and

$$\mathbf{E}|w(t)| = \sqrt{2t\pi}, \qquad \mathbf{E}|\xi| = \sqrt{2(k - nt)/(n\pi)},$$

and hence

$$\mathbf{E}T_k(t) \le t^{-\varepsilon}\frac{1}{2t}\mathbf{E}|\xi^2 - 2\xi w(t)| \le t^{-\varepsilon}\frac{1}{2t}(\mathbf{E}\xi^2 + 2\mathbf{E}|\xi|\mathbf{E}|w(t)|)$$
$$= t^{-\varepsilon}\frac{1}{2t}((k-nt)/n + 4\sqrt{t(k-nt)/n/\pi}) \le \frac{t^{-1-\varepsilon}}{2n} + \frac{2t^{-1/2-\varepsilon}}{\pi\sqrt{n}}.$$

It is easily seen that

$$\int_{1/n}^{1}\frac{dt}{t^{1+\varepsilon}} \le \frac{n^{\varepsilon}}{\varepsilon}, \qquad \int_{1/n}^{1}\frac{dt}{t^{(1+2\varepsilon)/2}} \le h_\varepsilon(n),$$

where

$$h_\varepsilon(n) = \begin{cases} n^{\varepsilon-1/2}/(\varepsilon - 1/2), & 1/2 < \varepsilon > 1, \\ \ln n, & \varepsilon = 1/2, \\ 1/(1/2 - \varepsilon), & 0 < \varepsilon < 1/2. \end{cases}$$

Hence it follows that

$$\mathbf{E}\sum_{k=2}^{n}\int_{(k-1)/n}^{k/n} T_k(t)dt \le \frac{1}{2n}\int_{1/n}^{1}\frac{dt}{t^{1+\varepsilon}} + \frac{2}{\pi\sqrt{n}}\int_{1/n}^{1}\frac{dt}{t^{(1/2+\varepsilon)}} \le f_\varepsilon(n), \tag{6}$$

where

$$f_\varepsilon(n) = \begin{cases} \left(\dfrac{1}{2\varepsilon} + \dfrac{4}{\pi(2\varepsilon - 1)}\right)\dfrac{1}{n^{1-\varepsilon}}, & 1/2 < \varepsilon < 1, \\ \dfrac{2\ln(n)}{\pi\sqrt{n}} + \dfrac{1}{\sqrt{n}}, & \varepsilon = 1/2, \\ \dfrac{4}{\pi(1-2\varepsilon)\sqrt{n}} + \dfrac{1}{2\varepsilon n^{1-\varepsilon}}, & 0 < \varepsilon < 1/2. \end{cases}$$

Let us estimate $R_k(t)$. Applying the Lagrange theorem to the function $f(x) = x^{-\varepsilon}$, we find that for some $0 \le \theta \le 1$

$$R_k(t) = \exp\{-S_k^2/(2nt)\}|t^{-\varepsilon} - (k/n)^{-\varepsilon}|$$
$$\le n^{\varepsilon}((nt)^{-\varepsilon} - k^{-\varepsilon}) = \varepsilon n^{\varepsilon}(k-\theta)^{-1-\varepsilon}(k-nt) \le \varepsilon n^{\varepsilon}(k-1)^{-1-\varepsilon}.$$

Using the inequality

$$\zeta(1+\varepsilon) \le \frac{1+\varepsilon}{\varepsilon}, \qquad \varepsilon > 0, \tag{7}$$

for the Riemann zeta function

$$\zeta(\alpha) = \sum_{k=1}^{\infty}\frac{1}{k^{\alpha}},$$

we obtain

$$\mathbf{E}\sum_{k=2}^{n}\int_{(k-1)/n}^{k/n} R_k(t)\,dt \le \frac{\varepsilon n^{\varepsilon}}{n}\sum_{k=2}^{n}\frac{1}{(k-1)^{1+\varepsilon}} \le \frac{\varepsilon}{n^{1-\varepsilon}}\zeta(1+\varepsilon) \le \frac{1+\varepsilon}{n^{1-\varepsilon}}. \tag{8}$$

Finally, we estimate $Q_k(t)$. By (5),

$$Q_k(t) = \left(\frac{n}{k}\right)^{\varepsilon} \left|\exp\{-S_k^2/(2nt)\} - \exp\{-S_k^2/(2k)\}\right| \le \left(\frac{n}{k}\right)^{\varepsilon} \left|\frac{S_k^2}{2}\right| \left|\frac{1}{nt} - \frac{1}{k}\right|.$$

Applying the Lagrange theorem to the function $f(x) = x^{-1}$, we see that for some $0 \le \theta \le 1$

$$Q_k(t) \le \left(\frac{n}{k}\right)^{\varepsilon} \left|\frac{S_k^2}{2}\right| \frac{k - nt}{(k-\theta)^2} \le \left(\frac{n}{k}\right)^{\varepsilon} \left|\frac{S_k^2}{2}\right| \frac{1}{(k-1)^2}.$$

Using (7), we obtain

$$\begin{aligned}
\mathbf{E}\sum_{k=2}^{n} \int_{(k-1)/n}^{k/n} Q_k(t)dt &\le \sum_{k=2}^{n} \frac{n^{\varepsilon}\mathbf{E}S_k^2}{2nk^{\varepsilon}(k-1)^2} \\
&= \frac{1}{2n^{1-\varepsilon}} \sum_{k=2}^{n} \frac{k^{1-\varepsilon}}{(k-1)^2} \le \frac{1}{2n^{1-\varepsilon}} \sum_{k=2}^{\infty} (1 + 1/k)^{1-\varepsilon} k^{-1-\varepsilon} \\
&\le \frac{2^{1-\varepsilon}\varepsilon}{2n^{1-\varepsilon}} \zeta(1+\varepsilon) \le \frac{(1+\varepsilon)\varepsilon}{\varepsilon 2^{\varepsilon} n^{1-\varepsilon}}
\end{aligned} \tag{9}$$

Combining (4), (6), (8), (9) and substituting them into (3), we obtain

$$\tau_1(V_\varepsilon, J_\varepsilon(n)) = \mathbf{E}\left|V_\varepsilon - J_\varepsilon(n)\right| \le g_\varepsilon(n),$$

where

$$g_\varepsilon(n) = \begin{cases}
\dfrac{4}{\pi(1-2\varepsilon)\sqrt{n}}(1 + \varphi_1(n)), & 0 < \varepsilon < 1/2, \\
\dfrac{2\ln n}{\pi\sqrt{n}}(1 + \varphi_2(n)), & \varepsilon = 1/2, \\
\left(\dfrac{1}{2\varepsilon} + \dfrac{4}{\pi(2\varepsilon - 1)} + \dfrac{1}{1-\varepsilon} + 1 + \dfrac{\varepsilon(1+\varepsilon)}{\varepsilon 2^{\varepsilon}}\right) \dfrac{1}{n^{1-\varepsilon}}, & 1/2 < \varepsilon < 1,
\end{cases}$$

where $\varphi_1(n)$ and $\varphi_2(n)$ are of the form given in the formulation of the theorem.

It is easily seen that the inequality

$$\left(\frac{1}{2\varepsilon} + \frac{4}{\pi(2\varepsilon - 1)} + \frac{1}{1-\varepsilon} + 1 + \frac{\varepsilon(1+\varepsilon)}{\varepsilon 2^{\varepsilon}}\right) \frac{1}{n^{1-\varepsilon}} \le \frac{4}{\pi(2\varepsilon - 1)} + \frac{1}{1-\varepsilon} + \frac{3}{2}(2+\sqrt{2})$$

holds, which completes the proof.

Note that the condition

$$C_0(\varepsilon) = \sup_{x\in\mathbf{R}} p_{V_\varepsilon}(x) < \infty$$

is typical of the analysis of the convergence rate of functionals (2) (see [5]). Therefore the theoretical proof of this condition is actual (by computer simulation, the value $C_0(1/2) \approx 1.8$ is obtained).

The author is greatly indebted to B. V. Ryazanov for suggesting the problem and many stimulating discussions.

REFERENCES

1. I. M. Arbekov, V. I. Vinokurov, and B. V. Ryazanov, To a random trajectory detection problem. *Theory Probab. Appl.* (1994) **39**, 669–980 (in Russian).

2. K. Sh. Zigangirov, Potential possibilities of random trajectory detection. *Probl. Inform. Trans.* (1977) **13**, No. 2, 72–82 (in Russian).

3. S. M. Ermakov, *Monte-Carlo Method and Related Problems*. Nauka, Moscow, 1975 (in Russian).

4. I. V. Kharlamov, A relation between uniform and compound mean probabilistic metrics. *These Proceedings*, pp. 275–277.

5. A. I. Sakhanenko, An estimate of the density function of integral-type functionals. In: *Limit Theorems in Probability Theory and Related Problems*. Nauka, Novosibirsk, 1982, pp. 180–186 (in Russian).

Probabilistic Methods in Discrete Mathematics, pp. 285–294
V. F. Kolchin et al. (Eds.)

ON THE ASYMPTOTIC BEHAVIOUR OF THE NUMBER OF HYPERFORESTS

A. V. KOLCHIN

Moscow

Abstract — We consider m-forests with n labelled vertices and T edges. Asymptotic expressions for the number of such hyperforests are obtained as m is fixed and $n, T \to \infty$ in such a manner that $T/n \to 1/(m(m-1))$ sufficiently fast; the case of fixed T is also considered.

This research was supported by the Russian Foundation for Basic Research, grant 96–15–96033.

1. INTRODUCTION

Let a set V and a family of its different subsets $E = (E_1, \ldots, E_T)$ be given. The pair (V, E) is called a hypergraph; V is the set of vertices, E is the set of edges. If $|E_t| = 2$, $t = 1, \ldots, T$, then the notion of a hypergraph coincides with the well-known notion of a graph. A hypertree is a connected hypergraph with no cycles. A hypertree is called an m-tree, if every its edge contains m vertices. A hypergraph whose all connected components are m-trees is called an m-forest.

We consider m-forests with n labelled vertices consisting of N hypertrees. Let T denote the number of edges in such a hyperforest. It is not difficult to see that in an m-forest with N trees

$$n = T(m-1) + N.$$

Let $\mathcal{F}_{n,N}^{(m)}$ denote the set of all such forests, and let $F_{n,N}^{(m)} = |\mathcal{F}_{n,N}^{(m)}|$ be the number of elements of this set.

For $m = 2$ the set $\mathcal{F}_{n,N}^{(m)}$ is the set of forests of N unrooted trees with n labelled vertices. The random forests with the uniform distribution given on $\mathcal{F}_{n,N}^{(2)}$ were thoroughly investigated by V. E. Britikov in [1–3]. The following exact formula for $F_{n,N}^{(2)}$ was obtained by A. Rényi [4]:

$$F_{n,N}^{(2)} = \frac{1}{N!} \sum_{j=0}^{N} \binom{N}{j} \left(-\frac{1}{2}\right)^j n^{n-N-j-1} \frac{n!\,(N+j)!}{(n-N-j)!}.$$

For an arbitrary $m \geq 2$, the exact formula for the number of m-forests was obtained by B. I. Selivanov [5–9]:

$$F_{n,N}^{(m)} = \frac{(T(m-1)+N)!}{N!\,((m-1)!)^T} \times \sum_{k=0}^{T} (-1)^{T-k} \binom{N}{T-k} \frac{(T(m-1)+N)^{k-1}}{k!} \times \left(\frac{m-1}{m}\right)^{T-k} ((T-k)(m-1)+N).$$

With $N = 1$, this formula gives us the number $b_n^{(m)}$ of unrooted m-trees with r edges and $n = r(m-1)+1$ vertices:

$$b_n^{(m)} = \frac{(r(m-1)+1)!\,(r(m-1)+1)^{r-2}}{r!\,((m-1)!)^r}, \qquad r = 0, 1, \ldots$$

Note that $b_n^{(m)}$ are defined above only for n of the form $r(m-1)+1$, $r = 0, 1, \ldots$; for the remaining values of n we set $b_n^{(m)} = 0$.

We introduce the generating function

$$B_m(x) = \sum_{r=0}^{\infty} \frac{b_n^{(m)} x^{r(m-1)+1}}{(r(m-1)+1)!} = \sum_{r=0}^{\infty} \frac{(r(m-1)+1)^{r-2} x^{r(m-1)+1}}{r!((m-1)!)^r}. \tag{1}$$

Note that the convergence radius of $B_m(x)$ is equal to $((m-2)!\,e^{-1})^{1/(m-1)}$.

Following [1, 2, 3, 10, 12], where the investigation of forests of unrooted trees was reduced to the investigation of sums of independent random variables, we apply the so-called generalized allocation scheme to the analysis of m-forests.

We introduce independent identically distributed random variables $\xi_1, \ldots, \xi_N$ with the distribution

$$\mathbf{P}\{\xi_1 = r(m-1)+1\} = \frac{b_{r(m-1)+1}^{(m)} x^{r(m-1)+1}}{(r(m-1)+1)!\,B_m(x)} = \frac{(r(m-1)+1)^{r-2} x^{r(m-1)+1}}{r!((m-1)!)^r B_m(x)}, \qquad r = 0, 1, \ldots, \tag{2}$$

where the normalizing function $B_m(x)$ is defined by (1) and the positive parameter x can be chosen arbitrarily from the convergence domain of the series $B_m(x)$.

It is not difficult to prove that

$$F_{n,N}^{(m)} = \frac{n!}{N!} \sum_{n_1+\ldots+n_N=n} \frac{b_{n_1}^{(m)} \ldots b_{n_N}^{(m)}}{n_1! \ldots n_N!} = \frac{n!\,(B_m(x))^N}{N!\,x^n} \mathbf{P}\{\xi_1 + \ldots + \xi_N = n\}, \tag{3}$$

where x is an arbitrary positive number from the convergence domain of series (1).

Thus, to deduce an asymptotic formula for $F_{n,N}^{(m)}$, it is sufficient to choose an appropriate value of the parameter x and prove a local limit theorem for the sum

$$\zeta_N = \xi_1 + \ldots + \xi_N,$$

where $\xi_1, \ldots, \xi_N$ are independent and identically distributed random variables with distribution (2).

2. LOCAL LIMIT THEOREMS

LEMMA 1. *If we choose*

$$x = \lambda \exp\left\{-\frac{\lambda^{m-1}}{(m-1)!}\right\}, \tag{4}$$

then

$$B_m(x) = \lambda\left(1 - \frac{\lambda^{m-1}}{m(m-2)!}\right). \tag{5}$$

PROOF. We introduce the function

$$\theta_m(z) = \sum_{r=0}^{\infty} \frac{(r(m-1)+1)^{r-1} z^{r(m-1)+1}}{r!\,((m-1)!)^r}.$$

In [5] it was proved that the function $\theta_m(z)$ satisfies the functional equation

$$\theta_m(z) = z \exp\left\{\frac{1}{(m-1)!}\theta_m^{m-1}(z)\right\}. \tag{6}$$

The equality

$$B_m(z) = \theta_m(z) - \frac{\theta_m^m(z)}{m(m-2)!} \tag{7}$$

holds. Indeed, differentiating (6), we obtain

$$\theta_m'(z)e^{-\theta_m^{m-1}(z)/(m-1)!} - \frac{\theta_m^{m-1}(z)\theta_m'(z)}{(m-2)!}e^{-\theta_m^{m-1}(z)/(m-1)!} = 1,$$

hence

$$\theta_m'(z) = \frac{1}{(1 - \theta_m^{m-1}(z)/(m-2)!)e^{-\theta_m^{m-1}(z)/(m-1)!}}.$$

To prove relation (7), we consider the function

$$u(z) = \theta_m(z) - \frac{\theta_m^m(z)}{m(m-2)!}.$$

It is easily seen that

$$u'(z) = \theta'_m(z) - \frac{\theta_m^{m-1}(z)\theta'_m(z)}{(m-2)!} = \theta'_m(z)\left(1 - \frac{\theta_m^{m-1}(z)}{(m-2)!}\right).$$

Applying (6), we obtain that

$$u'(z) = \frac{\theta_m(z)}{z} = \sum_{r=0}^{\infty} \frac{(r(m-1)+1)^{r-1} z^{r(m-1)}}{r!\,((m-1)!)^r}.$$

Since $u(0) = 0$, integrating this relation we arrive at

$$\int_0^z u'(t)\,dt = u(z) = \sum_{r=0}^{\infty} \frac{(r(m-1)+1)^{r-1}}{r!\,((m-1)!)^r} \int_0^z t^{r(m-1)}\,dt$$
$$= \sum_{r=0}^{\infty} \frac{(r(m-1)+1)^{r-2} z^{r(m-1)+1}}{r!\,((m-1)!)^r} = B_m(z).$$

Thus,

$$B_m(z) = \theta_m(z) - \frac{\theta_m^m(z)}{m(m-2)!};$$

now the assertion of the lemma for x of the form (4) follows from (6) and (7).

We choose λ in such a way that

$$\lambda^{m-1} = \frac{m!\,T}{n}, \tag{8}$$

where n is the number of vertices, T is the number of edges in a forest consisting of $N = n - T(m-1)$ trees.

To prove a local limit theorem for the sum of independent identically distributed summands, it is frequently convenient to transform the summands in such a way that the maximal span of the distribution of the summands becomes equal to one. We introduce the random variables

$$\xi_k^* = \frac{\xi_k - 1}{m-1}, \qquad k = 1, \dots, N,$$

with the maximal span of the distribution equal to one, where ξ_k, $k = 1, \dots, N$, have distribution (2) with the parameter x determined by (4) and (8). We denote $a^* = \mathbf{E}\xi_1^*$, $\sigma^{*2} = \mathbf{D}\xi_1^*$. It is not difficult to see that if x and λ are chosen according to (4) and (8), and $0 < \lambda^{m-1} < (m-2)!$, then

$$a^* = \mathbf{E}\xi_1^* = \frac{n-N}{(m-1)N}, \qquad (\sigma^*)^2 = \mathbf{D}\xi_1^* = \frac{Tn}{N^2(1 - m(m-1)T/n)}.$$

In [13], the following local limit theorem was proved.

THEOREM 1. *Let m be fixed, $n, T \to \infty$ and let there exist constants λ_0, λ_1 such that*

$$0 < \lambda_0 \le \lambda \le \lambda_1 < ((m-2)!)^{1/(m-1)}, \tag{9}$$

where λ is defined by relation (8). Then

$$\mathbf{P}\{\zeta_N^* = k\} = \frac{1}{\sigma^* \sqrt{2\pi N}} e^{-z^2/2}(1 + o(1))$$

uniformly with respect to the integers k such that

$$z = \frac{k - Na^*}{\sigma^* \sqrt{N}}$$

lies in any finite interval, and uniformly in λ lying in bounds (9).

Let us consider the case where $\lambda \to 0$. The following limit theorem is true.

THEOREM 2. *Let m be fixed, $n \to \infty$ and T be bounded by a constant. Then*

$$\mathbf{P}\{\zeta_N = n\} = \frac{e^{-(m-1)}(m-1)^{T(m-1)}}{(T(m-1))!}(1 + o(1)). \tag{10}$$

PROOF. We consider the characteristic function

$$\varphi(t) = \varphi(t, \lambda) = \mathbf{E}e^{it\xi_1} = \frac{B_m(xe^{it})}{B_m(x)}.$$

It can easily be seen that as $h \to 0$

$$B_m(he^{it}) = he^{it} + \frac{h^m}{m!}e^{itm} + O(h^{m+1})$$

uniformly in t. In addition, as $\lambda \to 0$

$$x = \lambda \exp\left\{-\frac{\lambda^{m-1}}{(m-1)!}\right\} = \lambda - \frac{\lambda^m}{(m-1)!} + O(\lambda^{m+1}).$$

Taking (5) into account, gathering the above estimates together, and choosing λ as in (8) (it is clear that under the hypotheses of the theorem $\lambda \to 0$), for any fixed t as $N \to \infty$ we obtain

$$\left(e^{-it}\varphi(t)\right)^N \to \exp\left\{e^{it(m-1)} - 1\right\},$$

i.e., the distribution of the random variable $\zeta_N - N$ weakly converges to the Poisson law with parameter $m - 1$. Therefore, as $N \to \infty$, for any fixed $k = 0, 1, \ldots$

$$\mathbf{P}\{\zeta_N - N = k\} \to \frac{e^{-(m-1)}(m-1)^k}{k!}.$$

Now, let us consider the critical case where the distribution of the summands of the sum ζ_N is fixed and is determined by the parameter $x = ((m-2)!\, e^{-1})^{1/(m-1)}$.

THEOREM 3. *Let m be fixed, $N \to \infty$, and let*

$$x = ((m-2)!\, e^{-1})^{1/(m-1)}.$$

Then

$$bN^{2/3}\mathbf{P}\{\zeta_N = k\} = p(z)(1 + o(1))$$

uniformly with respect to $k = N + l(m-1)$, where l are non-negative integers, such that

$$z = \frac{k - N(1 + 1/(m-1))}{bN^{2/3}(m-1)}$$

lies in any finite interval, where

$$b = \left(\frac{2m}{3(m-1)((m-2)!)^{1/(m-1)}}\right)^{2/3} \tag{11}$$

and $p(z)$ is the density of the stable law with the characteristic function $\varphi(t)$ such that

$$\ln \varphi(t) = -|t|^{3/2}\left\{1 + i\frac{3t}{2|t|}\tan\frac{3\pi}{4}\right\}. \tag{12}$$

PROOF. For the sake of convenience, we consider, as above, $\xi_k^* = (\xi_k - 1)/(m-1)$, $k = 1, \ldots, N$, with the maximal span of the distribution equal to one.

We investigate the asymptotic behaviour of the tail of the distribution of ξ_1^*. Note that under the hypotheses of the theorem

$$B_m(x) = B_m = \frac{(m-1)((m-2)!)^{1/(m-1)}}{m}, \qquad \mathbf{E}\xi_1^* = \frac{1}{(m-1)^2}.$$

It is not difficult to prove that as $y \to \infty$

$$\mathbf{P}\{\xi_1^* \geq y\} = \mathbf{P}\{\xi_1 \geq y(m-1) + 1\} = \sum_{r=y}^{\infty} \frac{(r(m-1)+1)^{r-2}x^{r(m-1)+1}}{r!((m-1)!)^r B_m(x)}$$
$$= \frac{2m(1 + o(1))}{3\sqrt{2\pi}y^{3/2}(m-1)((m-2)!)^{1/(m-1)}}.$$

Hence it follows that (see, e.g., [16, §35]) the distribution of ξ_1^* belongs to the domain of normal attraction of the stable law V with parameters $\gamma = 0$, $c = 1$, $\alpha = 3/2$, $\beta = -1$ in the Lévy–Khintchine parameterization [16, §34], i.e., for any fixed y

$$\lim_{N\to\infty} \mathbf{P}\left\{\frac{\zeta_N^* - Na^*}{bN^{2/3}} < y\right\} = V(y), \tag{13}$$

where

$$a^* = \mathbf{E}\xi_1^* = \frac{1}{(m-1)^2}, \qquad b = \left(\frac{2m}{3(m-1)((m-2)!)^{1/(m-1)}}\right)^{2/3},$$

and the characteristic function $\varphi(t)$ of the distribution V satisfies equality (12).

Since the maximal span of the distribution of the summands is equal to one, the validity of the integral limit theorem implies the validity of the local limit theorem (see [16, §50]), and from (13) it follows that

$$bN^{2/3}\mathbf{P}\{\zeta_N^* = k\} = p(z^*)(1 + o(1))$$

uniformly in the integers k such that $z^* = (k - N/(m-1)^2)/(bN^{2/3})$ lies in any finite interval, where $p(z)$ is the density of the stable distribution with the characteristic function $\varphi(t)$ satisfying (12). Returning to the random variable $\zeta_N = \zeta_N^*(m-1) + N$, we obtain that

$$bN^{2/3}\mathbf{P}\{\zeta_N = k\} = p(z)(1 + o(1))$$

uniformly in the integers k such that k can be represented as $k = N + l(m-1)$, where l are non-negative integers, and $z = (k - N(1 + 1/(m-1)))/(bN^{2/3}(m-1))$ lies in any finite interval.

The theorem is proved.

We note that the density $p(z)$ of the stable law with parameters $\alpha = 3/2$, $\beta = -1$ cannot be expressed in terms of elementary functions.

To deal with the domain of large deviations, we give here the local limit theorem due to S. G. Tkachuk [14], preserving our notation.

THEOREM 4. *Let η_k, $k = 1, 2, \ldots$, be independent identically distributed random variables whose distribution law belongs to the domain of normal attraction of the stable law with parameter α, $1 < \alpha < 2$; let $\mathbf{E}\eta_1 = 0$. Let η_1 be integer-valued and*

$$\mathbf{P}\{\eta_1 = l\} = cl^{-(1+\alpha)}(1 + o(1))$$

as $l \to \infty$, where c is a constant. Then

$$\mathbf{P}\left\{\sum_{\nu=1}^{N} \eta_\nu = k\right\} = N\,\mathbf{P}\{\eta_1 = k\}(1 + o(1)) = N^{-1/\alpha} p_\alpha(kN^{-1/\alpha})(1 + o(1))$$

as $kN^{-1/\alpha} \to \infty$. Here $p_\alpha(\cdot)$ is the density of the stable law with parameter α.

3. THE MAIN THEOREMS

The following assertion can be derived from Theorem 1.

THEOREM 5. *Let m be fixed, and let $n, T \to \infty$ in such a way that*

$$\lambda = \left(\frac{m!\,T}{n}\right)^{1/(m-1)}$$

lies in an arbitrary interval of the form

$$0 < \lambda_0 \le \lambda \le \lambda_1 < ((m-2)!)^{1/(m-1)},$$

where λ_0 and λ_1 are constants.

Then

$$F_{n,N}^{(m)} = \frac{n^{mT}}{(m!)^T T!}\sqrt{1 - \frac{m(m-1)T}{n}}(1+o(1))$$

uniformly in $\lambda \in [\lambda_0, \lambda_1]$.

A thorough proof can be found in [13].

Starting from Theorem 2, we arrive at the following assertion.

THEOREM 6. *Let m be fixed, $n \to \infty$ and T be bounded by a constant. Then*

$$F_{n,N}^{(m)} = \frac{e^{T-m+1}(m-1)^{T(m-1)}}{(m!)^T T^T (T(m-1))!} n^{Tm}(1+o(1)).$$

THEOREM 7. *Let m be fixed, and let $n, T \to \infty$ in such a way that $T/n \to 1/(m(m-1))$ and $(1 - Tm(m-1)/n)N^{1/3} \to bv(m-1)^3/m$, where $-\infty < v < \infty$. Then*

$$F_{n,N}^{(m)} = \frac{n!\,B_m^N}{bN^{2/3}N!\,x^n}p(-v)(1+o(1)), \tag{14}$$

where

$$B_m = \frac{(m-1)((m-2)!)^{1/(m-1)}}{m}, \qquad x = \left(\frac{(m-2)!}{e}\right)^{1/(m-1)},$$

$$b = \left(\frac{2m}{3(m-1)((m-2)!)^{1/(m-1)}}\right)^{2/3}$$

and $p(z)$ is the density of the stable law with the characteristic function $\varphi(t)$ such that

$$\ln \varphi(t) = -|t|^{3/2}\left\{1 + i\frac{3t}{2|t|}\tan\frac{3\pi}{4}\right\}.$$

PROOF. Under the hypotheses of the theorem, as is easily seen,

$$\begin{aligned} z &= \frac{n - N(1 + 1/(m-1))}{bN^{2/3}(m-1)} = \frac{n}{Nb(m-1)}N^{1/3}\left(1 - \frac{N(1+1/(m-1))}{n}\right) \\ &= \frac{n}{Nb(m-1)}N^{1/3}\left(1 - \frac{(n - T(m-1))(1 + 1/(m-1))}{n}\right) \\ &= -\frac{n}{bN(m-1)^2}N^{1/3}\left(1 - \frac{Tm(m-1)}{n}\right) \to -v. \end{aligned}$$

By virtue of Theorem 3 and the continuity of $p(z)$, the relation

$$\mathbf{P}\{\zeta_N = k\} = \frac{1}{bN^{2/3}}p(-v)(1 + o(1)),$$

holds, where $p(z)$ is the density of the stable distribution with the characteristic function $\varphi(t)$ satisfying (12) and b is determined by (11). To complete the proof, it remains to apply formula (3).

Finally, Tkachuk's theorem (Theorem 4) allows us to present the following assertion.

THEOREM 8. *Let m be fixed and $n, T \to \infty$ in such a way that $(Tm(m-1)/n-1)N^{1/3} \to \infty$. Then*

$$F_{n,N}^{(m)} = \frac{n!\, B_m^{N-1}}{(N-1)!\, x^n \sqrt{2\pi}(m-1)^2(Tm/(m-1) - (n-1)/(m-1)^2)^{5/2}}(1 + o(1)),$$

where

$$B_m = \frac{(m-1)((m-2)!)^{1/(m-1)}}{m}, \qquad x = \left(\frac{(m-2)!}{e}\right)^{1/(m-1)},$$

PROOF. The random variable $\hat{\xi}_1 = \xi_1^* - 1/(m-1)^2$ satisfies the hypotheses of Theorem 4 with $\alpha = 3/2$. It can easily be proved that the assumption $(Tm(m-1)/n-1)N^{1/3} \to \infty$ guarantees the validity of the asymptotic representation in Theorem 4. Then

$$\begin{aligned}\mathbf{P}\{\zeta_N = n\} &= N\,\mathbf{P}\{\xi_1^* = T + (1-N)/(m-1)^2\}(1 + o(1)) \\ &= \frac{N(1 + o(1))}{\sqrt{2\pi}(m-1)^2 B_m (Tm/(m-1) - (n-1)/(m-1)^2)^{5/2}}.\end{aligned}$$

It remains to substitute this expression into (3).

REMARK 1. The Stirling asymptotic formula can be applied to the factorials in (14), but in contrast to the case $m = 2$, this does not lead to significant simplification.

REMARK 2. For $m = 2$ these theorems were proved in [2, 11, 12].

REFERENCES

1. V. E. Britikov, Limit theorems for the maximum size of trees in a random forest of unrooted trees. In: *Probab. Problems in Discrete Math.* MIEM, Moscow, 1987, pp. 84–91 (in Russian).
2. V. E. Britikov, The asymptotics for the number of forests of non-rooted trees. *Math. Notes* (1988) **43**, 672–684.
3. V. E. Britikov, The asymptotic behaviour of the number of trees of a given size in a random forest of unrooted trees. In: *Probab. Processes Appl.* MIEM, Moscow, 1988, pp. 36–41 (in Russian).
4. A. Rényi, Some remarks on the theory of trees. *Magyar Tud. Akad. Mat. Kut. Int. Kőzl.* (1959) **4**, 73–85.

5. B. I. Selivanov, Enumeration of regular hypergraphs with simple cycle structure. *Combinatorial Analysis* (1972) No. 2, 60–67 (in Russian).
6. V. N. Lyamin and B. I. Selivanov, Enumeration problems for simple hypergraphs. *Combinatorial Analysis* (1974) No. 3, 57–67 (in Russian).
7. V. N. Lyamin and B. I. Selivanov, Hypertrees with a given numbers of end vertices and edges. *Combinatorial Analysis* (1974) No 3, 68–71 (in Russian).
8. V. N. Lyamin and B. I. Selivanov, On random hypertrees and hyperforests. *Math. Notes* (1974) **15**, 513–521.
9. B. I. Selivanov, On the branch size of random m-trees. *Discrete Math. Appl.* (1991) **1**, 97–103.
10. V. F. Kolchin, *Random Mappings*. Optimization Software, New York, 1986.
11. V. F. Kolchin, On the behaviour of a random graph near a critical point. *Theory Probab. Appl.* (1986) **31**, 439–451.
12. V. F. Kolchin, *Systems of Random Equations*. MIEM, Moscow, 1988 (in Russian).
13. A. V. Kolchin, On the number of hyperforests. *J. Math. Sci.* (1995) **76**, 2250–2258.
14. S. G. Tkachuk, Local limit theorems for large deviations in the case of stable limiting laws. *Izv. Akad. Nauk UzSSR Ser. Fiz.-Mat. Nauk* (1973) No. 2, 30–33 (in Russian).
15. V. N. Sachkov, *Combinatorial Methods in Discrete Mathematics*. Cambridge Univ. Press, Cambridge, 1996.
16. B. V. Gnedenko and A. N. Kolmogorov, *Limit Distributions for Sums of Independent Random Variables*. Addison–Wesley, Reading, MA, 1954.

Probabilistic Methods in Discrete Mathematics, pp. 295–304
V. F. Kolchin et al. (Eds.)

SYSTEMS OF RANDOM LINEAR EQUATIONS WITH SMALL NUMBER OF NON-ZERO COEFFICIENTS IN FINITE FIELDS

V. F. KOLCHIN

Steklov Institute of Mathematics

Abstract — We consider a system of equations of the form $AX = B$ over the finite prime field GF(p), where $X = (x_1, \ldots, x_n)$ is a column-vector of unknowns, $A = \|\alpha_{tj}\|$, $t = 1, \ldots, T$, $j = 1, \ldots, n$, is a matrix which consists of independent random variables taking values from GF(p) with probabilities

$$\mathbf{P}\{\alpha_{tj} = 0\} = 1 - \frac{\ln n + x}{n},$$
$$\mathbf{P}\{\alpha_{tj} = k\} = \frac{\ln n + x}{(p-1)n}, \qquad k = 1, \ldots, p-1,$$

x is a fixed number. We suppose that $b_1, \ldots, b_T$ are independent random variables which do not depend on the left-hand side of the system and take the value 0 with the same probability β. We consider the probability $P_{n,T}$ of consistency of the system and prove the following assertion.

If $n, T \to \infty$ in such a way that $T/n \to \alpha$, $0 < \alpha < 1$, then $P_{n,T} \to \exp\{-\alpha e^{-x}(1-\beta)\}$.

This research was supported by the Russian Foundation for Basic Research, grants 96–01–00338 and 96–15–96092.

1. INTRODUCTION

We consider a system of equations over the finite prime field GF(p)

$$AX = B, \tag{1}$$

where $X = (x_1, \ldots, x_n)$ is a column-vector of unknowns, $B = (b_1, \ldots, b_T)$ is a column-vector of right-hand side terms,

$$A = \|\alpha_{tj}\|, \quad t = 1, \ldots, T, \quad j = 1, \ldots, n,$$

is a matrix which consists of independent random variables taking values from GF(p) with probabilities

$$\mathbf{P}\{\alpha_{tj} = 0\} = 1 - \frac{\ln n + x}{n},$$
$$\mathbf{P}\{\alpha_{tj} = k\} = \frac{\ln n + x}{(p-1)n}, \qquad k = 1, \ldots, p-1,$$

x is a fixed number. We suppose that $b_1, \dots, b_T$ are independent random variables which do not depend on the left-hand side of the system and take the value 0 with the same probability

$$\mathbf{P}\{b_i = 0\} = \beta, \qquad i = 1, \dots, T.$$

We consider the probability $P_{n,T}$ of consistency of system (1) and prove the following theorem.

THEOREM 1. *If $n, T \to \infty$ in such a way that $T/n \to \alpha$, $0 < \alpha < 1$, then*

$$P_{n,T} \to \exp\{-\alpha e^{-x}(1 - \beta)\}.$$

Proving Theorem 1 we use the notion of a critical set introduced in [1, 2] (see also [3]). Let $s(A)$ be the maximal number of independent critical sets in the matrix A and let

$$S(A) = p^{s(A)} - 1$$

be the total number of the critical sets. We prove the following assertion.

THEOREM 2. *If $n, T \to \infty$ in such a way that $T/n \to \alpha$, $0 < \alpha < 1$, then the distribution of the maximal number of critical sets $s(A)$ converges to the Poisson distribution with parameter*

$$\lambda = \alpha e^{-x},$$

and if $T/n \to \alpha$, $\alpha > 1$, then

$$\mathbf{E}S(A) \to \infty.$$

Thus, $\alpha = 1$ is the threshold point in the behaviour of the number of critical sets and, consequently, of the rank of the matrix A.

The approach of this paper was used in [1] to prove Theorems 1 and 2 in GF(2). With the use of a different approach, Theorem 2 was proved by Balakin [4].

2. AUXILIARY RESULTS

We recall the definition of a critical set introduced in [1, 2]. We consider a $T \times n$ matrix $\|a_{tj}\|$ with elements from GF(p) and with rows

$$a_t = (a_{t1}, \dots, a_{tN}), \qquad t = 1, \dots, T.$$

The rows with numbers $t_1, \dots, t_m$ and weights $\varepsilon_1, \dots, \varepsilon_m$ form a critical set $B = \{t_1, \dots, t_m; \varepsilon_1, \dots, \varepsilon_m\}$ if the sum

$$\varepsilon_1 a_{t_1} + \dots + \varepsilon_m a_{t_m}$$

is the zero vector, here $\varepsilon_1, \dots, \varepsilon_m$ take non-zero values from GF(p) and the summation and multiplications are the operations in GF(p).

We define the multiplication of a critical set $B = \{t_1, \ldots, t_m; \varepsilon_1, \ldots, \varepsilon_m\}$ by a non-zero constant α, setting

$$\alpha B = \{t_1, \ldots, t_m; \alpha\varepsilon_1, \ldots, \alpha\varepsilon_m\},$$

and addition of critical sets

$$B_1 = \{t_1^{(1)}, \ldots, t_{m_1}^{(1)}; \varepsilon_1^{(1)}, \ldots, \varepsilon_{m_1}^{(1)}\}, \qquad B_2 = \{t_1^{(2)}, \ldots, t_{m_2}^{(2)}; \varepsilon_1^{(2)}, \ldots, \varepsilon_{m_2}^{(2)}\},$$

setting for non-zero constants α_1 and α_2

$$B = \alpha_1 B_1 + \alpha_2 B_2 = \{t_1, \ldots, t_m; \varepsilon_1, \ldots, \varepsilon_m\},$$

where a row t, $1 \le t \le T$, is included into B with weight $\varepsilon = \alpha_1\varepsilon_j^{(1)}$, if $t = t_j^{(1)}$ for some j and $t_j^{(1)} \notin \{t_1^{(2)}, \ldots, t_{m_2}^{(2)}\}$, a row t is included into B with weight $\varepsilon = \alpha_2 t_k^{(2)}$, if $t = t_k^{(2)}$ for some k and $t_k^{(2)} \notin \{t_1^{(1)}, \ldots, t_{m_1}^{(1)}\}$, and t is included into B with weight $\varepsilon = \alpha_1\varepsilon_j^{(1)} + \alpha_2\varepsilon_k^{(2)}$, if $t = t_j^{(1)} = t_k^{(2)} \in \{t_1^{(1)}, \ldots, t_{m_1}^{(1)}\} \cap \{t_1^{(2)}, \ldots, t_{m_2}^{(2)}\}$ and $\alpha_1\varepsilon_j^{(1)} + \alpha_2\varepsilon_k^{(2)} \neq 0$; otherwise t is not included into $\{t_1, \ldots, t_m\}$. Here the addition and multiplication of weights by a constant are fulfilled in GF(p).

It is not difficult to see that αB and $\alpha_1 B_1 + \alpha_2 B_2$ for $B_1 \neq B_2$ are critical sets.

Critical sets $B_1, \ldots, B_s$ are called independent if

$$\alpha_1 B_1 + \ldots + \alpha_s B_s = \varnothing$$

if and only if $\alpha_1 = \ldots = \alpha_s = 0$.

We denote by $s(A)$ the maximal number of critical sets of the matrix A. It is not difficult to show (see, for example, [1, 2]) that the rank $r(A)$ of the matrix A and $s(A)$ satisfy the relation

$$r(A) + s(A) = T. \tag{2}$$

We denote by $S(A)$ the total number of critical sets of the matrix A, it is clear that in GF(p)

$$S(A) = p^{s(A)} - 1.$$

Obviously, a zero row of the matrix A forms a critical set. We denote by $\xi_{n,T}$ the number of zero rows of the matrix A.

LEMMA 1. *If $n, T \to \infty$ in such a way that $T/n \to \alpha$, $0 < \alpha < \infty$, then for any fixed $k = 0, 1, \ldots$*

$$\mathbf{P}\{\xi_{n,T} = k\} \to \frac{\lambda^k e^{-\lambda}}{k!},$$

where

$$\lambda = \alpha e^{-x}.$$

PROOF. A row is the zero row with probability

$$q_n = \left(1 - \frac{\ln n + x}{n}\right)^n$$

and under the conditions of the lemma

$$q_n = \frac{1}{n}e^{-x}(1 + o(1)).$$

The random variable $\xi_{n,T}$ has the binomial distribution with parameters (T, q_n). Under the conditions of the lemma

$$Tq_n \to \alpha e^{-x}$$

and the binomial distribution converges to the Poisson distribution with parameter $\lambda = \alpha e^{-x}$.

Let $\xi_1, \dots, \xi_k$ be independent random variables which have the same distribution as α_{11} and let

$$S_k = \xi_1 + \dots + \xi_k.$$

We set

$$p_0 = 1, \quad p_k = \mathbf{P}\{S_k \equiv 0 \pmod p\}, \qquad k > 0.$$

LEMMA 2. *For any* $k \geq 0$

$$p_k = \frac{1}{p}(1 + (p-1)(q-r)^k),$$

where

$$q = \mathbf{P}\{\xi_1 = 0\}, \quad r = \frac{1-q}{1-p} = \mathbf{P}\{\xi_1 = i\}, \quad i = 1, \dots, p-1.$$

PROOF. We set

$$p_k(i) = \mathbf{P}\{S_k \equiv i \pmod p\}, \qquad i = 0, 1, \dots, p-1,$$

and

$$f(z) = \sum_{k=0}^{\infty} p_k z^k.$$

By the total probability formula, for $k > 1$

$$p_k = qp_{k-1} + rp_{k-1}(1) + \dots + rp_{k-1}(p-1) = (q-r)p_{k-1} + r.$$

It is easily seen that if we put $p_0 = 1$, then this equality is true for $k = 1$ as well. Multiplying the both sides of this equality by z^k and summing over k, we find that

$$f(z) - 1 = (q-r)zf(z) + \frac{rz}{1-z},$$

hence it follows that the generating function $f(z)$ is equal to

$$f(z) = \frac{1 - z + rz}{(1 - z)(1 - (q - r)z)}.$$

Expanding $f(z)$ into common fractions, we see that p_k is of the form given in the lemma.

LEMMA 3. *If $n, T \to \infty$ in such a way that $T/n \to \alpha$, $\alpha < 1$, then, with probability tending to one, the number of critical sets each of which contains a non-zero row is equal to zero.*

PROOF. By Lemma 2 the probability that the sum of the elements of one column of a submatrix with k rows for any non-zero weights $\varepsilon_1, \ldots, \varepsilon_k$ has the factor p is equal to

$$\frac{1}{p}(1 + (p - 1)(q - r)^k) = \frac{1}{p}\left(1 + (p - 1)\left(1 - \frac{p(\ln n + x)}{(p - 1)n}\right)^k\right)$$

and consequently, the probability that k rows form a critical set with weights $\varepsilon, \ldots, \varepsilon_k$ is equal to

$$\frac{1}{p^n}\left(1 + (p - 1)\left(1 - \frac{p(\ln n + x)}{(p - 1)n}\right)^k\right)^n.$$

Using the representation of the total number of critical sets as a sum of indicators, we obtain that the mathematical expectation of this number is equal to

$$\sum_{k=1}^{T}\binom{T}{k}\frac{(p - 1)^k}{p^n}\left(1 + (p - 1)\left(1 - \frac{p(\ln n + x)}{(p - 1)n}\right)^k\right)^n.$$

We have to subtract from this expression the mathematical expectation of the number of critical sets consisting of zero rows. This mean value is equal to

$$\sum_{k=1}^{T}\binom{T}{k}(p - 1)^k\left(1 - \frac{\ln n + x}{n}\right)^{kn}.$$

Thus, the mathematical expectation of the number of critical sets each of which contains a non-zero row can be written in the form

$$\sum_{k=0}^{T}\binom{T}{k}\frac{(p - 1)^k}{p^n}\left(1 + (p - 1)\left(1 - \frac{p(\ln n + x)}{(p - 1)n}\right)^k\right)^n$$
$$- \sum_{k=0}^{T}\binom{T}{k}(p - 1)^k\left(1 - \frac{\ln n + x}{n}\right)^{kn}.$$

First of all we note that

$$\sum_{k=0}^{T}\binom{T}{k}(p - 1)^k\left(1 - \frac{\ln n + x}{n}\right)^{kn} = \left(1 + (p - 1)\left(1 - \frac{\ln n + x}{n}\right)^n\right)^T.$$

Under the conditions of the lemma

$$\left(1+(p-1)\left(1-\frac{\ln n+x}{n}\right)^n\right)^T=\left(1+\frac{p-1}{n}e^{-x}+o\left(\frac{1}{n}\right)\right)^T$$
$$=\exp\{\alpha(p-1)e^{-x}\}(1+o(1)). \tag{3}$$

Now we consider the sum

$$\sum_{k=0}^{T}\binom{T}{k}\frac{(p-1)^k}{p^n}\left(1+(p-1)\left(1-\frac{p(\ln n+x)}{(p-1)n}\right)^k\right)^n$$
$$=\sum_{k=0}^{T}\binom{T}{k}\frac{(p-1)^k}{p^n}(1+(p-1)a^k)^n$$
$$=\sum_{k=0}^{T}\binom{T}{k}\frac{(p-1)^k}{p^n}\sum_{l=0}^{n}\binom{n}{l}(p-1)^l a^{kl}$$
$$=\sum_{l=0}^{n}\binom{n}{l}\frac{(p-1)^l}{p^n}\sum_{k=0}^{T}\binom{T}{k}(p-1)^k a^{lk}$$
$$=\sum_{l=0}^{n}\binom{n}{l}\frac{(p-1)^l}{p^n}(1+(p-1)a^l)^T,$$

where for the sake of brevity we set

$$a=1-\frac{p(\ln n+x)}{(p-1)n}.$$

Thus, we have to estimate the sum

$$S(n,T)=\sum_{k=0}^{n}\binom{n}{k}\frac{(p-1)^k}{p^n}(1-(p-1)a^k)^T.$$

We introduce the notation

$$r_k=1+(p-1)\left(1-\frac{p(\ln n+x)}{(p-1)n}\right)^k$$

and set

$$a_k=\binom{n}{k}\left(1-\frac{1}{p}\right)^k\left(\frac{1}{p}\right)^{n-k}r_k^T.$$

To estimate the sum $S(n,T)$ we divide it into five parts in such a way that

$$S(n,T)=S_1+\ldots+S_5,$$

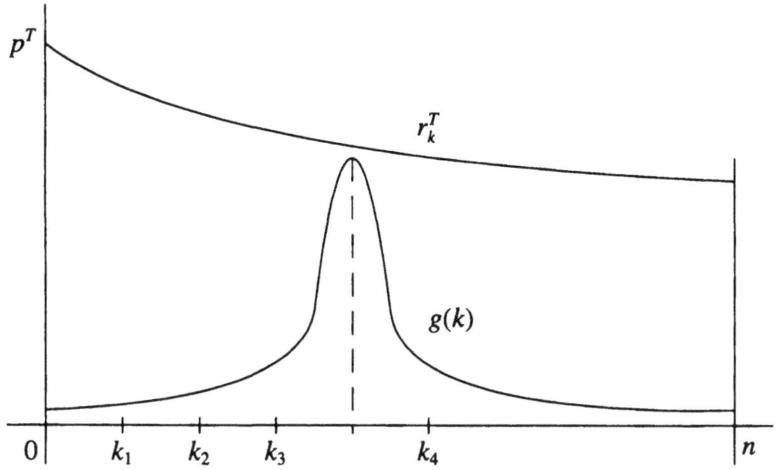

Figure 1.

where

$$S_1 = \sum_{0 \le k < k_1} a_k, \quad S_2 = \sum_{k_1 \le k < k_2} a_k, \quad S_3 = \sum_{k_2 \le k < k_3} a_k,$$
$$S_4 = \sum_{k_3 \le k < k_4} a_k, \qquad S_5 = \sum_{k_4 \le k \le n} a_k,$$

and

$$k_1 = \varepsilon n, \qquad k_2 = \frac{n(p-1)}{p}(1-\varepsilon),$$
$$k_3 = \frac{n(p-1)}{p} - \frac{n^{1/10}}{p}\sqrt{n(p-1)}, \qquad k_4 = \frac{n(p-1)}{p} + \frac{n^{1/10}}{p}\sqrt{n(p-1)}.$$

The value of $\varepsilon > 0$ will be chosen later.

Figure 1 contains the graphs of the functions

$$g(k) = \binom{n}{k}\left(1 - \frac{1}{p}\right)^k \left(\frac{1}{p}\right)^{n-k},$$
$$r_k^T = \left(1 + (p-1)\left(1 - \frac{p(\ln n + x)}{(p-1)n}\right)^k\right)^T.$$

If $\alpha < 1$, then the main contribution to the sum $S(n, T)$ is given by S_4. It is clear that

$$\left(1 - \frac{p(\ln n + x)}{(p-1)n}\right)^k = \exp\left\{\frac{pk}{(p-1)n}(\ln n + x)\right\}(1 + o(1)) = \frac{1}{n}e^{-x}(1 + o(1))$$

uniformly with respect to the integers

$$k = \frac{(p-1)n}{p} + \frac{u}{p}\sqrt{(p-1)n},$$

such that $|u| \le n^{1/10}$. Therefore

$$\begin{aligned} r_k^T &= \left(1 + (p-1)\left(1 - \frac{p(\ln n + x)}{(p-1)n}\right)^k\right)^T \\ &= \left(1 + \frac{p-1}{n}e^{-x} + o(\left(\frac{1}{n}\right))\right)^T \\ &= \exp\{\alpha(p-1)e^{-x}\}(1 + o(1)). \end{aligned}$$

uniformly with respect to k from the domain of summation in S_4.

Thus,

$$\begin{aligned} S_4 &= \sum_{k:\ |u| \le n^{1/10}} \binom{n}{k}\left(1 - \frac{1}{p}\right)^k \left(\frac{1}{p}\right)^{n-k} r_k^T \\ &= \sum_{k:\ |u| \le n^{1/10}} \binom{n}{k}\left(1 - \frac{1}{p}\right)^k \left(\frac{1}{p}\right)^{n-k} e^{\alpha(p-1)e^{-x}}(1 + o(1)) \\ &= e^{\alpha(p-1)e^{-x}} \sum_{k:\ |u| \le n^{1/10}} \binom{n}{k}\left(1 - \frac{1}{p}\right)^k \left(\frac{1}{p}\right)^{n-k} (1 + o(1)) \\ &= e^{\alpha(p-1)e^{-x}}(1 + o(1)). \end{aligned}$$

In the same way as in the case of GF(2) in [1], it can be shown that the remaining sums tend to zero under the conditions of the lemma. For example, using the monotonicity of r_k^T, we find that

$$\begin{aligned} S_5 &= \sum_{k:\ u \ge n^{1/10}} \binom{n}{k}\left(1 - \frac{1}{p}\right)^k \left(\frac{1}{p}\right)^{n-k} r_k^T \\ &\le r_{k_4}^T \sum_{k:\ u \ge n^{1/10}} \binom{n}{k}\left(1 - \frac{1}{p}\right)^k \left(\frac{1}{p}\right)^{n-k}, \end{aligned}$$

and $S_5 \to 0$ under the conditions of the lemma, since, as it was shown,

$$r_{k_4}^T \to e^{\alpha(p-1)e^{-x}}$$

and according to the Moivre–Laplace theorem

$$\sum_{k:\ u \ge n^{1/10}} \binom{n}{k}\left(1 - \frac{1}{p}\right)^k \left(\frac{1}{p}\right)^{n-k} \to 0.$$

Let us estimate

$$S_1 = \sum_{0 \le k < k_1} \binom{n}{k} \left(1 - \frac{1}{p}\right)^k \left(\frac{1}{p}\right)^{n-k} r_k^T.$$

Using the monotonicity of r_k, we find that for sufficiently small ε

$$\begin{aligned} S_1 &\le r_0^T \sum_{k<k_1} \binom{n}{k} \left(1 - \frac{1}{p}\right)^k \left(\frac{1}{p}\right)^{n-k} \le \frac{p^T (p-1)^{k_1}}{p^n} \sum_{k<k_1} \binom{n}{k} \\ &\le k_1 \binom{n}{k_1} \frac{p^T (p-1)^{k_1}}{p^n} \le \varepsilon n \frac{n^{\varepsilon n} p^T (p-1)^{\varepsilon n}}{(\varepsilon n)!\, p^n} \\ &\le \frac{\varepsilon n n^{\varepsilon n} p^T (p-1)^{\varepsilon n}}{(\varepsilon n)^{\varepsilon n} \sqrt{\varepsilon n} e^{-\varepsilon n} p^n} \le \sqrt{\varepsilon n} \left(\frac{p^{T/n} e^{\varepsilon} (p-1)^{\varepsilon}}{\varepsilon^{\varepsilon} p}\right)^n. \end{aligned}$$

If $T/n \to \alpha < 1$, then for sufficiently small ε the expression in the brackets is no greater that some constant $c < 1$. Therefore if $n \to \infty$ and $\varepsilon > 0$ is sufficiently small, then

$$S_1 \le \sqrt{\varepsilon n} c^n \to 0.$$

The remaining sums are estimated in the same way as in [1]. Note that the first term of the sum S_1 is p^{T-n} and if $T/n \to \alpha > 1$, then this term tends to infinity.

3. PROOFS OF THEOREMS 1 AND 2

Theorem 2 follows from Lemmas 1 and 3. Indeed, under the conditions of Theorem 2 according to Lemma 3

$$\mathbf{P}\{s(A) = \xi_{n,T}\} \to 1,$$

and therefore the assertion of the theorem follows from Lemma 1.

To prove Theorem 1, we note that according to Lemma 3 the probability of consistency of system (1) depends on the number $\xi_{n,T}$ of zero rows only, more precisely,

$$P_{n,T} = \sum_{k=0}^{\infty} \mathbf{P}\{\xi_{n,T} = k\} \beta^k + o(1).$$

Therefore under the conditions of Theorem 1 by virtue of Lemma 1

$$P_{n,T} \to \sum_{k=0}^{\infty} \frac{\lambda^k \beta^k}{k!} e^{-\lambda} = e^{-\lambda(1-\beta)} = \exp\{-\alpha e^{-x}(1-\beta)\}.$$

Note that the limit distribution of the rank of the matrix A is unknown even in GF(2) in the case where $n, T \to \infty$ and $|n - T|$ is bounded.

REFERENCES

1. V. F. Kolchin, *Systems of Random Equations*. MIEM, Moscow, 1988 (in Russian).
2. V. F. Kolchin and V. I. Khokhlov, A threshold effect for systems of random equations of a special form. *Discrete Math. Appl.* (1995) **5**, 425–436.
3. V. F. Kolchin, Random graphs and systems of linear equations in finite fields. *Random Structures and Algorithms* (1994) **5**, 135–146.
4. G. V. Balakin, On the distribution of the rank of a random matrix. *Theory Probab. Appl.* (1968) **13**, 631–641 (in Russian).

Probabilistic Methods in Discrete Mathematics, pp. 305–308
V. F. Kolchin et al. (Eds.)

ON THE ABSOLUTE CONSTANT IN THE REMAINDER TERM ESTIMATE IN THE CENTRAL LIMIT THEOREM FOR POISSON RANDOM SUMS

V. YU. KOROLEV and S. YA. SHORGIN

Moscow

Abstract — We show that if the remainder term estimate in the central limit theorem for Poisson random sums is expressed in terms of the non-central Lyapunov fraction, then the absolute constant in this estimate is not greater than that in the Berry–Esseen inequality for sums of identically distributed summands.
This work was supported in part by the Russian Foundation for Basic Research, grant 96–01–01919.

First of all introduce the notation.

The coincidence of the distributions of random variables X and Y is denoted as $\stackrel{d}{=}$. The distribution function and the characteristic function of a random variable X is denoted by $F_X(x)$ and $f_X(t)$ respectively. The generating function of a non-negative integer-valued random variable N is denoted by $\psi_N(s)$. We write $\{N, X\}$ for the random variable with the characteristic function $\psi_N(f_X(t))$. If random variables $N, X_1, X_2, \dots$ are defined on the same probability space so that $N, X_1, X_2, \dots$ are independent and $X_k \stackrel{d}{=} X$, $k \geq 1$, then

$$\{N, X\} \stackrel{d}{=} \sum_{j=1}^{N} X_j$$

(for definiteness, we assume $\sum_{j=1}^{0} = 0$). In this case the random variable $\{N, X\}$ is called a random sum. If a random variable X possesses two finite first moments, then the symbol X^* stands for the standardized random variable X defined as

$$X^* = (X - \mathbf{E}X)/\sqrt{\mathbf{D}X}.$$

For a random variable X with a finite third moment, $L(X) = \mathbf{E}|X|^3(\mathbf{E}|X|^2)^{-3/2}$ is called the (non-central) Lyapunov fraction (for definiteness, we assume that X is not degenerated at zero). If $\mathbf{E}X = 0$, then $L(X)$ coincides with the classical Lyapunov fraction $L_0(X) = L(X - \mathbf{E}X)$.

In what follows N_λ denotes the Poisson random variable with parameter λ. The subject of this note is to consider the estimate of the accuracy of the normal approximation to the distribution of the random variable

$$S = \{N_\lambda, X\} \tag{1}$$

with large λ. The random variable S defined in (1) is called the Poisson random sum and the distribution of S is called the generalized Poisson distribution.

Let $\Phi(x)$ be the standard normal distribution function. We set

$$\Delta = \sup_x |F_{S^*}(x) - \Phi(x)|.$$

The estimates for Δ obtained by the transfer method from the well-known convergence rate estimates for sums of a non-random number of independent summands with finite third moments usually contain two summands, one of which includes the classical Lyapunov fraction. For example, using the general techniques developed in [1], one can obtain the estimate

$$\sqrt{\lambda}\Delta \le C_0 + \inf_{0<q<1}\left(\frac{C_0}{\sqrt{q}}L_0(X) + Q(q)\right),$$

where

$$Q(q) = \max\left\{\frac{1}{1-q}, \frac{1}{\sqrt{2\pi e q}(1+\sqrt{q})}\right\},$$

C_0 is the absolute constant in the Berry–Esseen inequality. It is known that $(2\pi)^{-1/2} \le C_0 \le 0.7655$ (see [2]).

At the same time the analysis of asymptotic expansions of the distribution function of S^* in powers of $\sqrt{\lambda}$ shows that it is more natural to express the estimates of Δ in terms of the non-central Lyapunov fractions $L(X)$. Moreover, if $\mathbf{E}X \neq 0$, the Lyapunov fractions $L(X)$ are more favourable for convergence rate estimates due to the following properties proved in [3]:

(1) there exists a positive constant $A \le 2\sqrt{2}$ such that for any non-degenerate random variable X with a finite third moment we have $L(X) \le AL_0(X)$;

(2) there exists a sequence of random variables $\{Z_n\}_{n\ge1}$ with finite third moments such that $L(Z_n) \to 0$, as $n \to \infty$, while $L_0(Z_n) \ge \varepsilon > 0$ for all $n \ge 1$.

An estimate for Δ expressed via $L(X)$ is given in the following lemma.

LEMMA 1. *If the distribution of X is not concentrated in zero and X possesses a finite third moment, then there exists an absolute positive constant C_1 such that*

$$\sqrt{\lambda}\Delta \le C_1 L(X). \tag{2}$$

Apparently, this result was first presented in [4] (without any proof). It was stated in [4] that $C_1 \le 2.23$. We can prove that $C_1 \le 1.99$.

At first sight it seems that the right-hand side of (2) cannot be made better than the right-hand side of the Berry–Esseen inequality, since random summation contains an additional noise parameter, the random index. It is this conclusion the traditional proof of (2) based on the direct use of the smoothing inequality leads to. However as soon as we notice that Δ is the distance between two infinitely divisible distributions we appear at a somewhat unexpected result.

THEOREM 1. *Under the conditions of Lemma* 1,

$$C_1 \le C_0$$

where C_0 *in the constant in the Berry-Esseen inequality.*

PROOF. Let n be an arbitrary natural number. We represent $f_S(t)$ as

$$f_S(t) = \exp\{\lambda(f_X(t) - 1)\} = (\exp\{\mu(f_X(t) - 1)\})^n,$$

where $\mu = \lambda/n$. Hence,

$$S \stackrel{d}{=} \sum_{k=1}^{n} Z_k,$$

where the random variables $Z_1, Z_2, \ldots$ are independent and identically distributed, $Z_k \stackrel{d}{=} \{N_\mu, X\}$, $k \ge 1$. We set $Z = \{N_\mu, X\}$. Due to the arbitrariness of n, according to the Berry–Esseen inequality,

$$\Delta \le C_0 \inf_{n \ge 1} \frac{L_0(Z)}{\sqrt{n}}. \tag{3}$$

Without loss of generality we assume that $n > \lambda$, i.e., $\mu < 1$.

Let $\alpha = \mathbf{E}X$, $\sigma^2 = \mathbf{D}X$, $\beta = \mathbf{E}|X|^3$. Note that $\mathbf{E}Z = \mu\alpha$, $\mathbf{D}Z = \mu(\alpha^2 + \sigma^2)$. Consider $I = \mathbf{E}|Z - \mu\alpha|^3$. With the account of the distribution of Z, by the total probability formula we obtain

$$I \le e^{-\mu}\left(\mu^3|\alpha|^3 + \mu\mathbf{E}|X_1 - \mu\alpha|^3 + \sum_{k=2}^{\infty}\frac{\mu^k}{k!}\mathbf{E}|X_1 + \ldots + X_k - \mu\alpha|^3\right).$$

Using the inequalities

$$|\alpha|^3 \le |\alpha|\mathbf{E}X^2 \le \mathbf{E}|X|\mathbf{E}X^2 \le \beta,$$
$$(x_1 + \ldots + x_r)^3 \le r^2(x_1^3 + \ldots + x_r^3)$$

(the latter inequality holds for any $r \ge 1$ and any non-negative $x_1, \ldots, x_r$), we obtain

$$I \le \beta\left(\mu + 3\mu^2 + 4\mu^3 + \mu^4 + \sum_{k=2}^{\infty}\frac{\mu^k}{k!}(k+1)^2(k+\mu^3)\right) \le \beta(\mu + (8+M)\mu^2),$$

where

$$M = \sum_{k=2}^{\infty} \frac{(k+1)^3}{k!}.$$

Since

$$\frac{L_0(Z)}{\sqrt{n}} = \frac{l}{\sqrt{n}(\mu(\alpha^2+\sigma^2))^{3/2}} \le \frac{\beta(1+(8+M)\mu)}{\sqrt{\lambda}(\alpha^2+\sigma^2)^{3/2}} = \frac{L(X)}{\sqrt{\lambda}} + \frac{\beta\sqrt{\lambda}(8+M)}{n(\alpha^2+\sigma^2)^{3/2}},$$

by virtue of (3) and the arbitrariness of n we have

$$\Delta \le C_0 \frac{L(X)}{\sqrt{\lambda}}.$$

The theorem is proved.

REMARK 1. As we have already mentioned, it is well known that $C_0 \ge (2\pi)^{-1/2}$. But no lower bound for C_1 is still known.

In a similar way we can obtain non-uniform estimates of the accuracy of the normal approximation to the distributions of Poisson random sums from those for sums of a non-random number of independent identically distributed summands without any loss of accuracy due to the absolute constant. Namely, the following theorem holds.

For a natural n put $S_n = X_1 + \ldots + X_n$, where as before, $X_1, X_2, \ldots$ are independent random variables, $X_k \overset{d}{=} X$, $k \ge 1$.

THEOREM 2. *If the non-uniform estimate*

$$|F_{S_n^*}(x) - \Phi(x)| \le K \frac{L_0(X)}{(1+|x|^3)\sqrt{n}}, \quad x \in \mathbf{R},$$

holds with some absolute constant K, then for the Poisson random sum S defined in (1) *we have*

$$|F_{S_n^*}(x) - \Phi(x)| \le K \frac{L(X)}{(1+|x|^3)\sqrt{\lambda}}, \quad x \in \mathbf{R},$$

with the same constant.

The proof of this theorem is completely analogous to that of Theorem 1.

REFERENCES

1. V. M. Kruglov and V. Yu. Korolev, *Limit Theorems for Random Sums*. Moscow Univ. Press, Moscow, 1990 (in Russian).
2. I. S. Shiganov, Refinement of the upper bound of the constant in the central limit theorem. *J. Soviet Math.* (1986) **35**, 2545–2550.
3. S. Ya. Shorgin, On the estimate of the accuracy of the normal approximation to the distributions of random sums with infinitely divisible indices. *Theory Probab. Appl.* (1996) **41**, 920–926 (in Russian).
4. G. V. Rotar, On a problem of reserves control. *Econ. Math. Methods* (1976) **12**, 733–739 (in Russian).

Probabilistic Methods in Discrete Mathematics, pp. 309–312
V. F. Kolchin et al. (Eds.)

THE MEAN NUMBER OF SOLUTIONS OF A SYSTEM OF RANDOM CONGRUENCES

A. V. LAPSHIN

Moscow

Abstract — We consider the asymptotic behaviour of the mean value of the number of solutions of a system of random congruences as the number of unknowns and the number of equations tend to infinity.

Probabilistic characteristics of the solutions of systems of random equations whose unknowns and coefficients take values from sets with some algebraic structures were investigated in a number of papers. In [1–4], in particular, the systems of random equations were considered whose each equation contains exactly two unknowns. The aim of this paper is to find an asymptotic estimate of the mean value of the number of solutions of a system of random congruences of the form

$$x_{i_k} - x_{j_k} = 0, \qquad k = 1, 2, \ldots, n, \tag{1}$$

where the indices i_k, j_k, $k = 1, 2, \ldots, n$, are independent random variables taking values from the set $\{1, 2, \ldots, m\}$ with equal probabilities and the unknowns $x_1, x_2, \ldots, x_n$ take values from a finite additive Abelian group G.

If the equations of system (1) are written in the form

$$x_{i_k} = x_{j_k}, \qquad k = 1, 2, \ldots, n,$$

then the assumption that the unknowns take values from the group G can be removed; it is sufficient to assume that $x_1, \ldots, x_n$ take values from a finite set consisting of M distinct elements and we have a possibility to check whether or not two elements of this set are equal. For example, we can assume that the elements of the set above are of M distinct colours and the values of unknowns are the same if the corresponding colours are the same.

In the sequel, we consider a system of equations of the form (1) and denote by η the number of its solutions, in this case M is the order of the group G.

According to [1, 2], a random graph $G_{m,n}$ with m vertices and n edges can be put into correspondence to the random system (1). It is clear that the graph $G_{m,n}$ can have loops and multiple edges. The distribution of edges in the graph $G_{m,n}$ is determined by the distribution of the indices of unknowns in system (1). We denote by $G_{m,n}^{(0,1)}$ the graph which is obtained from $G_{m,n}$ if we remove the loops and replace each set of parallel edges joining any two vertices by one edge. It is clear that the pairs of indices of unknowns in the congruences of the system corresponding to the graph $G^{(0,1)}_{m,n}$ are chosen from the set of all $m(m-1)/2$ possible pairs by the simple sampling without replacement. The graphs without loops $G^{(0)}_{m,n}$ and without multiple edges $G^{(1)}_{m,n}$ will be considered as well. The random systems of congruences corresponding to the random graphs $G_{m,n}$, $G^{(0,1)}_{m,n}$, $G^{(0)}_{m,n}$, and $G^{(1)}_{m,n}$ and the numbers of their solutions are denoted by S, $S^{(0,1)}$, $S^{(0)}$, and $S^{(1)}$ and η, $\eta^{(0,1)}$, $\eta^{(0)}$, and $\eta^{(1)}$ respectively.

Note that in [5] the random graph $G^{(0,1)}_{m,n}$ was considered.

THEOREM 1. *The equality*

$$\mathbf{E}\eta = \sum_{s_1+\ldots+s_M=m} \frac{m!}{s_1!\ldots s_M!}\left(\sum_{i=1}^{M}\left(\frac{s_i}{m}\right)^2\right)^n \tag{2}$$

holds.

PROOF. Let a vector $x = (x_1, \ldots, x_m) \in G^m$ have s_i coordinates equal to g_i, $g_i \in G$, $i = 1, \ldots, M$. The vector x is a solution of system (1) if and only if each edge of the graph G_m connects vertices corresponding to unknowns both of which take an arbitrary but the same value g, $g \in G$. The probability of this event is

$$\left(\sum_{i=1}^{M}\left(\frac{s_i}{m}\right)^2\right)^n.$$

The number of vectors $x \in G^m$ that have s_i coordinates equal to g_i, $i = 1, \ldots, M$, is

$$\frac{m!}{s_1!\ldots s_M!},$$

where

$$g_i \in G, \quad s_i \geq 0, \quad i = 1, \ldots, M, \quad \sum_{i=1}^{M} s_i = m.$$

Formula (2) is thus validated.

The formulae for the mean number of solutions of the random systems $S^{(0)}$, $S^{(1)}$, and $S^{(0,1)}$ can be obtained in the same manner.

COROLLARY 1. *The equalities*

$$\mathbf{E}\eta^{(0)} = \sum_{s_1+\ldots+s_M=m} \frac{m!}{s_1!\ldots s_M!} \left(\sum_{i=1}^{M} \binom{s_i}{2} \binom{m}{n}^{-1} \right)^n,$$

$$\mathbf{E}\eta^{(1)} = \sum_{s_1+\ldots+s_M=m} \frac{m!}{s_1!\ldots s_M!} \binom{s_1^2+\ldots+s_M^2}{n} \binom{m^2}{n}^{-1},$$

$$\mathbf{E}\eta^{(0,1)} = \sum_{s_1+\ldots+s_M=m} \frac{m!}{s_1!\ldots s_M!} \binom{s_1(s_1-1)/2+\ldots+s_M(s_M-1)/2}{n} \binom{m(m-1)/2}{n}^{-1}$$

hold.

REMARK 1. In fact, the formula for $\mathbf{E}\eta^{(0,1)}$ given above is equivalent to Theorem 2.1 in [5], where Corollary 2.1 contains the expression for the generating function of the number of connected components of the random graph $G_{m,n}^{(0,1)}$. In a similar way, using the formulae for $\mathbf{E}\eta$, $\mathbf{E}\eta^{(0)}$, and $\mathbf{E}\eta^{(1)}$, we can obtain the generating functions for the numbers of connected components of the graphs $G_{m,n}$, $G_{m,n}^{(0)}$, and $G_{m,n}^{(1)}$ respectively.

THEOREM 2. *Let $m, n \to \infty$ in such a way that $n/m \to \rho$, where ρ is a constant. Then the mean number of the system S is equivalent to*

$$M^{m-n}(1-2\rho)^{-(M-1)/2}, \quad \rho < \rho_k, \tag{3}$$

$$M^{m-n-1}C \exp\left\{ m\left((\rho-1)\ln(\sigma M) + \ln(2\rho) - \frac{\sigma}{2\rho} u \ln u - \frac{\sigma}{2\rho}(M-1)\tau \ln \tau \right) \right\}, \quad \rho > \rho_k,$$

where the parameters u, τ, and σ are determined by the system of equations

$$u = \frac{2\rho}{\sigma M}(1+\sqrt{(M-1)(\sigma M-1)}),$$

$$\tau = \frac{2\rho}{\sigma M}(1-\sqrt{(\sigma M-1)/(M-1)}),$$

$$\frac{4\rho^2}{\sigma} = u^2 + (M-1)\tau^2, \quad 0 < \tau \le 1, \quad u \ge 1,$$

the value ρ_c is equal to

$$\rho_c = \rho(\sigma_c) = -\frac{\sigma_c}{2}\sqrt{\frac{M-1}{\sigma_c M-1}} \ln\left(1 - \frac{M\sqrt{\sigma_c M-1}}{\sqrt{M-1}+(M-1)\sqrt{\sigma_c M-1}}\right),$$

where σ_c is the root of the equation

$$\ln\left(\frac{1-\sqrt{(\sigma M-1)/(M-1)}}{1+\sqrt{(\sigma M-1)(M-1)}}\right)$$
$$\left(\frac{1+\sqrt{(\sigma M-1)(M-1)}}{M}-\frac{\sigma}{2}\sqrt{\frac{M-1}{\sigma M-1}}\ln(\sigma M)\right)$$
$$-\ln(1-\sqrt{(\sigma M-1)/(M-1)})=0,$$

and

$$C=\left(\frac{2\rho}{\sigma}\right)^{1/2}\left(u\tau(1-\tau)^{M-2}\left(\frac{(m-1)\tau+u}{u\tau}-M+\frac{\sigma(M-1)(u-\tau)^2}{2\rho^2}\right)\right)^{-1/2}.$$

The proof of Theorem 2 is similar to the proof of Theorem 3.1 in [5].

Under the conditions of Theorem 2 the asymptotic values of $\mathbf{E}\eta^{(0)}$, $\mathbf{E}\eta^{(1)}$, and $\mathbf{E}\eta^{(0,1)}$ differ from the asymptotic value of $\mathbf{E}\eta$ given in Theorem 2 only by a constant factors.

REFERENCES

1. G. V. Balakin, Graphs of systems of two-term equations with Boolean unknowns. *Theory Probab. Appl.* (1995) **40**, 241–259 (in Russian).
2. V. F. Kolchin, *Systems of Random Equations.* MIEM, Moscow, 1989 (in Russian).
3. V. F. Kolchin, Consistency of a system of random congruences. *Discrete Math. Appl.* (1993) **3**, 103–113.
4. V. F. Kolchin, A classification problem in the presence of measurement errors. *Discrete Math. Appl.* (1994) **4**, 19–30.
5. A. V. Lapshin, The property of phase transitions in random graphs. *Probabilistic Methods in Discrete Mathematics, Proc. III Intern. Petrozavodsk Conf.* 1993, VSP/TVP, Utrecht/Moscow, pp. 287–297.

Probabilistic Methods in Discrete Mathematics, pp. 313–319
V. F. Kolchin et al. (Eds.)

ESTIMATION OF THE DISTRIBUTION OF A SUMMAND ON THE BASE OF OBSERVATIONS OF SUMS OF TWO RANDOM SUMMANDS IN A FINITE ABELIAN GROUP

A. V. LAPSHIN

Moscow

Abstract — We investigate the estimate of a distribution of a summand obtained by the method of least squares on the base of observations of sums of this summand and independent summands taking values from a finite Abelian group. We suppose that each sum consists of two summands and the distributions of the second summands of the sums are known. The accuracy of the estimate is investigated.

Let G be a finite Abelian group and let α, $\beta_1, \dots, \beta_s$ be random variables with values in G. Let the probability distributions $\pi_j(g) = \mathbf{P}\{\beta_j = g\}$, $g \in G$, of the random variables β_j, $j = 1, \dots, s$, be known and the probability distribution $p(g) = \mathbf{P}\{\alpha = g\}$, $g \in G$, be unknown. The problem is to estimate the probabilities $p(g)$ on the base of the observations of the random variables

$$\alpha + \beta_j, \qquad j = 1, \dots, s.$$

We suppose that the random variables $\alpha\, \beta_1, \dots, \beta_s$ are independent and we have the results of n_j independent trials of the random variable $\alpha + \beta_j$, $j = 1, \dots, s$.

The case $s = 1$ or the problem of estimating the unknown distribution of the summand α on the base of independent observations of the random sum $\alpha + \beta$, where α and β are independent random variables on a finite cyclic group, was considered by some authors.

As an estimate $p^*(g)$ of the unknown distribution $p(g)$ we take the estimate obtained by the least squares method [1], that is, the function that provides the minimum

$$\min_{p(g)} \sum_{j=1}^{s} \|(p * \pi_j)(g) - P_j^*(g)\|^2,$$

where $(p * \pi_j)(g)$ is the convolution of the distributions $p(g)$ and $\pi_j(g)$, $j = 1, \dots, s$, which is determined by the equality

$$(p * \pi_j)(g) = \sum_{y \in G} p(y)\pi_j(g - y) = \sum_{y \in G} p(g - y)\pi_j(y),$$

and

$$P_j^*(g) = \frac{n_j(g)}{n_j}$$

is the frequency and $n_j(g)$ is the number of the event g in n_j independent realizations of the random variable $\alpha + \beta_j$,

$$\sum_{g\in G} n_j(g) = n_j, \qquad j = 1, \dots, s,$$

and, finally, for an arbitrary complex-valued function $f(g)$ of the variable $g \in G$,

$$\|f(g)\|^2 = \sum_{g\in G} |f(g)|^2.$$

We use the following notions and results of the paper [2]. There exists a one-to-one correspondence between the set of probability distributions of the random variables taking values from the group G and the set of their characteristic functions $\hat{p}(\chi)$ defined on the dual group $\hat{G}$, that is, on the group of characters of the group G. This correspondence is defined by the equalities

$$\hat{p}(\chi) = \sum_{g\in G} p(g)\chi(g), \qquad \chi(g) \in \hat{G},,$$

$$p(g) = \frac{1}{|G|} \sum_{\chi\in\hat{G}} \hat{p}(\chi)\bar{\chi}(g), \qquad g \in G, \tag{1}$$

where $\hat{p}(\chi)$ is the characteristic function of the probability distribution $p(g)$ on the group G. The characteristic function of the convolution of two probability distributions $p_1(g)$ and $p_2(g)$ on G is equal to the product of the characteristic functions of these distributions, that is, if $p(g) = (p_1 * p_2)(g)$, then

$$\hat{p}(\chi) = \hat{p}_1(\chi)\hat{p}_2(\chi).$$

Let us find the formula for the characteristic function $\hat{p}^*(\chi)$ of the least square estimate of the unknown distribution $p(g)$ of the random variable α.

THEOREM 1. *Let*

$$\sum_{j=1}^{s} |\hat{\pi}_j(\chi)| \neq 0, \qquad \chi \in \hat{G}.$$

Then

$$\hat{p}^*(\chi) = \sum_{j=1}^{s} \hat{P}_j^*(\chi)\bar{\hat{\pi}}_j(\chi) \left(\sum_{j=1}^{s} |\hat{\pi}_j(\chi)|^2 \right)^{-1}, \qquad \chi \in \hat{G}. \tag{2}$$

This estimate is an unbiased estimate of $p(g)$, $g \in G$.

PROOF. According to Theorem 1 of Section 12 in [3],

$$\|f(g)\|_G^2 = \|\hat{f}(\chi)\|_{\hat{G}}^2,$$

where the functions $f(g)$, $g \in G$, and $\hat{f}(\chi)$, $\chi \in \hat{G}$, are related by formulae (1), and

$$\|\hat{f}(\chi)\|_{\hat{G}}^2 = \frac{1}{|G|} \sum_{\chi \in \hat{G}} |\hat{f}(\chi)|^2.$$

Therefore

$$\min_{p(g)} \sum_{j=1}^{s} \|(p * \pi_j)(g) - P_j^*(g)\|_G^2 = \min_{\hat{p}(\chi)} \sum_{j=1}^{s} \|\hat{p}(\chi)\hat{\pi}_j(\chi) - \hat{P}_j^*(\chi)\|_{\hat{G}}^2. \tag{3}$$

We consider the right-hand side of (3) as a real-valued function of the arguments $\hat{p}_1(\chi)$ and $\hat{p}_2(\chi)$, where $\hat{p}_1(\chi) = \Re\hat{p}(\chi)$ and $\hat{p}_2(\chi) = \Im\hat{p}(\chi)$. Since

$$\hat{p}(1) = \hat{\pi}(1) = \hat{P}_j^*(1) = 1, \qquad j = 1, \dots, s,$$

the equality

$$\sum_{j=1}^{s} \|\hat{p}(\chi)\hat{\pi}_j(\chi) - \hat{P}_j^*(\chi)\|_{\hat{G}}^2 = \frac{1}{|G|} \sum_{j=1}^{s} \sum_{\chi \neq 1} |\hat{p}(\chi)\hat{\pi}(\chi) - \hat{P}_l^2(\chi)|^2$$

holds. Further, for

$$\sum_{j=1}^{s} |\hat{p}(\chi)\hat{\pi}_j(\chi) - \hat{P}_j^*(\chi)|^2,$$

we have

$$\begin{aligned}
\sum_{j=1}^{s} |\hat{p}(\chi)\hat{\pi}_j(\chi) - \hat{P}_j^*(\chi)|^2 &= (\hat{p}_1^2(\chi) + \hat{p}_2^2(\chi)) \sum_{j=1}^{s} |\hat{\pi}_j(\chi)|^2 + \sum_{j=1}^{s} |\hat{P}_j^*(\chi)|^2 \\
&\quad - 2\hat{p}_1(\chi) \sum_{j=1}^{s} (\hat{\pi}_{1j}(\chi)\hat{P}_{1j}^*(\chi) + \hat{\pi}_{2j}(\chi)\hat{P}_{2j}^*(\chi)) \\
&\quad + 2\hat{p}_1(\chi) \sum_{j=1}^{s} (\hat{\pi}_{2j}(\chi)\hat{P}_{1j}^*(\chi) - \hat{\pi}_{1j}(\chi)\hat{P}_{2j}^*(\chi)).
\end{aligned}$$

Now we find the point of the minimum of this function and, as the result, obtain $\hat{p}^*(\chi)$. Using the relations

$$\frac{\partial}{\partial \hat{p}_1(\chi)} \left(\sum_{j=1}^{s} |\hat{p}(\chi)\hat{\pi}_j(\chi) - \hat{P}_j^*(\chi)|^2 \right) = 0,$$

$$\frac{\partial}{\partial \hat{p}_2(\chi)} \left(\sum_{j=1}^{s} |\hat{p}(\chi)\hat{\pi}_j(\chi) - \hat{P}_j^*(\chi)|^2 \right) = 0,$$

we obtain

$$\hat{p}_1^*(\chi) = \frac{1}{\sum_{j=1}^s |\hat{\pi}_j(\chi)|^2} \sum_{j=1}^s (\hat{\pi}_{1j}(\chi)\hat{P}_{1j}^*(\chi) + \hat{\pi}_{2j}(\chi)\hat{P}_{2j}^*(\chi)),$$

$$\hat{p}_1^*(\chi) = \frac{1}{\sum_{j=1}^s |\hat{\pi}_j(\chi)|^2} \sum_{j=1}^s (\hat{\pi}_{1j}(\chi)\hat{P}_{2j}^*(\chi) - \hat{\pi}_{2j}(\chi)\hat{P}_{1j}^*(\chi)).$$

Here

$$\hat{\pi}_{1j}(\chi) = \Re\hat{\pi}_j(\chi), \qquad \hat{P}_{1j}^*(\chi) = \Re\hat{P}_j^*(\chi),$$
$$\hat{\pi}_{2j}(\chi) = \Im\hat{\pi}_j(\chi), \qquad \hat{P}_{2j}^*(\chi) = \Im\hat{P}_j^*(\chi).$$

Thus, we conclude that the point with coordinates (2) is the unique point of minimum of expression (3).

The last assertion of the theorem follows from the relations

$$\hat{P}_j^*(\chi) = \hat{p}_j^*(\chi)\hat{p}_j(\chi),$$
$$\mathbf{E}\hat{P}_j^*(\chi) = \hat{P}_j(\chi), \quad \chi \in \hat{G}, \quad j = 1, \ldots, s,$$

where $\mathbf{E}$ is the symbol of mathematical expectation and $P_j(g)$, $g \in G$, is the probability distribution of the random variable $\alpha + \beta_j$, $j = 1, \ldots, s$. The proof is completed.

The accuracy of the estimate $p^*(g)$ of the distribution $p(g)$, $g \in G$, is characterized by the value

$$\delta^2 = \|p^*(g) - p(g)\|_G^2. \tag{4}$$

We begin with the mean value of δ^2.

THEOREM 2. *The following equality holds:*

$$\mathbf{E}\delta^2 = \frac{1}{|g|} \sum_{j=1}^s \frac{1}{n_j} \sum_{\chi \neq 1} |\hat{\pi}_j(\chi)|^2 (1 - |\hat{p}(\chi)|^2 |\hat{\pi}_j(\chi)|^2) \left(\sum_{j=1}^s |\hat{\pi}_j(\chi)|^2 \right)^{-2}.$$

PROOF. In the same manner as we deduced (3) from (4), we obtain

$$\delta^2 = \frac{1}{|g|} \sum_{\chi \neq 1} |\hat{p}^*(\chi) - \hat{p}(\chi)|^2. \tag{5}$$

Using (2), we obtain

$$|\hat{p}^*(\chi)-\hat{p}(\chi)|^2 = \left(\sum_{j=1}^{s} |\hat{P}_j^*(\chi)|^2|\hat{\pi}_j(\chi)|^2 + \sum_{i\neq j} \hat{P}_j^*(\chi)\bar{\hat{\pi}}_j(\chi)\bar{\hat{P}}_i^*(\chi)\hat{\pi}_i(\chi)\right)\left(\sum_{j=1}^{s} |\hat{\pi}_j(\chi)|^2\right)^{-2}$$

$$- \bar{\hat{p}}(\chi)\sum_{j=1}^{s} \hat{P}_j^*(\chi)\bar{\hat{\pi}}_j(\chi)\left(\sum_{j=1}^{s} |\hat{\pi}_j(\chi)|^2\right)^{-1}$$

$$- \hat{p}(\chi)\sum_{j=1}^{s} \bar{\hat{P}}_j^*(\chi)\hat{\pi}_j(\chi)\left(\sum_{j=1}^{s} |\hat{\pi}_j(|\chi|^2)\right)^{-1} + |\hat{p}(\chi)|^2. \quad (6)$$

The characteristic function $\hat{P}^*(\chi)$ is a complex-valued random variable defined on the group $\hat{G}$. Since

$$\mathbf{E}P_j^*(g) = P(g),$$

where $P_j(g) = (p * \pi_j)(g)$, setting $\hat{P}_j^*(\chi) = \hat{P}_{1j}^* + i\hat{P}_{2j}^*(\chi)$, we see that the random variable $(\hat{P}_{1j}^*(\chi), \hat{P}_{2j}^*(\chi))$ has the mean vector $(\hat{P}_{1j}(\chi), \hat{P}_{2j}(\chi))$ and some covariance matrix

$$M(\hat{P}_j^*(\chi)) = \begin{pmatrix} \sigma^2(\hat{P}_{1j}^*(\chi)) & \rho(\hat{P}_j^*(\chi))\sigma(\hat{P}_{1j}^*(\chi))\sigma(\hat{P}_{2j}^*(\chi)) \\ \rho(\hat{P}_j^*(\chi))\sigma(\hat{P}_{1j}^*(\chi))\sigma(\hat{P}_{2j}^*(\chi)) & \sigma^2(\hat{P}_{2j}^*(\chi)) \end{pmatrix}$$

We set

$$\sigma^2(\hat{P}_j^*(\chi)) = \sigma^2(\hat{P}_{1j}^*(\chi)) + \sigma^2(\hat{P}_2 j^*(\chi)).$$

LEMMA 1. *For* $j = 1, \ldots, s$

$$\sigma^2(\hat{P}_j^*(\chi)) = \frac{1 - |\hat{P}_j(\chi)|^2}{n_j}. \quad (7)$$

PROOF. Using (1), we obtain

$$\begin{aligned}\mathbf{E}|\hat{P}_j^*(\chi) - \hat{P}_j(\chi)|^2 &= \mathbf{E}(\hat{P}_j^*(\chi) - \hat{P}_j(\chi))(\bar{\hat{P}}_j^*(\chi) - \bar{\hat{P}}_j(\chi)) \\ &= \mathbf{E}(|\hat{P}_j^*(\chi)|^2 - \hat{P}_j^*(\chi)\bar{\hat{P}}_j(\chi) - \bar{\hat{P}}_j^*(\chi)\hat{P}_j(\chi) + |\hat{P}_j(\chi)|^2) \\ &= \mathbf{E}|\hat{P}_j^*(\chi)|^2 - |\hat{P}_j(\chi)|^2 = \mathbf{E}(\hat{P}_{1j}^*(\chi))^2 + \mathbf{E}(\hat{P}_{2j}^*(\chi))^2 - \hat{P}_{1j}(\chi) - \hat{P}_{2j}(\chi) \\ &= \sigma^2(\hat{P}_{1j}^*(\chi)) + \sigma^2(\hat{P}_{2j}^*(\chi)) = \sigma^2(\hat{P}_j^*(\chi)),\end{aligned}$$

and on the other hand,

$$\begin{aligned}\mathbf{E}|\hat{P}_j^*(\chi) - \hat{P}_j(\chi)|^2 &= \mathbf{E}(\hat{P}_j^*(\chi) - \hat{P}_j(\chi))(\bar{\hat{P}}_j^*(\chi) - \bar{\hat{P}}_j(\chi)) \\ &= \mathbf{E}\sum_{g,g_1}(P_j^*(g) - P_j(g))(P_j^*(g_1) - P_j(g_1))\chi(g)\chi(-g_1) \\ &= \sum_{g,g_1}\mathbf{E}(P_j^*(g) - P_j(g))(P_j^*(g_1) - P_j(g_1))\chi(g)\chi(-g_1) \\ &= \frac{1}{n_j}\sum_{g,g_1}\delta_{g_1}^{g}\sqrt{P_j(g)P_j(g_1)}\chi(g)\chi(-g_1) - \frac{1}{n_j}\sum_{g,g_1}P_j(g)P_j(g_1)\chi(g)\chi(-g_1) \\ &= \frac{1}{n_j}\sum_{g}P_j(g)|\chi(g)|^2 - \frac{1}{n_j}|\hat{P}_j(\chi)|^2 = \frac{1}{n_j}(1 - |\hat{P}_j(\chi)|^2).\end{aligned}$$

Here $\delta_{g_1}^{g}$ is the Kronecker symbol and we used the equalities (see [2])

$$|\chi(g)| = 1, \qquad g \in G,$$

and the equality

$$\mathbf{E}(P_j^*(g) - P_j(g))(P_j^*(g_1) - P_j(g_1)) = \frac{1}{n_j}(\delta_{g_1}^{g}\sqrt{P_j(g)P_j(g_1)} - P_j(g)P(g_1)),$$

since $P_j^*(g)$, $g \in G$, is the polynomial distribution.

Now we return to the proof of Theorem 2. The random variables $\hat{P}_j^*(\chi)$ and $\hat{P}_l^*(\chi)$ are independent if $j \neq l$, therefore

$$\mathbf{E}\left(\sum_{j=1}^{s}|\hat{P}_j^*(\chi)|^2|\hat{\pi}_j(\chi)|^2 + \sum_{j\neq l}\hat{P}_j^*(\chi)\bar{\hat{\pi}}_j(\chi)\bar{\hat{P}}_l^*(\chi)\hat{\pi}_l(\chi)\right)$$
$$= \sum_{j=1}^{s}\sigma^2(\hat{P}_j^*(\chi))|\hat{\pi}_j(\chi)|^2 + \sum_{j=1}^{s}\hat{P}_j(\chi)\bar{\hat{\pi}}_j(\chi)\sum_{j=1}^{s}\bar{\hat{P}}_j(\chi)\hat{\pi}_j(\chi). \tag{8}$$

The assertion of Theorem 2 follows from (5)–(8).

COROLLARY 1. *If $n_j = n$, $j = 1, \dots, s$, then*

$$\mathbf{E}\delta^2 \le \frac{1}{n|G|}\sum_{\chi\neq 1}\left(\sum_{j=1}^{s}|\hat{\pi}_j(\chi)|^2\right)^{-1}.$$

PROOF. If $n_j = n$ for all $j = 1, \dots, s$, then the upper bound for $\mathbf{E}\delta^2$ can be obtained from Theorem 2 in the following way:

$$\begin{aligned}\mathbf{E}\delta^2 &\le \frac{1}{|G|}\sum_{j=1}^{s}\frac{1}{n}\sum_{\chi\neq 1}|\hat{\pi}_j(\chi)|^2\left(\sum_{l=1}^{s}|\hat{\pi}_l(\chi)|^2\right)^{-2} \\ &= \frac{1}{n|G|}\sum_{\chi\neq 1}\sum_{j=1}^{s}|\hat{\pi}_j(\chi)|^2\left(\sum_{l=1}^{s}\hat{\pi}_l(\chi)|^2\right)^{-2}.\end{aligned}$$

REFERENCES

1. W. Ledermann, *Handbook of Applied Mathematics. Vol. 6: Statistics.* Wiley, New York, 1984.
2. N. N. Vorobyev, The summation of two independent random variables on a finite Abelian group. *Soviet Math. Sb.* (1954) **34**, 89–126.
3. A. A. Kirillov, *Elements of Representation Theory.* Nauka, Moscow, 1978.

Probabilistic Methods in Discrete Mathematics, pp. 321–324
V. F. Kolchin et al. (Eds.)

THE LIMIT DISTRIBUTIONS OF THE MISES FUNCTIONAL OVER NON-EQUIPROBABLE BERNOULLI VECTORS

M. V. LEVASHOV

Moscow

Abstract — We describe all limit distributions of the Mises functional over non-equiprobable Bernoulli vectors in relation to the deviation of the probability of zero value of the coordinate of the vectors from 1/2.

Let $X_1, \dots, X_T$ be a sequence of independent identically distributed n-dimensional (0, 1)-vectors and f be a real-valued function defined on the set of such vectors.

Each vector consists of n independent identically distributed random coordinates taking the values 0 and 1 with probabilities $(1 + d)/2$ and $(1 - d)/2$, respectively, where $d > 0$.

We consider the Mises functional of the form

$$\xi_t = \sum_{1 \le t_1, \dots, t_m \le T} f(X_{t_1} \oplus \dots \oplus X_{t_m}),$$

where the summation is over all samples with replacement $(t_1, \dots, t_m)$ from the set $\{1, 2, \dots, T\}$ and $\oplus$ means the coordinate-wise addition modulo 2.

We use the Fourier series expansion of the function f of the form

$$f(x) = \sum_a \bar{f}(a)(-1)^{(a,x)},$$

where $a = (a_1, \dots, a_n)$, $a_1, \dots, a_n \in \{0, 1\}$,

$$(a, x) = \sum_{i=1}^{n} a_i x_i.$$

The Fourier coefficients are equal to

$$\bar{f}(a) = \frac{1}{2^n} \sum_x f(x)(-1)^{(a,x)}.$$

Using the Fourier series expansion, we can represent the Mises functional in the form

$$\begin{aligned}\xi_T &= \sum_{t_1,\dots,t_m}\sum_a \bar f(a)(-1)^{(X_{t_1}\oplus\dots\oplus X_{t_m},a)} = \sum_a \bar f(a)\left(\sum_{t=1}^T(-1)^{(X_t,a)}\right)^m \\ &= T^{m/2}\sum_a \bar f(a)\left(\frac{1}{\sqrt{T}}\sum_{t=1}^T(-1)^{(X_t,a)}\right)^m \\ &= T^{m/2}\sum_a \bar f(a)\left(\frac{1}{\sqrt{T}}\sum_{t=1}^T((-1)^{(X_t,a)}-d^{|a|})+\sqrt{T}d^{|a|}\right)^m, \qquad (1)\end{aligned}$$

where $|a|$ is the number of unities among the coordinates of the vector a.

We consider the (2^n-1)-dimensional vectors

$$\lambda_t = \left(\frac{1}{\sqrt{T}}((-1)^{(X_t,a)}-d^{|a|}),\ a\neq 0\right) = (\xi_t(a),\ a\neq 0), \qquad t=1,\dots,T,$$

and the sum of these vectors

$$\eta_T = \sum_{t=1}^T \lambda_t.$$

It is clear that $\mathbf{E}\lambda_t = 0$, and for any non-zero binary vectors a and b

$$\begin{aligned}&\mathbf{E}\frac{1}{\sqrt{T}}\sum_{t=1}^T((-1)^{(X_t,a)}-d^{|a|})\frac{1}{\sqrt{T}}\sum_{t=1}^T((-1)^{(X_t,b)}-d^{|b|}) \\ &\quad= \frac{1}{T}\sum_{t=1}^T(\mathbf{E}(-1)^{(X_t,a+b)}-d^{|a|+|b|}) + \frac{1}{T}\sum_{t_1\neq t_2}(\mathbf{E}(-1)^{(X_{t_1},a)+(X_{t_2},b)}-d^{|a|+|b|}) \\ &\quad= d^{|a+b|}-d^{|a|+|b|}.\end{aligned}$$

It is clear that the Lyapunov condition is satisfied for the sum η_T. Therefore, by the central limit theorem, the distribution of ζ_T converges weakly, as $T\to\infty$ and d and n are constants, to the (2^n-1)-dimensional normal distribution with zero mean and the matrix of the second moments

$$D = (d^{|a+b|}-d^{|a|+|b|},\ a,b\neq 0).$$

We return now to ξ_T. It is not difficult to see that, as $T\to\infty$,

$$\begin{aligned}\xi_T &= T^m\bar f(0) + T^{m/2}\sum_{a\neq 0}\bar f(a)\sum_{j=0}^m\binom{m}{j}\left(\frac{1}{\sqrt{T}}\sum_{t=1}^T((-1)^{(X_t,a)}-d^{|a|})\right)^j(\sqrt{T}d^{|a|})^{m-j} \\ &= T^m\bar f(0) + T^{m/2}\sum_{a\neq 0}\bar f(a)(\sqrt{T}d^{|a|})^m \\ &\quad + mT^{m/2}\sum_{a\neq 0}\bar f(a)(\sqrt{T}d^{|a|})^{m-1}\frac{1}{\sqrt{T}}\sum_{t=1}^T((-1)^{(X_t,a)}-d^{|a|}) \\ &\quad + T^{m/2}\sum_{a\neq 0}d^{|a|(m-2)}\zeta_T(a)O(T^{(m-2)/2}),\end{aligned}$$

where $\mathbf{E}|\zeta_T(a)|^3 \le C$ for $a \ne 0$.

Then

$$\frac{1}{m}T^{-m+1/2}\left(\xi_T - T^m\sum_a d^{m|a|}\bar{f}(a)\right)$$
$$= \sum_{a\ne 0} d^{|a|(m-1)}\bar{f}(a)\frac{1}{\sqrt{T}}\sum_{t=1}^{T}((-1)^{(X_t,a)} - d^{|a|})$$
$$+ \sum_{a\ne 0} d^{|a|(m-2)}\zeta_T(a)O\left(\frac{1}{\sqrt{T}}\right).$$

Let $(l_a,\ a \ne 0)$ be a random vector whose distribution coincides with the limit distribution of the vector η_T. Then under the conditions given above the distribution of the random variable

$$\frac{1}{m}T^{-m+1/2}\left(\xi_T - T^m\sum_a d^{m|a|}\bar{f}(a)\right)$$

converges weakly to the distribution of the random variable

$$\sum_{a\ne 0} d^{(m-1)|a|}\bar{f}(a)l_a.$$

Let $d = 0$. Then it follows from (1) that the distribution of the random variable

$$T^{-m/2}(\xi_T - T^m\bar{f}(0))$$

converges weakly to the distribution of the random variable

$$\sum_{a\ne 0}\bar{f}(a)\eta_a^m,$$

where η_a, $a \ne 0$, is a set of independent random variables with the standard normal distribution.

Let now $T \to \infty$ and $d \to 0$ in such a way that $\sqrt{T}d \to \infty$. Then

$$\xi_T = T^m\sum_a d^{m|a|}\bar{f}(a) + T^{m/2}\sum_{|a|=1}\bar{f}(a)m(\sqrt{T}d)^{m-1}\frac{1}{\sqrt{T}}\sum_{t=1}^{T}((-1)^{(X_t,a)} - d)$$
$$+ T^{m/2}\sum_{|a|=1}\sum_{j=2}^{m}\binom{m}{j}(\sqrt{T}d)^{m-j}\left(\frac{1}{\sqrt{T}}\sum_{t=1}^{T}((-1)^{(X_t,a)} - d)\right)^j$$
$$+ T^{m/2}\sum_{|a|\ge 2}\bar{f}(a)\sum_{j=1}^{m}\binom{m}{j}(\sqrt{T}d^{|a|})^{m-j}\left(\frac{1}{\sqrt{T}}\sum_{t=1}^{T}((-1)^{(X_t,a)} - d^{|a|})\right)^j$$
$$= T^m\sum_a d^{m|a|}\bar{f}(a) + T^{m/2}\sum_{|a|=1}\bar{f}(a)m(\sqrt{T}d)^{m-1}\frac{1}{\sqrt{T}}\sum_{t=1}^{T}((-1)^{(X_t,a)} - d)$$
$$+ T^{m/2}\sum_{|a|=1}\sum_{j=2}^{m}\binom{m}{j}(\sqrt{T}d)^{m-j}\eta_T^j(a) + T^{m/2}\sum_{|a|\ge 2}\bar{f}(a)\sum_{j=1}^{m}(\sqrt{T}d^{|a|})^{m-j}\eta_T^j(a).$$

Hence it follows that the random variable

$$\frac{1}{md^{m-1}} T^{-m+1/2} \left(\sum_a \bar{f}^2(a) \right)^{-1/2} \left(\xi_T - T^m \sum_a d^{m|a|} \bar{f}(a) \right)$$

has in limit the standard normal distribution.

Let $T \to \infty$ and $d \to 0$ in such a way that $\sqrt{T}d \to \beta$. Then in the same way as in the case $d = 0$, it follows from representation (1) that the distribution of the random variable

$$T^{-m/2}(\xi_T - T^m \bar{f}(0))$$

converges weakly to the distribution of the random variable

$$\sum_{|a|=1} \bar{f}(a)(\eta_a + \beta)^m + \sum_{|a|\geq 2} \bar{f}(a)\eta_a^m .$$

Probabilistic Methods in Discrete Mathematics, pp. 325–331
V. F. Kolchin et al. (Eds.)

ON A CLASS OF DISTRIBUTIONS CONNECTED WITH A NON-HOMOGENEOUS RANDOM WALK ON A FINITE ABELIAN GROUP

YU. I. MAKSIMOV

Moscow

Abstract — We consider an approach intended for the extension of statements of some problems on classical random walks on the integer lattice to random walks on a finite abelian group, and for the generation of new distributions on the group. The statements of the problems and the results have certain relations to the problems on automata in a random environment, mathematical genetics and crystallography.

Let G be a finite abelian group, H be its subgroup, and let m and m_0 be the numbers of elements of G and H respectively. We suppose that a two-parameter family of independent variables

$$\xi_t(x), \quad x \in G, \quad t = 0, 1, 2, \ldots,$$

taking values from G is given. We define a sequence of random elements y_t, $t = 0, 1, 2, \ldots$, on G setting

$$y_{t+1} = y_t + \xi_t(y_t), \tag{1}$$

where + means the operation in G.

Suppose that the distributions of the random variables $\xi_t(x)$ are the same for all $x \in H$, $t = 0, 1, 2, \ldots$, and coincide with some distribution $p_0 = \{p_0(g),\ g \in G\}$ on G. All the distributions of $\xi_t(x)$ for $x \notin H$, $t = 0, 1, 2, \ldots$, are also the same and coincide with some distribution $p_1 = \{p_1(g),\ g \in G\}$. If $p_0 = p_1$, then we obtain the usual homogeneous random walk on G. We denote by π the stationary distribution of the Markov chain defined by (1) supposing that this distribution exists and is unique. Let Γ be the group of characters of G. We denote by $\langle g, \gamma\rangle$ the value of the character $\gamma \in \Gamma$ at the element $g \in G$. If p is a distribution on G, then we denote its characteristic function by $\hat{p} = (\hat{p}(\gamma),\ \gamma \in \Gamma)$.

Using (1) and turning to the characteristic functions, we obtain

$$\begin{aligned}\hat{\pi} &= \mathbf{E}\langle y_{t+1}, f\rangle = \sum_g \pi(g)\mathbf{E}\langle g + \xi_t(g), \gamma\rangle \\ &= \hat{p}_0(\gamma)\sum_{h\in H}\pi(h)\langle h,\gamma\rangle + \hat{p}_1(\gamma)\sum_{h\notin H}\pi(h)\langle h,\gamma\rangle \\ &= \frac{1}{m}\sum_{\beta\in\Gamma}\hat{\pi}(\beta)\left(\hat{p}_0(\gamma)\sum_{h\in H}\langle h,\gamma-\beta\rangle + \hat{p}_1(\gamma)\sum_{h\notin H}\langle h,\gamma-\beta\rangle\right) \\ &= \frac{1}{m}\sum_{\beta\in\Gamma}\hat{\pi}(\beta)\left(\hat{p}_1(\gamma)\sum_{g\in G}\langle g,\gamma-\beta\rangle + (\hat{p}_0(\gamma)-\hat{p}_1(\gamma))\sum_{h\in H}\langle h,\gamma-\beta\rangle\right)\end{aligned}$$

Let Λ be the annihilator of H in Γ. Then

$$\begin{aligned}\pi(\gamma) &= \frac{m_0}{m}(\hat{p}_0(\gamma)-\hat{p}_1(\gamma))\delta_\Lambda(\gamma) \\ &\quad + \frac{1}{m}\sum_{\beta\neq 0}\hat{\pi}(\beta)(m\hat{p}_1(\gamma)\delta_0(\gamma-\beta) + m_0(\hat{p}_0(\gamma)-\hat{p}_1(\gamma))\delta_\Lambda(\gamma-\beta)) \\ &= \frac{m_0}{m}(\hat{p}_0(\gamma)-\hat{p}_1(\gamma))\delta_\Lambda(\gamma) + \hat{p}_1(\gamma)\hat{\pi}(\gamma) \\ &\quad + \frac{m_0}{m}(\hat{p}_0(\gamma)-\hat{p}_1(\gamma))\sum_{\beta\neq 0}\hat{\pi}(\beta)\delta_\Lambda(\gamma-\beta). \end{aligned} \tag{2}$$

Here δ_0 and δ_Λ are the corresponding indicators.

We introduce a distribution

$$\varphi_H(p_0, p_1) = (\varphi_H(p_0, p_1, g),\ g\in G)$$

on G by means of its characteristic function $\hat{\varphi}(p_0, p_1, \gamma)$ setting

$$\hat{\varphi}_H(p_0,p_1,\gamma) = \begin{cases} 1, & \gamma = 0, \\ \dfrac{m_0}{m}\dfrac{\hat{p}_0(\gamma)-\hat{p}_1(\gamma)}{1-\hat{p}_1(\gamma)}\left(1 - \sum\limits_{\beta\in\Lambda\setminus 0}\dfrac{m_0}{m}\dfrac{\hat{p}_0(\beta)-\hat{p}_1(\beta)}{1-\hat{p}_1(\beta)}\right)^{-1}, & \gamma\in\Lambda\setminus 0, \\ 0, & \gamma\notin\Lambda. \end{cases} \tag{3}$$

Then equality (2) implies the following result.

PROPOSITION 1. *If the stationary distribution of the Markov chain defined by* (1) *is unique, then its characteristic function is equal to* $\hat{\varphi}(p_0, p_1, \gamma)$ *provided the right-hand side of* (3) *is meaningful.*

If $\hat{\varphi}(p_0, p_1, \gamma)$ *in* (3) *is determined for all* $\gamma\in\Gamma$, *then it is the characteristic function of some distribution on the group* G.

For the case where the subgroup H consists of zero only, we rewrite formula (3) omitting the index H:

$$\hat{\varphi}(p_0, p_1, \gamma) = \begin{cases} 1, & \gamma = 0, \\ \dfrac{\hat{p}_0(\gamma) - \hat{p}_1(\gamma)}{m(1 - \hat{p}_1(\gamma))} \left(1 - \sum_{\beta \neq 0} \dfrac{\hat{p}_0(\beta) - \hat{p}_1(\beta)}{m(1 - \hat{p}_1(\beta))}\right)^{-1}, & \gamma \neq 0. \end{cases} \tag{4}$$

We give several examples of distributions on G obtained by constructions (3) and (4). We denote by k_φ the normalizing constant which is in the brackets in (3) and (4).

EXAMPLE 1. Let p be a distribution on G. We set in (4) $p_0 = p^{*2}$ and $p_1 = p$, where $*$ denotes the operation of convolution. Then

$$\frac{\hat{p}_0(\gamma) - \hat{p}_1(\gamma)}{m(1 - \hat{p}_1(\gamma))} = -\frac{1}{m}\hat{p}(\gamma),$$

$$k_\varphi = 1 - \frac{1}{m} + p(0),$$

and inverting the characteristic function, we obtain

$$\varphi(p_0, p_1, g) = \frac{1}{m}\frac{1 + p(0) - p(g)}{1 - 1/m + p(0)}.$$

EXAMPLE 2. Let now, in the dual manner, $p_0 = p$ and $p_1 = p^{*2}$ for some distribution p on G. In this case,

$$\frac{\hat{p}_0(\gamma) - \hat{p}_1(\gamma)}{m(1 - \hat{p}_1(\gamma))} = \frac{1}{m}\left(1 - \frac{1}{1 + \hat{p}(\gamma)}\right),$$

$$k_\varphi = 1 + \sum_{k=1}^{\infty}(-1)^k(p^{*k}(0) - 1/m)$$

and

$$\varphi(p_0, p_1, g) = \frac{1}{mk_\varphi}\left(1 + \sum_{k=1}^{\infty}(-1)^k(p^{*k}(0) - p^{*k}(g))\right). \tag{5}$$

EXAMPLE 3. Let $p_0 = p$ and $p_1 = \omega$, where ω is the uniform distribution on G. We set $\hat{p}^{(1)} = \varphi(p, \omega)$ and define the characteristic functions $\hat{p}^{(k)}$ by the recurrence relation

$$\hat{p}^{(k)} = \hat{\varphi}(\hat{p}^{(k-1)}, \omega), \qquad k = 2, 3, \ldots$$

By induction we obtain that for $\gamma \neq 0$

$$\hat{p}^{(k)}(\gamma) = \frac{1}{m^k}\hat{p}(\gamma)\left(1 - \frac{1}{m-1}\left(1 - \frac{1}{m^k}\right)\sum_{\beta \neq 0}\hat{p}(\beta)\right)^{-1},$$

and if for some $\gamma \neq 0$ we have $|\hat{p}(\gamma)| < 1$, then the sequence of distributions corresponding to the sequence of the characteristic functions $\hat{p}^{(k)}$ converges to the uniform distribution on G.

In the examples given below, we consider random walks on the group G of residues modulo m or walks on the m points of a circle which divide the circle into m equal arcs. We use below some relations containing the roots of the unity. Namely, let $\varepsilon = e^{2\pi i/m}$ be a primitive root of degree m of the unity and let $w(x, m)$ be the smallest non-negative residue of the integer x modulo m. Then

$$\sum_{\gamma\neq 0} \frac{\varepsilon^{-x\gamma}}{1-\varepsilon^{\gamma}} = \frac{m}{2} - \frac{1}{2} - w(x,m), \tag{6}$$

$$\sum_{\gamma\neq 0} \frac{\varepsilon^{-x\gamma}}{1-\rho\varepsilon^{\gamma}} = \frac{m\rho^{w(x,m)}}{1-\rho^m} - \frac{1}{1-\rho}, \tag{7}$$

$$\sum_{\gamma\neq 0} \frac{\varepsilon^{-x\gamma}}{(1-\varepsilon^{\gamma})^2} = \frac{1-m^2}{12} + w(-x-1,m)(m-w(-x-1,m)), \tag{8}$$

$$\sum_{\gamma\neq 0} \frac{\varepsilon^{-x\gamma}}{(1-\rho\varepsilon^{\gamma})^2} = \frac{m}{1-\rho^m}((w(x,m)+1)\rho^{w(x,m)} + (m-1-w(x,m))\rho^{m-w(x,m)}) \tag{9}$$

(in (7) and (9) we suppose that $1-\rho^m \neq 0$).

These relations can be proved with the use of the corresponding properties of the roots of the unity. Relation (6) for $x = 0$ can be found in [5], and (8) can be deduced from one of the results in [6]. Further, as above, ω is the uniform distribution on G and χ_x is the degenerate distribution concentrated in x.

EXAMPLE 4. Let $p_0 = \omega$ and $p_1 = \chi_x$. In this case,

$$\hat{\varphi}(p_0, p_1, \gamma) = -\frac{2\varepsilon^{\gamma}}{(m+1)(1-\varepsilon^{\gamma})}, \qquad \gamma \neq 0,$$

that corresponds to the distribution

$$\varphi(p_0, p_1, g) = \frac{2}{m(m+1)} w(g-1, m)$$

where the inversion can be realized with the use of (6).

EXAMPLE 5. Let $p_0 = \chi_{m/2}$, $p_1 = \chi_1$, and m is even. Then

$$\hat{\varphi}(p_0, p_1, \gamma) = \begin{cases} \dfrac{2}{m+2}, & \gamma \equiv 0 \pmod 2, \quad \gamma \neq 0, \\ -\dfrac{2}{m+2}\dfrac{1+\varepsilon^{\gamma}}{1-\varepsilon^{\gamma}}, & \gamma \equiv 1 \pmod 2, \end{cases}$$

and inverting $\hat{\varphi}$ with the use of (7), we obtain the uniform distribution on the discrete points of the lower half of the circle including 0 and $m/2$.

EXAMPLE 6. Let m be divided by 4 and

$$p_0 = \frac{1}{2}\chi_1 + \frac{1}{2}\chi_{-1}, \qquad p_1 = \frac{1}{2}\chi_{m/2+1} + \frac{1}{2}\chi_{m/2-1}.$$

Then

$$\frac{\hat{p}_0(\gamma) - \hat{p}_1(\gamma)}{m(1 - \hat{p}_1(\gamma))} = \begin{cases} 0, & \gamma \equiv 0 \pmod 2, \quad \gamma \neq 0, \\ \dfrac{2}{m}\left(1 - \dfrac{2\varepsilon^\gamma}{(1+\varepsilon^\gamma)^2}\right), & \gamma \equiv 1 \pmod 2. \end{cases}$$

Using (9) and taking into account that $m/2$ is even, we obtain $k_\varphi = m/4$ and

$$\varphi(p_0, p_1, x) = \frac{1}{m}\left(1 + \frac{4\delta_{m/2}(x)}{m} - \frac{4\delta_0(x)}{m} + \frac{(-1)^{w(x-1,m)+v(w(x-1,m),m/2)}}{m}(m - 4 - 4w(x-1, m/2))\right).$$

Here we used an additional notation $v(r, s)$ which means the quotient of division of integer r by the integer s. Note that the presence of the factor $(-1)^{w(x-1,m)}$ in the right-hand side of the last equality leads to appearance of a high-frequency component (of period 2) in the distribution.

EXAMPLE 7. Let $p_0 = \chi_{-1}$ and $p_1 = \chi_1/2 + \chi_{-1}/2$. Then

$$\frac{\hat{p}_0(\gamma) - \hat{p}_1(\gamma)}{m(1 - \hat{p}_1(\gamma))} = \frac{1}{m}\left(-1 - \frac{2\varepsilon^\gamma}{1 - \varepsilon^\gamma}\right)$$

and using (6), we obtain $k_\varphi = 1$ and

$$\varphi(p_0, p_1, g) = \frac{1}{m}\left(\frac{2}{m}w(g-1, m) + \frac{2}{m} - \delta_0(g)\right).$$

If, following the example in [7] concerning the classical random walk, we replace the distribution p_1 by the distribution $\chi_2/2 + \chi_1/2$, then the calculations become more complicated, since then

$$\frac{\hat{p}_0(\gamma) - \hat{p}_1(\gamma)}{m(1 - \hat{p}_1(\gamma))} = \frac{1}{m}\left(1 + \frac{2}{\varepsilon^{2\gamma} + \varepsilon^\gamma - 1}\right).$$

Nevertheless, we trace the dependence of probabilistic characteristics of the walk, which exists in the classical case, on the roots of the equation $z^2 + z + 1 = 0$, that is, on $\rho_1 = -(\sqrt{5} + 1)/2$ and the golden section $\rho_2 = (\sqrt{5} - 1)/2$. Namely, up to small values in comparison with $1/m$, we have

$$\varphi(p_0, p_1, 0) = \frac{1}{m}\left(2 + \frac{1}{2}\rho_1(2\rho_1 + 1)\right),$$

$$\varphi(p_0, p_1, -1) = \frac{1}{m}\left(1 + \frac{\rho_1(2\rho_1 + 1)}{2\rho_2 + 1}\right),$$

$$\varphi(p_0, p_1, 1) = \frac{1}{m}\left(1 + \frac{1}{\rho_1}\right).$$

EXAMPLE 8. Let $H = (0, m/2)$, where m is even, and $p_0 = \omega$, $p_1 = \chi_1$. Then the annihilator Λ consists of all even residues and for $\gamma \in \Lambda$, $\gamma \neq 0$,

$$\frac{2(\hat{p}_0(\gamma) - \hat{p}_1(\gamma))}{m(1 - \hat{p}_1(\gamma))} = -\frac{2\varepsilon^\gamma}{m(1 - \varepsilon^\gamma)}.$$

Using (6), we obtain

$$k_\varphi = \frac{1}{2} + \frac{1}{m},$$
$$\varphi(p_0, p_1, g) = \frac{1}{m(1 + 2/m)} \left(\frac{4}{m} w(g, m/2) + 2\delta_0(g) + 2\delta_{m/2}(g) \right).$$

If, in the dual manner, H consists of even residues modulo m, p_0 and p_1 are arbitrary distributions for which (3) is meaningful, then $\Lambda = (0, m/2)$ and setting

$$q_j = \sum_{x \in H} p_j(x) - \sum_{x \notin H} p_j(x), \qquad j = 0, 1,$$

we have

$$k_\varphi = 1 - \frac{q_0 - q_1}{2(1 - q_1)}$$

and

$$\varphi_H(p_0, p_1, g) = \frac{1 - (q_0 + q_1)/2 + (-1)^g (q_0 - q_1)/2}{m(1 - (q_0 + q_1)/2)}.$$

EXAMPLE 9. Let $G = G_m \times G_{m'}$, where G_m and $G_{m'}$ are the groups of residues modulo m and m', respectively, $H = \{(0, x), x \in G_{m'}\}$, and $p_0 = \omega$, $p_1 = \chi_{(1,1)}$. Then for $\tilde{\gamma} = (\gamma, \gamma')$ the characteristic function given by (3) is equal to

$$\hat{\varphi}_H(p_0, p_1, \tilde{\gamma}) = \begin{cases} -\dfrac{2\varepsilon^\gamma}{(m+1)(1 - \varepsilon^\gamma)}, & \gamma \neq 0, \quad \gamma' = 0, \\ 0, & \gamma' \neq 0. \end{cases}$$

This characteristic function corresponds to the distribution

$$\varphi_H(p_0, p_1, \tilde{x}) = \frac{2}{m(m+1)m'} (w(x - 1, m) + 1), \qquad \tilde{x} = (x, x'),$$

whose second component is the uniform distribution and the first component coincides with the distribution from Example 4.

Let now $m = m'$, that is, $G = G_m \times G_m$, and

$$H = \{(x, x), x \in G_m\}, \qquad p_0 = \omega, \qquad p_1 = \frac{1}{2}\chi_{(1,0)} + \frac{1}{2}\chi_{(0,1)}.$$

Then $\Lambda = \{(\gamma, -\gamma), \gamma \in \Gamma_m\}$ and for $\tilde{\gamma} = (\gamma, \gamma')$ the characteristic function given by (3) is of the form

$$\hat{\varphi}_H(p_0, p_1, \tilde{\gamma}) = \begin{cases} \dfrac{1}{mk_\varphi}\left(1 + \dfrac{2\varepsilon^\gamma}{(1-\varepsilon^\gamma)^2}\right), & \gamma + \gamma' = 0, \quad \gamma \neq 0, \\ 0, & \gamma + \gamma' \neq 0. \end{cases}$$

Using (8), we find that

$$k_\varphi = \frac{m}{6} + \frac{5}{6m}$$

and for $\tilde{x} = (x, x')$,

$$\varphi_H(p_0, p_1, \tilde{x}) = \frac{1}{m^2(1 + 5/m^2)}\left(\frac{6}{m}w(x - x', m)\left(1 - \frac{1}{m}w(x - x', m)\right) + \frac{1}{m}\delta_0(x - x')\right).$$

Since this distribution depends on the difference $(x - x')$ (mod m), it is conditionally uniform on the cosets of G with respect to H and in a sense it is a discrete analogue of the distribution from the class of beta distributions on the unit square.

REFERENCES

1. F. Spitzer, *Principles of Random Walk.* Van Norstand, Princeton, 1964.
2. V. I. Varshavskii, *Collective Behaviour of Automata.* Nauka, Moscow, 1972.
3. P. Moran, *The Stochastic Processes of Evolutionary Theory.* Oxford Univ. Press, London, 1962.
4. A. B. Montroll, Statistics of lattices. In: *Applied Combinatorics.* Mir, Moscow, 1968, 9–60.
5. P. R. Halmos, *A Hilbert Space Problem Book.* Van Nostrand, Princeton, 1965.
6. T. Hoholdt and H. Jensen, Determination of the merit factor of Legendre sequences. *IEEE Trans. Inf. Theory* (1988) **34**, 161–164.
7. G. J. Székely, *Paradoxes in Probability and Mathematical Statistics.* Acad. Kiado, Budapest, 1986.

Probabilistic Methods in Discrete Mathematics, pp. 333–342
V. F. Kolchin et al. (Eds.)

THE NUMBER OF SOLUTIONS OF SYSTEMS OF RANDOM MONOMIAL AND BINOMIAL LINEAR EQUATIONS

A. V. SHAPOVALOV

Moscow

Abstract — We obtain the limit distributions of the numbers of solutions of systems of random monomial equations over a finite ring and of binomial linear Boolean equations. Previously these problems were considered by Balakin and Kolchin for a priori solvable random systems with the limit Poisson distribution of the number of isolated vertices in the associated graphs. We study the case where the random graphs associated with random and a priori solvable random systems of binomial linear equations are at a subcritical stage of their evolution.

1. INTRODUCTION

The paper is devoted to random systems of monomial equations of the form

$$x_{i_k} = a_k, \qquad k = 1, \ldots, M,$$

over the finite ring $\mathbf{Z}/z$ and random systems of binomial Boolean equations of the form

$$x_{i_k} \oplus x_{j_k} = a_k, \qquad k = 1, \ldots, M.$$

The symbol $\oplus$ means the addition modulo 2, $z \geq 2$ is a natural number. The random systems under consideration depend on n unknowns and consist of M equations. Throughout the paper we consider all asymptotic formulae under condition that $n \to \infty$ and $M = M(n)$ is a function of n.

We introduce some definitions (see [2, 6, 12, 14]).

A random hypergraph (as well as a random system of equations) is a random element taking values from some set of hypergraphs (systems of equations) according to some probability measure. We use the term a random system of equations for a random system with equations independent of each other and with independent the right-hand and left-hand sides of equations [2, 6, 7, 14]. A priori solvable random system of equations is a random system of equations whose right-hand side is a result of substitution of some fixed vector into the left-hand side [2, 6]. Each realization of such a system has at least one solution.

We say that almost all realizations of a random hypergraph possess property Q if the probability $P(Q)$ that the random hypergraph possesses the property Q is such that $P(Q) = 1+o(1)$ (see, for example [9]). We say that almost all realizations of some system S_0 of random equations are consistent or inconsistent if the probability of consistency $P_c(S_0)$ of system S_0 is such that $P_c(S_0) = 1 + o(1)$ or $P_c(S_0) = o(1)$ respectively. The additional notations of graph theory used below can be found in [12] and [14].

The probability of consistency and the number of solutions of random systems of linear equations with the same fixed number of unknowns in each equation are studied in [1–7, 9, 11, 14].

Asymptotic estimates of the probability of consistency of a random system and of the probability of uniqueness of solution of a priori consistent random system with one unknown in each equation are obtained in [6] with the use of results on of random allocations of particles [8].

The uniqueness of solution of a priori solvable random system of binomial Boolean linear equations is determined by connectedness of the random graph associated with the random system. The limiting result is the same for sampling unknowns in each equation with replacement and without replacement [5, 6, 11]. Such systems are studied in [1] and [3] provided that the right-hand sides of equations are equal only to one or only to zero.

The probability of consistency of random binomial systems of linear equations is determined by the number of cycles of a specific form in the associated random graph with marked edges [6, 14]. If the sampling of unknowns is equiprobable, then for $M = o(n)$, almost all realizations of random system are consistent and for $M \sim cn$, $c \geq 1/2$, almost all realizations are inconsistent. If $M \sim cn$, $0 < c < 1/2$, then the probability of consistency decreases to zero with growth of c from zero to 1/2. The probability of consistency for sampling with replacement [6] and sampling without replacement [13] differs only by a factor defined by the distribution of the number of loops in the associated random graph.

In [4] and [9], estimates for the mean value of the random variable η which is the number of linearly dependent equations in a random system of equations over simple finite field with $m \geq 3$ unknowns in each equation and sampling with replacement are obtained. For each m, there exists a constant c_m such that $\mathbf{E}\eta = o(1)$ for $M \sim cn$, $0 < c < c_m$, and $\mathbf{E}\eta \to \infty$ for $M \sim cn$, $c > c_m$. Thus, almost all realizations of the random system are consistent if $M \sim cn$, $0 < c < c_m$.

In this paper, we obtain limit theorems for the number of solutions of random linear systems with one or two unknowns in each of the equations which are obtained by equiprobable sampling without replacement. In Section 2 the complete variety of the limit distributions is obtained for the numbers of solutions of a priori solvable random system of monomial equations and for a random monomial system of equations with the non-equiprobable right-hand side over the finite ring $\mathbf{Z}/z$. Section 3 is devoted to the systems of binomial linear Boolean equations. We study the numbers of solutions of a priori solvable random system and of a random system with non-equiprobable right-hand side. We consider the case of sampling without replacement. The number of equations in the systems is $M = o(n)$ and $M \sim cn$, $0 < c < 1/2$. This case corresponds to the subcritical stage of evolution of the associated random graphs [6, 13, 14].

All results are obtained on the basis of the investigations of structural properties of random hypergraphs fulfilled in [13, 14] and on the results on the equiprobable scheme of allocating particles [8]. Under the conditions given above the structure of the random graphs associated with the random systems is rather simple, so that we can obtain the limit distributions of the number of components which determine the distributions of the numbers of solutions of the random systems under consideration.

We denote the multivariate Poisson distribution with independent components and mean values $\lambda_1, \dots, \lambda_k$ by $P(\lambda_1, \dots, \lambda_k)$.

2. RANDOM SYSTEMS OF MONOMIAL EQUATIONS

We consider a random system of equations S_1 over the finite ring $\mathbf{Z}/z$ with unknowns $x_1, x_2, \dots, x_n$ of the form

$$x_{s_i} = a_i, \qquad i = 1, \dots, M.$$

where the indices $s_1, \dots, s_M$ are independent identically distributed random variables taking the values $1, 2, \dots, n$ with equal probabilities, the right-hand sides $a_1, \dots, a_M$ are independent identically distributed random variables independent of the left-hand side, and taking the values $0, 1, \dots, z-1$ with probabilities $p_0, p_1, \dots, p_{z-1}$. Here $z \geq 2$ and $p_0, p_1, \dots, p_{z-1}$ are constants not equal to one. Note that the right-hand sides of the system contain α_j values of $j, j = 0, 1, \dots, z-1$, with probability equal to $p_0^{\alpha_0} \dots p_{z-1}^{\alpha_{z-1}} n^{-M}$.

Let us define a priori solvable system of equations $S_1^{(c)}$. It is similar to the random system of equations S_1, but its right-hand sides depend on the left-hand side and are the result of the substitution of some vector $b^0 = (b_1^0, \dots, b_n^0)$ with the components from $\mathbf{Z}/z$ into the left-hand side, so that

$$a_i = b_{s_i}^0, \qquad i = 1, \dots, M.$$

All realizations of the random system $S_1^{(c)}$ are equiprobable.

The case $p_i = 1$ for some $i \leq z-1$ is excluded from the definition of the random system S_1. If this case is realized, then the random system S_1 is a priori solvable random system $S_1^{(c)}$ with $b^0 = (i, \dots, i)$.

The one-to-one correspondence between the random systems of equations considered in this paper and random graphs is described in [3, 6], and [14].

The random hypergraph $MG(S_1)$ with n vertices and M ordered edges corresponds to random system S_1. Realizations of this hypergraph are 1-graphs with marks of edges belonging to the set $\{0, 1, \dots, z-1\}$. The probability of each hypergraph H with α_j marks $j, j = 0, 1, \dots, z-1$, is equal to $n^{-M}\varphi(H)$, where

$$\varphi(H) = p_0 \dots p_{z-1}$$

is the product of the probabilities of marks of edges of the hypergraph H (see [14]).

It follows from results of [8] and [14] that for $M = o(\sqrt{n})$ almost all realizations of $MG(S_1)$ consist of vertices of degrees zero or one. For $M \sim c\sqrt{n}$ the vertices of degree two appear. The number of these vertices has the Poisson limit distribution. Almost all realizations of the random hypergraph have no vertices of degrees three and more. Let ζ_1 be the number of solutions of the random system S_1 and ξ_0, ξ_1 be the numbers of vertices of $MG(S_1)$ with two equal and two different marks respectively.

The weak convergence of the probability distribution of the bivariate random variable (ξ_0, ξ_1) to the bivariate Poisson distribution with independent components follows from Theorem 2.1 in [14].

THEOREM 1. *If*

$$M \sim c\sqrt{n}, \tag{1}$$

and c is a positive constant, then the joint distribution of (ξ_0, ξ_1) *converges to* $P(\lambda_0, \lambda_1)$, *where*

$$\lambda_0 = \frac{c^2}{2}\sum_{i=0}^{z-1} p_i^2, \qquad \lambda_1 = \frac{c^2}{2} - \lambda_0.$$

The structure of the random hypergraph $MG(S_1)$ described above leads to the following assertion.

THEOREM 2. *If condition* (1) *is true, then*

$$\lim_{n\to\infty} \mathbf{P}\{\zeta_1 - z^{n-M+\xi_0} = 0\} = \lim_{n\to\infty} \mathbf{P}\{\xi_1 = 0\}$$

and

$$\lim_{n\to\infty} \mathbf{P}\{\zeta_1 = 0\} = \lim_{n\to\infty} \mathbf{P}\{\xi_1 \geq 1\}.$$

By the use of Theorems 1 and 2 we obtain the limit distribution of the number of solutions of the random systems of equations S_1.

THEOREM 3. *If* $M = o(n)$, *then*

$$\lim_{n\to\infty} \mathbf{P}\{\zeta_1 - z^{n-M} = 0\} = 1.$$

If condition (1) *is true, then*

$$\lim_{n\to\infty} \mathbf{P}\{\zeta_1 - z^{n-M+k} = 0\} = \frac{1}{k!}\left(\frac{c^2}{2}\sum_{i=0}^{z-1} p_i^2\right)^k \exp\left\{\frac{c^2}{2}\right\}, \qquad k = 0, 1, \ldots,$$

and

$$\lim_{n\to\infty} \mathbf{P}\{\zeta_1 = 0\} = 1 - \exp\left\{\frac{c^2}{2}\left(\sum_{i=0}^{z-1} p_i^2 - 1\right)\right\}.$$

If $M/\sqrt{n} \to \infty$, *then*

$$\mathbf{P}\{\zeta_1 = 0\} = 1.$$

The asymptotic behaviour of the probability of consistency of the random system of equations S_1 was obtained by Kolchin in [6] for $z = 2$ and $p_0 = p_1 = 1/2$.

Let $\zeta_1^{(c)}$ be the number of solutions of a priori solvable random system of equations $S_1^{(c)}$. The following obvious formula makes it possible to obtain the limit distributions of the random variable $\zeta_1^{(c)}$:

$$\zeta_1^{(c)} = z^{\mu_0(M,n)}, \tag{2}$$

where $\mu_0(M, n)$ is the number of empty cells in the equiprobable scheme of allocating M particles to n cells [8]. Hence, the probability distributions of the random variable $\zeta_1^{(c)}$ are determined by the probability distributions of the random variable $\mu_0(M, n)$. In particular, the limit distribution of the number of solutions of the random system of equations $S_1^{(c)}$ for $M \sim c\sqrt{n}$ can be obtained from the results about the limit distribution of the number of empty cells in the left 2-domain in the scheme of equiprobable allocations [8]. This result can be proved also with the use of Theorem 1.

THEOREM 4. *Under condition* (1)

$$\lim_{n\to\infty} \mathbf{P}\{\zeta_1^{(c)} - z^{n-M+k} = 0\} = \frac{(c^2/2)^k}{k!} \exp\{-c^2/2\}, \qquad k = 0, 1, 2, \ldots,$$

If $M = o(\sqrt{n})$, *then*

$$\lim_{n\to\infty} \mathbf{P}\{\zeta_1^{(c)} - z^{n-M}\} = 1.$$

We prove the following theorem.

THEOREM 5. *Let* $n \to \infty$ *and one of the following three conditions be satisfied:*

(1) $c_1 \le M/n \le c_2$, *where* c_1 *and* c_2 *are positive constants,*

(2) $M/\sqrt{n} \to \infty$ *and* $M/n \to 0$,

(3) $M/n \to \infty$ *and* $M = n \log n + xn + o(n)$, *where* $x \to -\infty$.

Then

$$\lim_{n\to\infty} \mathbf{P}\left\{ \frac{\log_z \zeta_1^{(c)} - ne^{-c}}{\sqrt{ne^{-c}(1-(1+c)e^{-c})}} \le u \right\} = \frac{1}{\sqrt{2\pi}} \int_{-\infty}^{u} e^{-u^2/2} du.$$

PROOF. It follows from (2) that

$$\frac{\mu_0(M,n) - \mathbf{E}\mu_0(M,n)}{\sqrt{\mathbf{D}\mu_0(M,n)}} = \frac{\log_z \zeta_1^{(c)} - \mathbf{E}\mu_0(M,n)}{\sqrt{\mathbf{D}\mu_0(M,n)}}.$$

The conditions of the theorem define, respectively, the central, the left-hand intermediate and the right-hand intermediate domains in the scheme of random allocations respectively [8]. Under the assumptions of the theorem the limit distribution of the random variable in the left-hand side of the last equality is the standard normal distribution. Hence, the limit distribution of the random variable in the right-hand side of the equation is the standard normal distribution as well. Using the formulas [8] for the mean value and the variance of the random variable $\mu_0(M, n)$, we complete the proof of the theorem.

The remaining domain of the parameter M is

$$M = n \log n + xn + o(n), \tag{3}$$

where x is a constant or $x \to \infty$ as $n \to \infty$.

The case $z = 2$ was considered by Kolchin in [6]. If x is a constant, then using the reasoning from [6] or (2) and [5], it is easy to see that for any fixed $k = 0, 1, \ldots,$

$$\lim \mathbf{P}\{\zeta_1^{(c)} = z^k\} = \frac{e^{-kx}}{k!} \exp\{-e^{-x}\}.$$

Evidently, under condition (3) if $n \to \infty$ and $x \to \infty$, then

$$\lim_{n\to\infty} \mathbf{P}\{\zeta_1^{(c)} = 1\} = 1.$$

3. RANDOM SYSTEMS OF LINEAR BINOMIAL EQUATIONS

In this section we consider random systems of binomial linear Boolean equations with unknowns $x_1, \ldots, x_n$. Let the realizations of the random system S_2 be of the form

$$x_{i_k} \oplus x_{j_k} = a_k, \qquad k = 1, \ldots, M,$$

where $a_1, \ldots, a_M$ take the value 0 and 1. The equations of the system are independent, the right-hand sides are independent from left-hand side and take the value 0 and 1 with probabilities p_0 and p_1, where $p_1 > 0$. The choice of indices of unknowns from the set $\{1, \ldots, n\}$ for each equation is random and carried out by equiprobable sampling without replacement, i.e., the probability to choose any pair (i_k, j_k) is equal to $(n(n-1)/2)^{-1}$

The a priori solvable random system of equations $S_2^{(c)}$ is defined in a similar way, but the right-hand sides of equations depend on the left-hand sides and are the result of substitution of some Boolean vector $b^1 = (b_1^1, \ldots, b_n^1)$ into the left-hand side of the system. All realizations of the random system $S_2^{(c)}$ have equal probabilities.

Defining the random system S_2 we excluded the case $p_0 = 1$ from consideration. In this case the system S_2 is a priori solvable random system of equations $S_2^{(c)}$ with $b^1 = (0, \ldots, 0)$.

A random graph $MG(S_2)$ with n vertices and M ordered edges corresponds to the random system S_2 (see [6, 7, 14]). The edges of realizations of this hypergraph can be multiple. The vertices of edges are unordered, loops are absent. Every edge has the mark 0 or 1, which is equal to the value of the right-hand side of the corresponding equation. The probability that $MG(S_2)$ is equal to a graph H is

$$\mathbf{P}\{MG(S_2) = H\} = \binom{n}{2}^{-M} \varphi(H),$$

where $\varphi(H)$ is the product of the probabilities of marks of the edges of H. The random graph $MG(S_2^{(c)})$ is associated to the random system $S_2^{(c)}$in a similar way. It differs only by absence of marks of the edges. The probability of any realization of $MG(S_2^{(c)})$ is equal to $(n(n-1)/2)^{-M}$. It is easy to see that in other notation the random graph $MG(S_2^{(c)})$ is the random graph $MG_{n,M}^{(2)}$ which is a special case of the random graph $MG_{n,M}^{(m)}$ with $m = 2$ studied in [12, 14].

Let us introduce an equivalence relation on the realizations of $MG(S_2)$. Two graphs are equivalent if and only if they differ only by the marks of their edges. The probability of an equivalence class is equal to the sum of probabilities of realizations included in this class. In such a way we obtain the above-mentioned random graph $MG_{n,M}^{(2)}$. Hence, the topological structures of the random graphs $MG_{n,M}^{(2)}$ and $MG(S_2)$ are the same.

Let $\zeta_2^{(c)}$ be the number of solutions of the random system $S_2^{(c)}$, and $\psi_{n,M}$ is the number of components of the random graphs $MG(S_2^{(c)})$ and $MG(S_2)$.

THEOREM 6. *If* $M = o(n)$, *then*

$$\lim_{n\to\infty} \mathbf{P}\{\zeta_2^{(c)} - 2^{n-M} = 0\} = 1.$$

If

$$M \sim cn, \qquad 0 < c < 1/2, \tag{4}$$

where c *is a constant, then for any fixed* $k = 0, 1, 2, \ldots$

$$\mathbf{P}\{\zeta_2^{(c)} - 2^{n-M+k} = 0\} = \frac{e^c\sqrt{1-2c}}{k!}\left(-\frac{1}{2}\log(1-2c) - c\right)^k.$$

PROOF. It is easy to see that

$$\zeta_2^{(c)} = 2^{\psi_{n,M}}.$$

Therefore the assertion of the theorem follows from Theorem 3.4 in [12], where the limit distribution of the random variable $\psi_{n,M}$ under condition (4) is given.

The limit distribution of the number of solutions of the random system S_2 can be obtained in a similar way, but it is necessary to take into account that some realizations of S_2 have no solution. In accordance with [6] and [14] the realizations of S_2 can have contradictory subsystems associated only to cycles with odd sums of the marks of edges. We call such cycles contradictory and all other cycles non-contradictory. Let ζ_2 be the number of solutions of the system S_2 and let $C^{(1)}$, $C^{(0)}$ be the numbers of contradictory and non-contradictory cycles, respectively, in $MG(S_2)$. Hence $C^{(1)}$ and $C^{(0)}$ are the numbers of all cycles with odd and even sums of the marks of edges, respectively, in $MG(S_2)$.

THEOREM 7. *Under condition* (4)

$$\lim_{n\to\infty} \mathbf{P}\{\zeta_2 - 2^{n-M+C^{(0)}} = 0\} = \lim_{n\to\infty} \mathbf{P}\{C^{(1)} = 0\},$$

$$\lim_{n\to\infty} \mathbf{P}\{\zeta_2 = 0\} = \lim_{n\to\infty} \mathbf{P}\{C^{(1)} > 0\}.$$

PROOF. If a realization of the random system has at least one contradictory cycle, then this realization is contradictory and it has no solutions. This proves the second assertion of the theorem. If a realization of the random system has no contradictory cycles, then the number of its solutions is determined by the number of components of the associated graph. If the associated graph has s components, then this realization of the random system has 2^s solutions. Therefore under the condition $\{C^{(1)} = 0\}$ the formula $\zeta_2 = 2^{\psi_{n,M}}$ is valid. Thus, the first assertion of the theorem follows from formula (3.3) and Theorem 3.3 in [13].

Now we study the joint distribution of the numbers of contradictory and non-contradictory cycles. Let $C_l^{(1)}$ and $C_l^{(0)}$ be the numbers of cycles of length l with odd and even sums of the marks, respectively, in $MG(S_2)$. The following theorem generalizes formula (4.1) in the proof of Theorem 4.12 in [14].

THEOREM 8. *Let $l_1, \dots, l_s$ and $k_1, \dots, k_t$ be two sets each of which consists of distinct natural numbers no less than 2.*

Then under condition (4) *the joint distribution of*

$$(C_{l_1}^{(1)}, \dots, C_{l_s}^{(1)}, C_{k_1}^{(0)}, \dots, C_{k_t}^{(0)})$$

converges to the Poisson distribution

$$P(\lambda_{l_1}^{(1)}, \dots, \lambda_{l_s}^{(1)}, \lambda_{k_1}^{(0)}, \dots, \lambda_{k_t}^{(0)}),$$

where

$$\lambda_{l_i}^{(1)} = \frac{1}{4l_i}(2c)^{l_i}(1 - (p_0 - p_1)^{l_i}), \qquad i = 1, \dots, s,$$

$$\lambda_{k_j}^{(0)} = \frac{1}{4k_j}(2c)^{k_j}(1 + (p_0 - p_1)^{k_j}), \qquad j = 1, \dots, t.$$

PROOF. The convergence of the joint distribution of the multivariate random variable to the multivariate Poisson distribution follows from Theorem 3.6 in [14]. The estimates of the mean values of the random variables $C_{l_i}^{(1)}$ are obtained in the proof of Theorem 4.12 in [14]. The estimates of the mean values of the random variables $C_{l_i}^{(1)}$ follow from the relation

$$\lambda_k^{(0)} + \lambda_k^{(1)} = \frac{(2c)^k}{2k},$$

where $(2c)^k/(2k)$ is the limit of the mean value of the number of all cycles of length k in $MG(S_2)$ and $MG_{n,M}^{(2)}$ (see Theorem 2.3 in [13]).

By means of Theorem 8 and the bivariate version of Lemma 3 in [10], we obtain the joint limit distribution of the total numbers of contradictory and non-contradictory cycles of all lengths in the random graph $MG(S_2)$.

THEOREM 9. *Under condition* (4) *the joint distribution of* $(C^{(1)}, C^{(0)})$ *converges to* $P(\lambda^{(1)}, \lambda^{(0)})$, *where*

$$\lambda^{(1)} = -\frac{1}{4}\log\left(\frac{1-2c}{1-2c(p_0-p_1)}\right) - cp_1,$$

$$\lambda^{(0)} = -\frac{1}{4}\log((1-2c)(1-2c(p_0-p_1))) - cp_0.$$

The limit distribution of the number of solutions of the random system of equations is easily obtained from Theorems 7 and 9.

THEOREM 10. *Under condition* (4) *for any fixed* $k = 0, 1, 2, \ldots$

$$\lim_{n\to\infty} \mathbf{P}\{\zeta_2 - 2^{n-M+k} = 0) = \frac{\sqrt{1-2c}}{4^k k!} e^c(-\log((1-2c)(1-2c(p_0-p_1))) - cp_0)^k,$$

$$\lim_{n\to\infty} \mathbf{P}\{\zeta_2 = 0\} = 1 - \left(\frac{1-2c}{1-2c(p_0-p_1)}\right)^{1/4} e^{cp_1}.$$

If $M = o(n)$, *then*

$$\lim_{n\to\infty} \mathbf{P}\{\zeta_2 - 2^{n-M} = 0\} = 1.$$

If $\lim_{n\to\infty} M/n \geq 1/2$, *then*

$$\lim_{n\to\infty} \mathbf{P}\{\zeta_2 = 0\} = 1.$$

Note that Theorem 4.1 of [14] follows from Theorem 10. By a different approach the probability of consistency or $\mathbf{P}\{\zeta_2 = 0\}$ was obtained in [7].

REFERENCES

1. G. V. Balakin, A priori solvable systems of random equations over a finite field. In: *Abstr. First Petrozavodsk Conference on Probabilistic Methods in Discrete Mathematics*. Karelian Branch Acad. Sci. USSR, Petrozavodsk, 1973, pp. 8–10 (in Russian).
2. G. V. Balakin, On the number of solutions of systems of pseudo-Boolean random equations. In: *Probabilistic Methods in Discrete Mathematics, Proc. Third Intern. Petrozavodsk Conference*. VSP/TVP, Utrecht/Moscow, 1993, pp. 71–98.
3. G. V. Balakin, Graphs of systems of two-term equations with Boolean unknowns. *Theory Probab. Appl.* (1995) **4**, 241–259 (in Russian).
4. G. V. Balakin, V. F. Kolchin, and V. I. Khokhlov, Hypercycles in random hypergraph. *Discrete Math. Appl.* (1992) **2**, 563–570.
5. G. I. Ivchenko and Yu. I. Medvedev, On the probability of connectedness of a class of random graph. In: *Proc. Seminar on Combinatorial Math.*, Moscow Univ. Press, Moscow, 1973, pp. 164–185 (in Russian).
6. V. F. Kolchin, *Systems of Random Equations*. MIEM, Moscow, 1988 (in Russian).
7. V. F. Kolchin, Consistency of a system of random congruences. *Discrete Math. Appl.* (1993) **3**, 103–113.
8. V. F. Kolchin, B. A. Sevastyanov, and V. P. Chistyakov, *Random Allocations*. Wiley, New York, 1978.

9. V. F. Kolchin and V. I. Khokhlov, A threshold effect for systems of random equations of a special form. *Discrete Math. Appl.* (1995) **5**, 425–436.

10. V. F. Kolchin and V. I. Khokhlov, On the number of cycles in a random non-equiprobable graph. *Discrete Math. Appl.* (1992) **2**, 109–118.

11. V. A. Kopytzev, On the distribution of the number of solutions of a priori solvable systems of equations. *Theory Probab. Appl.* (1995) **4**, 430–437 (in Russian).

12. A. V. Shapovalov, On the connectedness and threshold functions of subgraphs of regular random hypergraphs. *Diskretnaya Matematika* (1993) **5**, No. 3, 105–115 (in Russian).

13. A. V. Shapovalov, On the number of strictly balanced subgraphs of regular random hypergraphs. *Diskretnaya Matematika* (1993) **5**, No. 4, 131–142 (in Russian).

14. A. V. Shapovalov, The probability of consistency of systems of random Boolean equations. *Discrete Math. Appl.* (1995) **5**, 173–185.

Probabilistic Methods in Discrete Mathematics, pp. 343–347
V. F. Kolchin et al. (Eds.)

ON A CONDITION OF EXISTENCE OF INTEGER-VALUED RANDOM VARIABLES WITH GIVEN TWO MOMENTS

S. YA. SHORGIN

Moscow

Abstract — We consider conditions of existence of integer-valued random variables with given mathematical expectation and variance.

In analysis of different counting processes in actual practice with restricted information on their distributions, the problem of development of sufficiently simple and natural models of such processes arises. For example, when risk processes in actuarial mathematics are analyzed,we often know only several first moments of the number of insurance contracts which are concluded during certain time or of the number of claims occurring during some period. Analogous problems may arise in queueing theory (in particular, when complex queueing networks are modelled) and in all cases when the arrangement of certain object in space or time is investigated (biology, nuclear physics and so on).

In connection with the aforesaid, the problem of development of such models which guarantee, for example, given values of two first moments has both theoretical and practical significance. We consider the following probabilistic problem.

Let Λ and M be non-negative numbers. The question is whether an integer-valued random variable N exist whose two first moments are

$$\mathbf{E}N = \Lambda, \tag{1}$$

$$\mathbf{D}N = M^2, \tag{2}$$

First of all we give a necessary condition for existence of such random variables.

THEOREM 1. *If an integer-valued random variable N satisfies* (1)–(2), *then Λ and M satisfy the inequality*

$$M^2 \geq \{\Lambda\}(1 - \{\Lambda\}), \tag{3}$$

where $\{a\}$ is the fractional part of a.

PROOF. If Λ is an integer, then the assertion of the theorem is trivial.

Let Λ be not an integral number, that is, $\Lambda = k + h$, where $k = [\Lambda]$ and $h = \{\Lambda\}$ are the integral part and the fractional parts of Λ respectively. Let us consider the random variable $Z = N - k$. Obviously, $\mathbf{E}Z = h$ and $\mathbf{D}Z = M^2$. Thus,

$$\mathbf{E}Z^2 = \sum_i i^2 \mathbf{P}\{Z = i\} \geq \sum_i i\mathbf{P}\{Z = i\} = h$$

since $i^2 \geq i$ for all positive and negative integers i. Therefore

$$M^2 = \mathbf{D}Z = \mathbf{E}Z^2 - h^2 \geq h - h^2.$$

The theorem is thus proved.

It can be proved that condition (3) is both necessary and sufficient condition for existence of an integer-valued random variable N satisfying (1)-(2). Moreover, if (3) is fulfilled and $\Lambda > 0$, then examples of such random variables N can be found within the following classes of random variables.

The distribution of the number of successes in N' Bernoulli trials with different probabilities of success (we denote by s_i the probability of success in the ith trial) is called the generalized binomial distribution with parameters $(N'; s_1, \ldots. s_{N'})$. The class of all random variables which have the generalized binomial distribution is denoted by K_1.

Let K_2 be the class of all random variables with the generalized Poisson distribution, i.e., the class of random variables with the generating function

$$\exp\{u(f(z) - 1\}, \qquad u > 0,$$

where u is a parameter and $f(z)$ is a generating function of some non-negative integer-valued random variable.

Note that the class K_1 is the class of random variables whose distributions coincide with one-dimensional distributions of superpositions of counting processes each of which includes no more than one event. In the example concerning the risk process, this class obviously corresponds to the case where there exists a bounded number of possible (potential) insureds (clients of an insurer), each of this possible insureds may conclude (with certain probability) an insurance contract with the insurer during some period, and may not conclude the contract (with the supplementary probability). This model may be interpreted analogously for other applications. Thus the random variables from the class K_1 provide a sufficiently natural model for the processes discussed in the beginning of the paper.

The class K_2 is the class of random variables whose distributions coincide with the one-dimensional distributions of infinitely divisible counting processes; this is a quite natural model of the processes in actual practice as well.

THEOREM 2. *Let condition* (3) *be satisfied. In this case,*
if $M^2 < \Lambda$, *then there exists a random variable* $N \in K_1$ *which satisfies* (1)–(2);

if $M^2 \geq \Lambda > 0$, *then there exists a random variable* $N \in K_2$ *which satisfies* (1)–(2);
if $\Lambda = 0$, *then there exists a symmetric random variable* $N \in K_1$ *which satisfies* (1)–(2).

Note that the case $\Lambda = 0$ is considered for the sake of completeness, since this case is of no practical importance in comparison with the case $\Lambda > 0$.

Before proving the theorem, we have to prove the following simple assertion.

LEMMA 1. *Let* $a > 0$ *and* $b > 0$, *and let* n *be an arbitrary positive integer, then the system of equations*

$$x_1 + \ldots + x_n = a, \tag{4}$$

$$x_1^2 + \ldots + x_n^2 = b^2 \tag{5}$$

has at least one non-negative solution if and only if

$$a^2/n \leq b^2 \leq a^2. \tag{6}$$

PROOF. Obviously, for any non-negative $x_1, \ldots, x_n$

$$\frac{1}{n}(x_1 + \ldots + x_n)^2 \leq x_1^2 + \ldots + x_n^2 \leq (x_1 + \ldots + x_n)^2.$$

Hence it follows that if system (4)–(5) has a non-negative solution, then inequality (6) holds.

Let us suppose that (6) is satisfied, and consider the continuous function of n variables

$$f(x_1, \ldots, x_n) = (x_1 + \ldots + x_n)^2$$

on the connected set

$$A = \{x_1, \ldots, x_n : x_1 \geq 0, \ldots, x_n \geq 0,\ x_1 + \ldots + x_n = a\}.$$

The function f has its maximum a^2 over A at the point $\{x_1 = a, x_2 = \ldots = x_n = 0\}$, and its minimum a^2/n at the point $\{x_1 = \ldots = x_n = a/n\}$, therefore by virtue of the theorem on intermediate values of a continuous function, the function f passes all intermediate values on the set A including the value b^2. The lemma is proved.

PROOF OF THEOREM 2. Suppose that $M^2 < \Lambda$. We begin first with the case where $M^2 < \{\Lambda\}$. By virtue of (3), we have

$$M^2 \geq \{\Lambda\}(1 - \{\Lambda\}).$$

Let Λ be represented in the form $\Lambda = k + h$, where $k = [\Lambda], h = \{\Lambda\}$. Let n be an arbitrary integer such that

$$n > \frac{h^2}{h - M^2},$$

and let N be a random variable which has the generalized binomial distribution with parameters $(N'; s_1, \ldots, s_{N'})$, where $N' = n + k$ and $s_{n+1}, \ldots. s_{N'} = 1$. Thus the random variable N is represented as $N = S + k$, where S has the generalized binomial distribution with parameters $(n; s_1, \ldots, s_n)$. It is clear that

$$\mathbf{E}S = h, \qquad \mathbf{D}S = M^2.$$

The parameters $s_1, \ldots, s_n$ satisfy the system of equations

$$s_1 + \ldots + s_n = h, \tag{7}$$

$$s_1^2 + \ldots + s_n^2 = h - M^2. \tag{8}$$

It is obvious that

$$h^2/n \le h - M^2 \le h^2,$$

thus, due to Lemma 1 there exist non-negative $s_1, \ldots, s_n$ which are the solution of (7)–(8). Since $h < 1$, we see that $s_i < 1$, $i = 1, \ldots, n$. Thus, a random variable N with the generalized binomial distribution and the given two moments is constructed.

Let now $M^2 \ge \{\Lambda\}$. By virtue of the condition $M^2 < \Lambda$, we have $\{\Lambda\} < \Lambda$, hence $\Lambda \ge 1$.

We set $\Lambda = k + h$, where $k = [\Lambda] \ge 1$ and $h = \{\Lambda\}$. Let l be the minimal integer which satisfies the condition $M^2 < l + h$. It is clear that $1 \le l \le k$. Let N be a random variable which has the generalized binomial distribution with parameters $(N'; s_1, \ldots, s_{N'})$. We set

$$N' = 2n + k - l,$$

$$s_1 = \ldots = s_n = s',$$

$$s_{n+1} = \ldots = s_{2n} = s'',$$

$$s_{2n+1} = \ldots = s_{2n+k-l} = 1$$

(the integer n will be determined below, and we will show that the numbers s' and s'' with necessary properties exist). Thus, the random variable N is represented as $N = S + k - l$, where S has the generalized binomial distribution with parameters $(2n; s_1, \ldots, s_{2n})$. It is clear that

$$\mathbf{E}S = l + h, \qquad \mathbf{D}S = M^2,$$

and $l + h > M^2$. Relations (1)–(2) are reduced to

$$s' + s'' = a, \tag{9}$$

$$(s')^2 + (s'')^2 = b^2, \tag{10}$$

where $a = (l + h)/n$, $b^2 = (l + h - M^2)/n$. We set

$$n = [(l + h)^2/(l + h - M^2)].$$

It is clear that n satisfies the conditions

$$\frac{(l+h)^2}{2(l+h-M^2)} \le n \le \frac{(l+h)^2}{l+h-M^2},$$

since

$$(l+h)^2 \geq l+h \geq l+h-M^2.$$

Hence it follows that $a^2/2 \leq b^2 \leq a^2$, and by virtue of Lemma 2, there exist non-negative s', s'' which are the solution of system (9)–(10). Since

$$l+h-M^2 \leq l \leq [(l+h)^2/(l+h-M^2)] = n,$$

we have $b \leq 1$ and hence $s' \leq 1$ and $s'' \leq 1$.

Suppose that $M^2 \geq \Lambda > 0$.

If $M^2 = \Lambda$, then we can take as N an ordinary Poisson distributed random variable with parameter Λ. This random variable of course belongs to the class K_2.

Suppose that $M^2 > \Lambda > 0$. We denote by W an arbitrary integers which is no less than M^2/Λ. Let a random variable N have the generating function $\exp(u(f(z) - 1)$, where $u > 0$ is the parameter we have to determine and $f(z)$ is the generating function of the random variable S with the distribution determined as follows:

$$S = \begin{cases} 0 & \text{with probability } 1 - p' - p'', \\ 1 & \text{with probability } p', \\ W & \text{with probability } p''. \end{cases}$$

We set

$$p' = \frac{W\Lambda - M^2}{(W-1)u}, p'' = \frac{M^2 - \Lambda}{(W^2 - W)u},$$

and choose u large enough to satisfy the condition $p' + p'' \leq 1$. It is clear that the first two moments of the random variable N satisfy (1)–(2).

Finally, suppose that $\Lambda = 0$. Then as N we can take a random variable with the symmetric distribution

$$N = \begin{cases} -W & \text{with probability } p, \\ 0 & \text{with probability } 1 - 2p, \\ W & \text{with probability } p, \end{cases}$$

where $p = M^2/(2W^2)$ and W is an arbitrary integer which is no less than M.

Theorem 2 is thus proved.

Probabilistic Methods in Discrete Mathematics, pp. 349–351
V. F. Kolchin et al. (Eds.)

ON THE PROBLEM OF OPTIMAL STACK CONTROL IN TWO-LEVEL MEMORY

A. V. SOKOLOV

Petrozavodsk

Abstract — The paper concerns the application of probabilistic methods to a problem of the theory of programming related to the construction of mathematical model and optimal algorithms of stack control in two-level memory.

This research was supported by the Russian Foundation for Basic Research, grant 95–01–00600, and by grant INTAS–93–0893.

The problem considered in this paper was originally posed by Knuth and formulated as follows. Let two stacks grow and collide inside a shared memory of volume m. In this case the location of their collision may be considered as a random variable. It is known that at each stage an element may be purged from the stack with the probability p and some information may be pushed into one of the stacks with the probability $1 - p$. Let $M(m, p)$ denote the expectation of the random variable $\max(k_1, k_2)$, where k_1 and k_2 are the heights of the stacks collided. Knuth [1] posed the problem to construct the mathematical model of this process and to find the form of the function $M(m, p)$.

In [2] the mathematical model of the process was constructed as the two-dimensional random walk in a triangle with two reflecting boundaries snd one absorbing boundary. The algorithm of computing $M(m, p)$ was proposed for the fixed parameter m. In order to solve the problem, the Markov chain theory was used.

In [3] $M(m, p)$ was studied as $m \to \infty$ and $0 < p < 0.5$.

In [4–6] the asymptotic behaviour of the stack sizes at the instant of the overflow, as well as the time till the memory overflow for $0.5 \leq p < 1$ were investigated.

In [7] the same problem was solved assuming that the probabilities of deletion and insertion of information depend on the stack sizes.

In the present paper we consider a generalization of Knuth's problem. We suppose that in the case of memory overflow the process does not terminate. Instead, swapping with the second level memory occurs, and tops of stacks in the fast memory transfer into some state which becomes the starting position of subsequent computing. Our task is to find this state. The tendency is to have as few swapping as possible but the swappinga does occur when the memory is overflowed and the tops of the stacks are emptied. Hence, we have to search for the state where the mean time till swapping would be maximal provided the process starts from this state.

Let x_1 and x_2 denote the current stack heights. Then the 2-dimensional random walk in the integer lattice space, where $x_1 \geq 0$, $x_1 + x_2 \leq m$, is used as the model. The probability of moving out of the position (x_1, x_2) to the left is $p_1(x_1)$, to the right, $p_2(x_1)$, downwards, $q_1(x_1)$, upwards, $q_2(x_2)$, and

$$p_1(x_1) + p_2(x_1) + q_1(x_2) + q_2(x_2) = 1.$$

The objective of our study is to find the state $s_0 = (x_1^{(0)}, x_2^{(0)})$ where the mean time of walk to the absorption at the lines $x_1 + x_2 = m + 1$, $x_1 = -1$, $x_2 = -1$ would be maximal, provided the process starts from the state s_0.

This problem is solved using the Markov absorbing chain theory. The algorithm of numbering the states is proposed and realized as a computer program. This algorithm makes it possible to estimate the transition matrix corresponding to the Markov chain. Since the transposition of an $m \times 4$ matrix is necessary, special methods of storing sparse matrices are involved. In the case of $m = 37$ we can solve the problem using the IBM PC with the 8 MB RAM.

Let us consider the special case of the problem where there is only a single stack. Then the problem is reduced to the search for the initial state x_0 of the 1-dimensional random walk which maximizes the expectation of the walk time till absorption at the states -1 and $+1$. The model is reduced to the gambler's ruin problem, where the common capital of the gamblers is $m + 1$ and the capital of the first gambler is $x_0 + 1$. In the case of fixed probabilities of insertion and deletion of information the analytical solution of the corresponding difference equation for the game time is well known [8], and the optimal value x_0 may be derived:

$$x_0 = \begin{cases} \dfrac{m}{2}, & p = q = \dfrac{1}{2}, \\ \log_{q/p} \dfrac{\left((q/p)^{m+2} - 1\right) p}{(m+2) q \log(q/p)}, & p \neq q. \end{cases}$$

In the case where the probabilities depend on the stack height, the calculations were carried out using MATHCAD 5.0 PLUS. If the dependence $p(x)$ is linear and it results from turning the line $p(x)$ about the point $(m/2, 1/2)$, we obtain the results given below as an illustration. Intuitively, the value x_0 seems to be equal to $m/2$, but it appears not to be so.

Let

$$p(x) = \frac{1 - 2c}{m} x + c.$$

The functionce $x_0(c)$ is studied, where c varies between 0 to 1. For example, for $m = 50$ the results of calculations are the following.

c	0	0.1	0.2	0.3	0.4	0.5	0.6	0.7	0.8	0.9	1
x_0	26	26	26	26	26	25	22	20	19	17	0

In the case of two stacks, the calculations were carried out for various kinds of dependencies between the probabilities and stack heights. In the symmetric case

$$p_1(x_1) = p_2(x_1) = q_1(x_2) = q_2(x_2) = \frac{1}{4}$$

it was found that $x_1^{(0)} \approx x_2^{(0)} \approx m/3$.

In the case of fixed probabilities the problem was considered for 3 stacks. Then we have a 3-dimensional random walk in a prism as a model.

REFERENCES

1. D. E. Knuth, *The Art of Computer Programming, Vol.* 1. Addison–Wesley, Reading, MA, 1976.
2. A. Sokolov, About storage allocation for implementing two stacks. In: *Automation of Experiment and Data Processing*. Petrozavodsk, 1980, pp. 65–71 (in Russian).
3. A. C. Yao, An analysis of a memory allocation scheme for implementing stacks. *SIAM J. Computing* (1981) **10**, 398–403.
4. P. Flajolet, The evolution of two stacks in bounded space and random walks in a triangle. In: *Lect. Notes Computer Sci.* (1986) **223**, 325–340.
5. G. Louchard and R. Schott, Probabilistic analysis of some distributed algorithms. In: *Lect. Notes Computer Sci.* (1990) **431**, 239–253.
6. G. Louchard, R. Schott, M. Tolley, and P. Zimmermann, Random walks, heat equation and distributed algorithms. *J. Comput. Appl. Math.* (1994) **53**, 243–274.
7. R. S. Maier, Colliding stacks: a large deviations analysis. *Random Structures and Algorithms* (1991) **2**, 379–421.
8. W. Feller. *An Introduction to Probability Theory and its Applications, Vol.* 1. Wiley, New York, 1964.

Probabilistic Methods in Discrete Mathematics, pp. 353–357
V. F. Kolchin et al. (Eds.)

LOCAL THEOREMS ON LARGE DEVIATIONS IN THE INVERSE ALLOCATION PROBLEM

A. N. TIMASHOV

Moscow

Abstract — In the equiprobable scheme of allocating particles to N cells, we obtain local limit theorems on large deviations for the random variable $\nu_k(N) - k - 1$, where $\nu_k(N)$ is the number of particles allocated until for the first time exactly $k+1$ cells are occupied, $k = 1, 2, \dots, N-1$.

We consider the waiting time $\nu_k(N)$ which is the number of particles allocated until exactly $k+1$ cells are occupied in the equiprobable scheme of allocating particles to N cells, $k = 1, \dots, N-1$. We write $\nu(N)$ for $\nu_{N-1}(N)$. The random variable $\nu(N)$ is the number of particles allocated until all cells are occupied for the first time. We consider the equiprobable allocation scheme where the particles are allocated independently of each other and occupy each of the N cells with equal probabilities [1]. It is clear that $\nu_k(N) \geq k+1$ and therefore $\nu_k(N) - k - 1$ can be regarded as the extra number of particles needed to fill $k+1$ cells. In [2], it is shown that for $n = 0, 1, 2, \dots$,

$$\mathbf{P}\{\nu_k(N) - k - 1 = n\} = \binom{N-1}{k} N^{-k-n} \Delta^k 0^{k+n}, \tag{1}$$

where $\Delta_k 0^{k+n}$ are Morgan numbers.

In [2], it is noted that this formula gives a possibility to analyze the local probabilities with the use of the known results on the asymptotic behaviour of Morgan numbers for the various relations between the parameters n and k and that this possibility has not been realized.

In this paper we realize the possibility pointed out in [2]. To this end, we use results on the asymptotic behaviour of Stirling numbers of the second kind $\sigma(k+n, k)$, since

$$\Delta^k 0^{k+n} = k!\, \sigma(k+n, k).$$

The relations obtained for the probabilities $\mathbf{P}\{\nu_k(N) - k - 1 = n\}$ are of the form of local limit theorems on large deviations and let us write these probabilities (up to equivalence) as the exponent function of convergent series which can be regarded as asymptotic expansions. In particular, if $k = N - 1$, we obtain the corresponding asymptotic expansions for the probability $\mathbf{P}\{\nu(N) - N = n\}$.

To obtain those relations, we use the representation

$$\mathbf{P}\{\nu_k(N) - k - 1 = n\} = \frac{(N-1)_k}{N^{k+n}} \sigma(k+n, k), \tag{2}$$

which is equivalent to (1). In (2) $(N-1)_0 = 1$ and for $k = 1, \dots, N-1$,

$$(N-1)_k = k! \binom{N-1}{k} = (N-1)(N-2)\dots(N-k).$$

THEOREM 1. *Let $n, N, k \to \infty$ in such a way that*

$$\frac{n}{k} - \ln k \to \infty, \qquad N \geq k+1.$$

Then

$$\mathbf{P}\{\nu_k(N) - k - 1 = n\} = \binom{N-1}{k} \left(\frac{k}{N}\right)^{k+n} (1 + o(1)). \tag{3}$$

PROOF. To prove (3), we use (2) and the following estimate of Stirling numbers given in [3]. If $n, k \to \infty$ in such a way that $k/n \to 0$, then

$$\sigma(k+n, k) = \frac{k^{k+n}}{k!} \exp\{-ke^{-(k+n)/k}\}(1 + o(1)). \tag{4}$$

It follows from (2) and (4) that

$$\mathbf{P}\{\nu_k(N) - k - 1 = n\} = \binom{N-1}{k} \left(\frac{k}{N}\right)^{k+n} \exp\{-ke^{-(k+n)/k}\}(1 + o(1)). \tag{5}$$

Under the conditions of the theorem,

$$\exp\{-ke^{-(k+n)/k}\} = 1 + o(1)),$$

therefore (5) implies (3).

Theorem 1 is thus proved.

COROLLARY 1. *Let the conditions of Theorem 1 be satisfied and $k/N \to 0$. Then*

$$\mathbf{P}\{\nu_k(N) - k - 1 = n\} = \frac{e^k}{\sqrt{2\pi k}} \left(\frac{k}{N}\right)^n \exp\left\{-\sum_{m=1}^{\infty} \frac{k^{m+1}}{m(m+1)N^m}\right\} (1 + o(1)).$$

PROOF. Indeed, if $k, N \to \infty$ and $k/N \to 0$, then

$$(N-1)_k = \left(1 - \frac{k}{N}\right)(N)_k = (N)_k(1 + o(1)),$$

$$(N)_k = N^k \left(1 - \frac{k}{N}\right)^{-(n-k)} e^{-k}(1 + o(1)).$$

The assertion of the corollary follows from (3) and the relation

$$\left(1-\frac{k}{N}\right)^{-(N-k)} e^{-k} = \exp\left\{-\sum_{m=1}^{\infty} \frac{k^{m+1}}{m(m+1)N^m}\right\}$$

with the use of the Stirling formula.

COROLLARY 2. *Let $n, N \to \infty$ in such a way that*

$$\frac{n}{N} - \ln N \to \infty$$

and let $k = N - s$, where s is a fixed positive integer. Then

$$\mathbf{P}\{\nu_k(N) - k - 1 = n\} = \frac{N^{s-1}}{(s-1)!} \exp\left\{-s - n\sum_{m=1}^{\infty} \frac{s^m}{mN^m}\right\}(1 + o(1)).$$

In particular, for $s = 1$,

$$\mathbf{P}\{\nu_k(N) - N = n\} = \exp\left\{-1 - n\sum_{m=1}^{\infty} \frac{1}{mN^m}\right\}(1 + o(1)).$$

Note that the series under the exp sign converge in all cases considered above, and these series are asymptotic expansions that let us replace the infinite sum of the series by a finite sum with the remainder term of the corresponding order. For example, if the conditions of Corollary 2 are satisfied and $n/N^2 \to 0$, then

$$\mathbf{P}\{\nu_k(N) - k - 1 = n\} = \frac{N^{s-1}}{(s-1)!} \exp\left\{-s - \frac{ns}{N}\right\}(1 + o(1)).$$

More generally, if under the same conditions $n/N^{M+1} \to 0$, where $M = 1, 2, \ldots$, then

$$\mathbf{P}\{\nu_k(N) - k - 1 = n\} = \frac{N^{s-1}}{(s-1)!} \exp\left\{-s - n\sum_{m=1}^{M} \frac{s^m}{mN^m}\right\}(1 + o(1)).$$

Similarly, if the conditions of Corollary 1 are satisfied and $k/N(M+1)/(M+2) \to 0$, where $M = 1, 2, \ldots$, then

$$\mathbf{P}\{\nu_k(N) - k - 1 = n\} = \frac{e^k}{\sqrt{2\pi k}} \left(\frac{k}{N}\right)^n \exp\left\{-\sum_{m=1}^{M} \frac{k^{m+1}}{m(m+1)N^m}\right\}(1 + o(1)).$$

In particular, if under the same conditions $k/N^{2/3} \to 0$, then

$$\mathbf{P}\{\nu_k(N) - k - 1 = n\} = \frac{e^k}{\sqrt{2\pi k}} \left(\frac{k}{N}\right)^n \exp\left\{-\frac{k^2}{2N}\right\}(1 + o(1)).$$

THEOREM 2. *Let $N, k \to \infty$, $N \geq k+1$, and $n^2/k \to 0$. Then*

$$\mathbf{P}\{\nu_k(N) - k - 1 = n\} = \frac{(N-1)_k k^{2n}}{N^{k+n} 2^n n!}(1 + o(1)). \tag{6}$$

PROOF. To prove (6), we use the following estimate which can be found in [4]. If $k \to \infty$ and $n^2/k \to 0$, then

$$\sigma(k+n, k) = \frac{(k+n)^{2n}}{2^n n!}(1 + o(1)). \tag{7}$$

Since

$$(k+n)^{2n} = k^{2n}(1 + o(1)),$$

as $k \to \infty$ and $n^2/k \to 0$, estimate (6) follows from (7) and (2).

The theorem is thus proved.

As a corollary to Theorem 2 we obtain the following assertion. If, in addition to the conditions of Theorem 2, $k/N \to 0$, then

$$\mathbf{P}\{\nu_k(N) - k - 1 = n\} = \frac{k^{2n}}{2^n n!\, N^n} \exp\left\{-\frac{k^2}{2N} - \sum_{m=2}^{\infty} \frac{k^{m+1}}{m(m+1)N^m}\right\}(1 + o(1)). \tag{8}$$

In particular, if the conditions of Theorem 2 are satisfied and $k^2/(2N) \to \lambda > 0$, then we obtain the well-known result [1] on the convergence of the distribution of $\nu_k(N) - k - 1$ to the Poisson distribution with parameter λ.

One more corollary to Theorem 2 is obtained if we assume that $N \to \infty$, $n^2/N \to 0$, and $k = N - s$, where s is a fixed positive integer. In this case,

$$\mathbf{P}\{\nu_k(N) - k - 1 = n\} = \frac{\sqrt{2\pi} N^{n+s-1/2}}{(s-1)!\, 2^n n!} e^{-N}(1 + o(1)). \tag{9}$$

Formulae (8) and (9) can be obtained by the use of (6) in the same way as Corollaries 1 and 2.

THEOREM 3. *Let $n, k \to \infty$ and there exist constants α_0 and α_1 such that*

$$0 < \alpha_0 \leq \alpha = \frac{n}{k} \leq \alpha_1,$$

let also $N \to \infty$ and $N \geq k+1$. Then

$$\mathbf{P}\{\nu_k(N) - k - 1 = n\}$$
$$= \binom{N-1}{k} \left(\frac{(1+\alpha)k}{Ne}\right)^{(1+\alpha)k} \frac{1}{z_0^{\alpha k}(1 + \alpha - z_0)^k \sqrt{z_0 - \alpha}}(1 + O(1/k)), \tag{10}$$

where $z = z_0$ is the unique positive root of the equation

$$\frac{z}{1 - e^{-z}} = 1 + \alpha. \tag{11}$$

PROOF. To prove (10), we use (2) and the well-known integral representation of Stirling numbers of the second kind

$$\sigma(k+n,k) = \frac{(k+n)!}{k!\,2\pi i}\int \frac{(e^z-1)^k}{z^{k+n+1}}dz, \tag{12}$$

where the integral is taken along the circle with the centre at zero in the clockwise direction. To estimate the integral, we can use the saddle point method. It follows from [6] that under the conditions of Theorem 3

$$\frac{1}{2\pi i}\int \frac{(e^z-1)^k}{z^{k+n+1}}\,dz = \frac{1}{z^{\alpha k}(1+\alpha-z_0)^k\sqrt{2\pi k(1+\alpha)(z_0-\alpha)}}(1+O(1/k)). \tag{13}$$

Substituting (13) into (12), using the Stirling formula, and taking into account (2), we obtain (10), where $z = z_0$ is the unique positive root of equation (11).

The theorem is thus proved.

In a similar way, the following theorem can be proved.

THEOREM 4. *Let $n, N \to \infty$ and there exist constants α_0 and α_1 such that*

$$0 < \alpha_0 \le \alpha = \frac{N}{N} \le \alpha_1$$

and let $k = N - s$, where s is a fixed positive integer. Then

$$\mathbf{P}\{\nu_k(N) - k - 1 = n\} = \frac{N^{s-1}(1+\alpha)^{(1+\alpha)N-s}}{(s-1)!z_0^{\alpha N}(1+\alpha-z_0)^{N-s}\sqrt{z_0-\alpha}}e^{-(1+\alpha)N}(1+O(1/N)). \tag{14}$$

In particular, for $s = 1$,

$$\mathbf{P}\{\nu_k(N) - N = n\} = \frac{1-z_0(1+\alpha)^{-1}}{\sqrt{z_0-\alpha}}\left(\frac{(1+\alpha)^{1+\alpha}}{z_0^{\alpha}(1+\alpha-z_0)}\right)^N e^{-(1+\alpha)N}(1+O(1/N)).$$

REFERENCES

1. V. F. Kolchin, B. A. Sevastyanov, and V. P. Chistyakov, *Random Allocations.* Wiley, New York, 1978.
2. G. I. Ivchenko and Yu. I. Medvedev, Allocations with reflection; waiting time. *Discrete Math. Appl.* (1993) **3**, 221–228.
3. Yu. I. Medvedev and G. I. Ivchenko, Asymptotic representations of finite differences of the power function at an arbitrary point. *Theory Probab. Appl.* (1965) **10**, 151–156 (in Russian).
4. A. D. Korshunov, On the asymptotic behaviour of Stirling numbers of the second kind. *Methods of Discrete Analysis in Investigations of Extreme Structures* (1983) **39**, 24–41 (in Russian).
5. A. N. Timashov, Multivariate local limit theorems on large deviations in the equiprobable allocation scheme. *Diskretnaya Matematika* (1990) **2**, No. 2, 145–149 (in Russian).
6. I. J. Good, An asymptotic formula for the differences of the powers at zero. *Ann. Math. Statist.* (1961) **32**, 249–256.

Probabilistic Methods in Discrete Mathematics, pp. 359–365
V. F. Kolchin et al. (Eds.)

ASYMPTOTIC EXPANSIONS IN LOCAL THEOREMS ON LARGE DEVIATIONS IN THE EQUIPROBABLE ALLOCATION SCHEME

A. N. TIMASHOV

Moscow

Abstract — We consider the equiprobable scheme of allocating n particles to N cells. Let $\mu_r(n,N)$ be the number of cells each of which contains exactly r particles. We obtain asymptotic expansions of the distributions of the random variables

$$\mu_0(n,N), \quad \mu_0(n,N) - N + n, \quad \frac{1}{2}(n - \mu_1(n,N))$$

as $n, N \to \infty$ in such a way that either $N/n \to 0$ or $N/n \to \infty$.

We consider the equiprobable scheme of allocating n particles to N cells [1]. Each of the n particles is allocated to the N cells independently of each other and occupies each of the N cells with the same probability equal to $1/N$. Let $\mu_r(n,N)$ be the random variable equal to the number of cells each of which contains exactly r particles, $r = 0, 1, \ldots$ Multivariate local theorems on large deviations for the random variables $\mu_r(n,N)$ in the central domain [1] of variation of the parameters n and N were obtained by the author in [2]. In this paper, for the particular cases $r = 0$ and $r = 1$, we obtain asymptotic expansions for the probabilities $\mathbf{P}\{\mu_0(n,N) = k\}$ as $n, N \to \infty$ in such a way that $N/n \to 0$ and for the probabilities $\mathbf{P}\{\mu_0(n,N) - N + n = k\}$ and $\mathbf{P}\{\mu_1(n,N) = n - 2k\}$ as $n, N \to \infty$ in such a way that $n/N \to 0$. These expansions are valid in the domain of attraction of the Poisson distribution with parameter λ which is the limit distribution for the random variable $\mu_0(n,N)$ in the right-hand domain [1] of variation of n and N, where $n, N \to \infty$ in such a way that

$$Ne^{-n/N} \to \lambda > 0,$$

and in the left-hand domain [1] of variation of n and N, where $n, N \to \infty$ in such a way that

$$\frac{n^2}{2N} \to \lambda > 0.$$

The results proved in this paper were partially announced by the author in [3].

THEOREM 1. *Let $n, N \to \infty$ in such a way that $N/n \to 0$. Then for any fixed $k = 0, 1, 2, \dots$*

$$\mathsf{P}\{\mu_0(n,N) = k\} = \frac{1}{k!}(Ne^{-n/N})^k \exp\left\{-Ne^{-n/(N-k)} - n\sum_{m=2}^{\infty} \frac{k^m}{mN^m}\right\}(1+o(1)). \quad (1)$$

Expansion (1) *is asymptotic in the sense that for any fixed $k = 0, 1, 2, \dots$ and $M = 2, 3, \dots$,*

$$\mathsf{P}\{\mu_0(n,N) = k\} =$$

$$\frac{1}{k!}(Ne^{-n/N})^k \exp\left\{-Ne^{-n/(N-k)} - n\sum_{m=2}^{M} \frac{k^m}{mN^m} + O\left(\frac{n}{N^{M+1}}\right)\right\}(1+o(1)). \quad (2)$$

PROOF. To prove (1), we use the well-known equalities

$$\mathsf{P}\{\mu_0(n,N) = k\} = \frac{N!}{k!\,N^n}\sigma(n, N-k), \quad k = 0, 1, \dots, N-1, \quad (3)$$

where $\sigma(n, N-k)$ are Stirling numbers of the second kind.

Under the conditions of the theorem,

$$N - k = o(n), \qquad 0 \le k \le N-1,$$

and therefore (see [4])

$$\sigma(n, N-k) = \frac{(N-k)^n}{(N-k)!}\exp\left\{-(N-k)e^{-n/(N-k)}\right\}(1+o(1)) \quad (4)$$

uniformly in $k \in \{0, 1, \dots, N-1\}$.

It follows from (3) and (4) that

$$\mathsf{P}\{\mu_0(n,N) = k\} = \binom{N}{k}\left(1 - \frac{k}{N}\right)^n \exp\left\{-(N-k)e^{-n/(N-k)}\right\}(1+o(1)) \quad (5)$$

uniformly in $k \in \{0, 1, \dots, N-1\}$.

For any fixed $k = 0, 1, 2, \dots$,

$$\binom{N}{k} = \frac{N^k}{k!}(1+o(1))$$

and

$$\left(1 - \frac{k}{N}\right)^n = \exp\left\{-\frac{kn}{N} - n\sum_{m=2}^{\infty}\frac{k^m}{mN^m}\right\}.$$

Moreover, under the conditions of the theorem,

$$\exp\left\{-(N-k)e^{-n/(N-k)}\right\} = \exp\left\{-Ne^{-n/(N-k)}\right\}(1+o(1)).$$

Expansion (1) follows from (5) and the relations obtained above. Expansion (2) follows from (1) and the Taylor formula for $\ln(1 - k/N)$ with the remainder term of the form $O(n^{-M-1})$ for any fixed $M = 2, 3, \dots$

The theorem is thus proved.

Setting $M = 2$ in (2), we obtain the following assertion.

If $n, N \to \infty$ in such a way that $N/n \to 0$ and $N^3/n \to \infty$, then for any fixed $k = 0, 1, 2, \ldots$,

$$\mathbf{P}\{\mu_0(n,N) = k\} = \frac{(Ne^{-n/N})^k}{k!} \exp\left\{-Ne^{-n/(N-k)} - \frac{k^2 n}{2N}\right\}(1 + o(1)). \tag{6}$$

If $n, N, k \to \infty$ in such a way that $N/n \to 0$ and $k^2/N \to 0$, then

$$\mathbf{P}\{\mu_0(n,N) = k\} = \frac{(Ne^{-n/N})^k}{k!} \exp\left\{-(N-k)e^{-n/(N-k)} - n\sum_{m=2}^{\infty} \frac{k^m}{mN^M}\right\}(1 + o(1)).$$

This assertion follows from estimate (5), since if $N \to \infty$ in such a way that k^2/N, then

$$\binom{N}{k} = \frac{N^k}{k!}(1 + o(1)).$$

Expansion (6) is asymptotic expansion in the same sense as (2).

Under more general conditions, if $n, N, k \to \infty$ in such a way that $N/n \to 0$ and $k/N \to 0$, we have

$$\mathbf{P}\{\mu_0(n,N) = k\}$$
$$= \frac{(Ne^{-n/N})^k}{k!} \exp\left\{-(N-k)e^{-n/(N-k)} - \frac{k^2}{N} - (n-N+k)\sum_{m=2}^{\infty} \frac{k^m}{nN^m}\right\}(1 + o(1)). \tag{7}$$

To obtain (7), it is sufficient to make use of (5) and the equality

$$\binom{N}{k} = \frac{N^k}{k!}\left(1 - \frac{k}{N}\right)^{-N+k} e^{-k}(1 + o(1)),$$

which follows from the Stirling formula under the conditions above.

THEOREM 2. *Let $n, N \to \infty$ in such a way that $n/N \to 0$. Then for any fixed $k = 0, 1, 2, \ldots$*

$$\mathbf{P}\{\mu_0(n,N) - (N-n) = k\} = \frac{1}{k!}\left(\frac{n^2}{2N}\right)^k \exp\left\{-\sum_{m=1}^{\infty} \frac{n^{m+1}}{m(m+1)N^m}\right\}(1 + o(1)). \tag{8}$$

Expansion (8) is asymptotic expansion in the sense that for any fixed $k = 0, 1, 2, \ldots$ and $M = 1, 2, \ldots$

$$\mathbf{P}\{\mu_0(n,N) - (N-n) = k\}$$
$$= \frac{1}{k!}\left(\frac{n^2}{2N}\right)^k \exp\left\{-\sum_{m=1}^{M} \frac{n^{m+1}}{m(m+1)N^m} + O\left(\frac{n^{M+2}}{N^{M+1}}\right)\right\}(1 + o(1)). \tag{9}$$

PROOF. It is well known [1] that for all k, $0 \le k \le N$,

$$\mathbf{P}\{\mu_0(n,N) = k\} = \binom{N}{k} \frac{n!}{N^n 2\pi i} \int \frac{(e^z - 1)^{N-k}}{z^{n+1}} dz, \tag{10}$$

where the integral is taken around the circle with centre at zero in the clockwise direction.

Substituting $N - n + k$ for k in (10), we obtain for k such that $n - N \le k \le n$,

$$\mathbf{P}\{\mu_0(n,N) = k\} = \binom{N}{n-k} \frac{n!}{N^n 2\pi i} \int \frac{(e^z - 1)^{n-k}}{z^{n+1}} dz. \tag{11}$$

If $n, N \to \infty$ in such a way that $n^2/(2N) \to \lambda > 0$, then by the theorem proved by Bekessy [1], for any fixed $k = 0, 1, 2 \ldots$

$$\mathbf{P}\{\mu_0(n,N) - (N - n) = k\} \to \frac{\lambda^k}{k!} e^{-\lambda}.$$

Let $\lambda = 1/2$ and $N = n^2$. Then, as $n \to \infty$,

$$\mathbf{P}\{\mu_0(n,n^2) - n(n-1) = k\} \to \frac{1}{k! 2^k} e^{-1/2}$$

for any fixed $k = 0, 1, 2, \ldots$

On the other hand, it follows from (11) that under the same conditions,

$$\mathbf{P}\{\mu_0(n,n^2) - n(n-1) = k\} = \frac{n^2}{n-k} \frac{n!}{n^{2n} 2\pi i} \int \frac{(e^z - 1)}{z^{n+1}} dz.$$

Comparing the two last relations and using the Stirling's formula, we see that for any fixed $k = 0, 1, 2, \ldots$, as $n \to \infty$,

$$\frac{1}{2\pi i} \int \frac{(e^z - 1)^{n-k}}{z^{n+1}} dz = \frac{n^k}{2^k k!} (1 + o(1)). \tag{12}$$

Substituting (12) into (11) and using the Stirling formula once more, we obtain from (11) that under the conditions of Theorem 2 for any fixed $k = 0, 1, 2, \ldots$

$$\mathbf{P}\{\mu_0(n,N) - N + n = k\} = \frac{1}{k!} \left(\frac{n^2}{2N}\right)^k \exp\{-n - (N - n)\ln(1 - n/N)\}(1 + o(1)). \tag{13}$$

Expanding the logarithm in the series, we obtain (8) from (13). Estimate (9) follows from (8) and the Taylor formula for $\ln(1 - n/N)$ with the remainder term of the form $O((n/N)^{M+2})$.

Theorem 2 is thus proved.

Setting $M = 1$ in (9), we obtain the following assertion.

If $n, N \to \infty$ in such a way that $n^3/N^2 \to 0$, then for any fixed $k = 0, 1, 2, \ldots,$

$$\mathbf{P}\{\mu_0(n,N) - N + n = k\} = \frac{1}{k!}\left(\frac{n^2}{2N}\right)^k \exp\left\{-\frac{n^2}{2N}\right\}(1 + o(1)). \tag{14}$$

Note that estimates (8) and (9) remain true in the case where $n < N, k \to \infty$ in such a way that

$$n/N \to 0, \qquad k = o(\min(\sqrt{n}, N/n)).$$

To justify this remark, it is sufficient to use the relation

$$\mathbf{P}\{\mu_0(n,N) - N + n = k\} = \frac{N}{(N-n+k)!N^n}\sigma(n, n-k)$$

which is valid for any k, the well-known [5] asymptotics of Stirling numbers of the second kind, namely,

$$\sigma(n, n-k) = \frac{n^{2k}}{2^k k!}(1 + o(1))$$

for $n \to \infty$ and $k = o(\sqrt{n})$, and the Stirling formula.

In particular, estimate (14) is also valid in the case where $n < N, k \to \infty$ in such a way that

$$n^3/N^2 \to 0, \qquad k^2/n \to 0.$$

We have preferred another method to prove (8) and (9), since it can be used in the case $r = 1$ as well.

Note also that estimates (11) and (12) can be used to analyze the asymptotic behaviour of the probability $\mathbf{P}\{\mu_0(n,N) - N + n = k\}$ in the central domain. If $n, N \to \infty$ in such a way that

$$0 < \alpha_0 \le \alpha = \frac{n}{N} \le \alpha_1 < 1,$$

where α_0 and α_1 are constants, then for any fixed $k = 0, 1, 2, \ldots$, uniformly in $\alpha \in [\alpha_0, \alpha_1]$,

$$\mathbf{P}\{\mu_0(n,N) - N + n = k\} = \frac{\alpha^{2k}N^k}{2^k(1-\alpha)^k k!}\exp\{-\alpha N - ((1-\alpha)N + 1/2)\ln(1-\alpha)\}(1 + o(1)).$$

If $\alpha = 1$ and $n = N \to \infty$, then for any fixed $k = 0, 1, 2, \ldots,$

$$\mathbf{P}\{\mu_0(N,N) = k\} = \frac{N^{2k}}{2^k(k!)^2\sqrt{2\pi N}e^{-N}}(1 + o(1)).$$

The two last relations are obtained from (11) and (12) by the use of the Stirling formula.

Estimates (11) and (12) imply a more general result than (13). Namely, if $n, N \to \infty$ in such a way that $N - n \to \infty$, then for any fixed $k = 0, 1, 2, \ldots,$

$$\mathbf{P}\{\mu_0(n,N) - N + n = k\}$$
$$= \frac{1}{k!}\left(\frac{n^2}{2N}\right)^k \exp\{-n - (N - n + k + 1/2)\ln(1 - n/N)\}(1 + o(1)). \tag{15}$$

Estimate (15) can be obtained from (11) and (12) with the use of the Stirling formula.

Now we turn to the case $r = 1$. It is known [1] that for $k = 1, \dots, N$

$$\mathbf{P}\{\mu_1(n,N) = k\} = \binom{N}{k} \frac{n!}{N^n 2\pi i} \int \frac{(e^z - z)^{N-k}}{z^{n-k+1}} dz, \tag{16}$$

where the integration is taken along the same path as in (10).

Substituting $n - 2k$ for k in (16), we obtain that for any k, $(n - N)/2 \le k \le [n/2]$,

$$\mathbf{P}\{\mu_1(n,N) = n - 2k\} = \binom{N}{n-2k} \frac{n!}{N^n 2\pi i} \int \frac{(e^z - z)^{N-n+2k}}{z^{2k+1}} dz. \tag{17}$$

We set $m = N - n$. If $N, m \to \infty$ and $n = [\sqrt{m}]$, then

$$\frac{n^2}{2N} \to \frac{1}{2}.$$

In this case, according to [1], for any fixed $k = 0, 1, 2, \dots$,

$$\mathbf{P}\{\mu_1(n,N) = n - 2k\} = \frac{1}{2^k k!} e^{-1/2}(1 + o(1)).$$

Comparing this relation and (17) and setting $n = [\sqrt{m}]$, $N = m + [\sqrt{m}]$ in (17), we conclude that for any fixed $k = 0, 1, 2, \dots$

$$\frac{1}{2\pi i} \int \frac{(e^z - z)^{m+2k}}{z^{2k+1}} dz = \frac{m^k}{2^k k!}(1 + o(1)). \tag{18}$$

THEOREM 3. *Let $n, N \to \infty$ in such a way that $n/N \to 0$. Then for any fixed $k = 0, 1, 2, \dots$*

$$\mathbf{P}\{\mu_1(n,N) = n - 2k\} = \frac{1}{k!}\left(\frac{n^2}{2N}\right)^k \exp\left\{-\sum_{m=1}^{\infty} \frac{n^{m+1}}{m(m+1)N^m}\right\}(1 + o(1)). \tag{19}$$

Expansion (19) *is an asymptotic expansion in the same sense as (8).*

PROOF. We set $m = N - n$ in (18) and use (17) and the Stirling formula. Reasoning as in the proof of Theorem 2, we obtain (19).

Theorem 3 is thus proved.

In particular, if $n, N \to \infty$ in such a way that $n^3/N^2 \to 0$, then we obtain, in a similar manner as (14), that for any fixed $k = 0, 1, 2, \dots$,

$$\mathbf{P}\{\mu_1(n,N) = n - 2k\} = \frac{1}{k!}\left(\frac{n^2}{2N}\right)^k \exp\left\{-\frac{n^2}{2N}\right\}(1 + o(1)).$$

If $n, N \to \infty$ in such a way that there exist constants α_0 and α_1 such that

$$0 < \alpha_0 \le \alpha = \frac{n}{N} \le \alpha_1 < 1,$$

then for any fixed $k = 0, 1, 2\ldots,$

$$\mathbf{P}\{\mu_1(n, N) = n - 2k\} = \frac{\alpha^{2k} N^k}{2^k (1-\alpha)^k k!} \exp\{-\alpha N - ((1-\alpha)N + 1/2)\ln(1-\alpha)\}(1 + o(1))$$

uniformly in $\alpha \in [\alpha_0, \alpha_1]$.

If $\alpha = 1$ and $n = N \to \infty$, then for any fixed $k = 0, 1, 2, \ldots$

$$\mathbf{P}\{\mu_1(N, N) = N - 2k\} = \frac{c_k N^{2k}}{(2k)!}\sqrt{2\pi N} e^{-N}(1 + o(1)),$$

where

$$c_k = \frac{1}{2\pi i}\int \frac{(e^z - z)^{2k}}{z^{2k+1}} dz.$$

The two last relations are obtained from (17) and (18) with the use of the Stirling formula.

If $n, N, N - n \to \infty$, then the equality which is obtained from (15) by the substitution of the probability $\mathbf{P}\{\mu_1(n, N) = n - 2k\}$ for the left-hand side of (15) holds. This assertion can be obtained from (17) and (18) by the use of the Stirling formula. In the same way it follows from (17) and (18) that if $N \to \infty$ and n is a constant, then

$$\mathbf{P}\{\mu_1(n, N) = n - 2k\} = \frac{n!}{(n-2k)!\, 2^k k!\, N^k}(1 + o(1))$$

uniformly in $k \in \{0, 1, \ldots, [n/2]\}$.

REFERENCES

1. V. F. Kolchin, B. A. Sevastyanov, and V. P. Chistyakov, *Random Allocations*. Wiley, New York, 1978.
2. A. N. Timashov, Multivariable local limit theorems on large deviations in the equiprobable allocation scheme. *Diskretnaya Matematika* (1990) **2**, No. 2, 145–149 (in Russian).
3. A. N. Timashov, Local limit theorems on large deviations in the equiprobable allocation scheme. *Abstr. 5th Intern. Vilnius Conf. on Probab. Theory and Math. Statist.*, Vol. 4. Vilnius, 1989, pp. 273–274 (in Russian).
4. Yu. I. Medvedev and G. I. Ivchenko, Asymptotic representations of the finite differences of the power function at an arbitrary point. *Theory Probab. Appl.* (1965) **10**, No. 1, 151–156 (in Russian).
5. A. D. Korshunov, On the asymptotic behaviour of Stirling numbers of the second kind. *Methods of Discrete Analysis in Investigations of Extreme Structures* (1983) **39**, 24–41 (in Russian).

LIST OF CONTRIBUTORS

ORGANIZING COMMITTEE

V. Ya. Kozlov, Chairman (Moscow)
V. F. Kolchin, Vice-Chairman (Moscow)
A. D. Sorokin, Vice-Chairman (Petrozavodsk)

A. S. Fomin (Petrozavodsk)
G. I. Ivchenko (Moscow)
V. I. Khokhlov (Moscow)
Yu. I. Medvedev (Moscow)
V. G. Mikhailov (Moscow)
Yu. L. Pavlov (Petrozavodsk)
Yu. V. Prokhorov (Moscow)
B. A. Sevastyanov (Moscow)
V. A. Vatutin (Moscow)
A. B. Zhizhchenko (Moscow)
A. M. Zubkov (Moscow)

Bei Fragen zur Produktsicherheit wenden Sie sich bitte an:
If you have any questions regarding product safety,
please contact:

Walter de Gruyter GmbH
Genthiner Straße 13
10785 Berlin
productsafety@degruyterbrill.com